MICROCYSTIS AERUGINOSA REMOVAL BY DISSOLVED AIR FLOTATION (DAF)

Microcystis aeruginosa Removal by Dissolved Air Flotation (DAF)

Options for Enhanced Process Operation and Kinetic Modelling

DISSERTATION
Submitted in fulfilment of the requirements of
the Board for the Doctorates of Delft University of Technology
and the Academic Board of the International Institute for Infrastructural,
Hydraulic and Environmental Engineering for the Degree of DOCTOR
to be defended in public
on Monday, 26 January 1998 at 13:30 h

by
ALEKSANDAR VLAŠKI
Master of Science in Sanitary Engineering (IHE-Delft)
born in Skopje, Macedonia

CRC Press
Taylor & Francis Group
Boca Raton London New York

CRC Press is an imprint of the
Taylor & Francis Group, an **informa** business
A BALKEMA BOOK

First published 1998 by CRC Press/Balkema
4 Park Square, Milton Park, Abingdon, Oxon, OX14 4RN

and by CRC Press/Balkema
2385 NW Executive Center Drive, Suite 320, Boca Raton FL 33431

First issued in hardback 2018

CRC Press/Balkema is an imprint of the Taylor & Francis Group, an informa business

ISBN-13: 978-1-138-47475-8 (hbk)
ISBN-13: 978-90-5410-410-0 (pbk)

Visit the Taylor & Francis Web site at
http://www.taylorandfrancis.com

and the CRC Press Web site at
http://www.crcpress.com

Contents

Acknowledgements

My deepest gratitude goes for ir. A.N. van Breemen and Prof. Dr. ir. G.J. Alaerts for entrusting me the opportunity to work on this project. Without the involvement of my mentor ir. A.N. van Breemen who meticulously worked on initiating and organising the research, this project would have never been carried out. Together with my promoter Prof. G.J. Alaerts they formed a team of wide vision, immense theoretical knowledge and broad research experience, giving me the opportunity to learn, develop, direct and critically assess my own research skills and knowledge. They helped me to develop a systematical and analytical approach to technical problems solutions, an invaluable achievement for my future engineering career.

I am most sincerely indebted to the Dutch companies which financially supported this research. I gratefully acknowledge the support of: Amsterdam Water Supply (GWA), North Holland Water Supply Companies (PWN), Rhine-Kennemerland Water Transport Company (WRK), Dune Water Supply Company South Holland (DZH), Energy and Water Supply Company Rhineland (EWR), Water Supply Association Overijssel-Regio Oost (WMO, formerly Water Supply Company Oost-Twente-WOT), Groningen Water Supply Company (GWG), and KIWA Research and Consultancy. I also wish to express my gratefulness to the representatives of these companies who participated on the Advisory Committee which guided and critically assessed the research in its progress. My thanks go to: Dr. ir. A. Graveland and ir. E.T. Baars (Gemeentewaterleidingen - Amsterdam), Dr. P.G.M. Stocks, ing. R.A. Wyatt and ing. N. Koelman (N.V.Watertransportmaatschappij Rijn-Kennemerland), Dr. W. Hoogenboezem (N.V.Provinciaal Waterleidingbedrijf van Noord-Holland), the late ir. J. van Puffelen and ing. P.L. Buijinck (N.V. Duinwaterbedrijf Zuid Holland), Mr. J.H. Heijnen (N.V. Waterleiding Maatschappij Overijssel-Regio Oost), Drs. A. de Ruyter and ing. W. Oorthuizen (N.V. Duinwaterbedrijf Zuid Holland, formerly N.V. Energie en Watervoerziening Rijnland), ir. A.I.A. Soppe and ir. R.T. van der Velde (Gemeente Waterbedrijf Groningen), ir. P.Buys and ir. H. Koppers (N.V. KIWA Onderzoek en Advies), and Drs. G. Bolier (Technische Universiteit-Delft). Their criticism and comments resulted in a praiseworthy quality of the research by combining the theory with the practical aspects of drinking water production.

I also wish to thank Prof. Dr. K.J. Ives (University College London) and Prof. Dr. J.C. Schippers (IHE-Delft) for their involvement, constructive criticism and useful comments on the draft version of the thesis.

I sincerely appreciate the hospitality of the Laboratory for Sanitary Engineering (Faculty of Civil Engineering, Delft University of Technology) which provided the facilities for carrying out the experimental part of the research. The invaluable help of the laboratory staff including Mr. C. Boeter, Mr. T. Schuit and others, in setting-up the experimental facilities and the analytical work and support is highly appreciated.

My equally sincere appreciations and gratefulness goes for the Laboratory of IHE-Delft. I thank Mr. F. Kruis, Mr. D. Lindenburg, Mr. C. Bik and all the other laboratory staff for their help in the analytical work, but also their kindness and hospitality. My sincere thanks go to Mrs. J. Van Hooijdonk for her priceless contribution in culturing the often unpredictable cyanobacteria used in the research.

I also have to express my gratefulness to the Rhine-Kennemerland Water Transport Company (WRK III, Andijk) for the hospitality in allowing me to carry out the pilot plant research on their premises. The always friendly atmosphere and the cooperativeness of the treatment plant and laboratory personnel will remain in pleasant memories. There I sincerely thank ing. R. Wyatt, ing. N. Koelman, Mr. W.M. Groeneveld, and all the others who helped carry out this not always easy task.

My endless appreciation goes to the enthusiastic and praiseworthy work of the four M.Sc. fellows who participated in the project : Mr. L.O. Karimu, Mr. A.A. Daniel, Mr. F. Merid and Mr. J.S. Hwang. Their commitment and hard work resulted in a valuable contribution to the research.

I would also like to thank to all my colleagues at IHE who provided the solid friendly and scientific background for my stay and research. I shall always praise and treasure the time spent in this environment.

Finally, I thank my wife Valentina and our children Ivana and Josif for centring their lives around my career. I apologise for all the late dinners, raised voices and nervous attitudes. Your support was priceless and shall always be treasured.

Abstract

Vlaški A., (1997). ***Microcystis aeruginosa Removal by Dissolved Air Flotation (DAF) - Options for Enhanced Process Operation and Kinetic Modelling.*** **Doctoral thesis, International Institute for Infrastructure, Hydraulic and Environmental Engineering (IHE)/Delft University of Technology (TUD), Delft, the Netherlands, 253 pages.**

Seven Dutch surface water treatment companies share the problem of periodical troublesome algal blooms in their raw water sources and reservoirs, and participated in this research : Amsterdam Water Supply (GWA), North Holland Water Supply Companies (PWN), Rhine-Kennemerland Water Transport Company (WRK), Dune Water Supply Company South Holland (DZH), Energy and Water Supply Company Rhineland (EWR), Water Supply Association Overijssel - Regio Oost (WMO, formerly Water Supply Company Oost-Twente - WOT), and Groningen Water Supply Company (GWG). KIWA Research and Consultancy also participated actively in the research.

The eutrophication status of the raw water utilised by the water treatment companies varies substantially. Phosphorus was found to be the limiting nutrient; concentrations of >0.05 mg P/L were recognised as potentially troublesome in terms of algal growth (>10 µg chlorophyll-*a*/L). The predominant algal species depend on a range of additional factors, including pH, temperature, combined ratio of other nutrients, light availability, etc. Different sets of problems are experienced due to the resultant seasonal algal population fluctuations (from 8.4 µg/L chlorophyll-*a* for Loenderveen Lake to 238 µg/L chlorophyll-*a* for the PWN reservoir, on a yearly average maximum basis). In all cases however, the periodical peak concentrations of algae cause substantial treatment problems. Cyanobacteria in general and the *Microcystis aeruginosa* and the *Oscillatoria agardhii* species in particular, proved to be the most persistent and associated with most treatment problems. The Dutch experiences are generally representative of industrialised countries where increased eutrophication results commonly in algal blooms and related treatment problems. These included an up to 100% higher coagulant dose, substantially increased filter backwash quantities, trihalomethane formation (during short periods of intermittent chlorination for mussels growth control), (surface) clogging of filters, passing of algae through treatment in objectionable quantities, increase of MFI (modified fouling index) and AOC (assimilable organic carbon) values, etc. The spherical *M. aeruginosa* single cells form is also representative of particles in the size range that are typically difficult to remove in treatment (3-10 µm). Its size and shape make it also a suitable representative of the pathogenic *Cryptosporidium* oocysts and *Giardia* cysts.

This study targeted the removal of the cyanobacteria *M. aeruginosa* by conventional and advanced treatment technologies, that are available or considered feasible under raw water quality and process circumstances that are typical for the Netherlands. The ultimate barrier preventing them, and their products, from penetrating into the distribution system is rapid or/and slow sand filtration. Pre-treatment modes practised in the involved companies include sedimentation-filtration and more advanced dissolved air flotation (DAF)-filtration. The coagulation/flocculation process is of critical importance regardless of the down-stream treatment; however, the down-stream treatment itself is defined by the coagulation/flocculation. The option of algae conditioning by oxidants, e.g. ozone or $KMnO_4$, although not actively practised in the Netherlands, was considered attractive to (periodically or continuously) increase the algae removal efficiency. Consequently, this study aimed to investigate a range of treatment options for the removal of *M. aeruginosa*, with an emphasis on DAF and different modes for its enhancement. The evaluation of process efficiency was to be complemented with insight into the process mechanisms, and to serve as a basis for the evaluation of process kinetics.

For this purpose two experimental modes were used: a bench-scale modified jar test filtration apparatus with facilities for both DAF and sedimentation, and a commercial DAF pilot plant (Purac, Sweden). The former utilised model water prepared by spiking water originating from the Biesbosch reservoirs (the Netherlands) with laboratory cultured *M. aeruginosa* to a standard initial concentration of ≈ 10,000 ±cells/mL. The standard experimental temperature was 20°C and coagulation G was 1,000 s^{-1} for 30 s at pH 8. Other coagulation pH conditions were also tested (pH 4, 6, 7, 8, and 9). The flocculation G and time were varied (G_f=10, 30, 50, 70, 100 and 120 s^{-1} and t_f=5, 10, 15, 25, 30 and 35 min). The coagulant ($FeCl_3 \cdot 6 H_2O$) doses were 0 - 15 mg Fe(III)/l, while the cationic polyelectrolytes Superfloc C-573 and Wisprofloc-P, the non ionic Wisprofloc-N, and the anionic Superfloc A-100 were tested as sole coagulants and as coagulant aids. Ozone was used as an algae conditioner in the dose range of 0.48-2.16 mg O_3/L, or 0.2-0.9 mg O_3/mg TOC at pH 7.5. $KMnO_4$ conditioning was tested in the dose range of 0-2 mg/L at pH 8, the optimal dose being based on the visual determination technique. The $KMnO_4$ contact time was varied from 0-30 min. The DAF pilot plant was (operated at a flow of Q=4-6 m^3/h) followed by a multi media rapid sand up-flow filtration (Q=0.4 m^3/h, v=10 m/h) and situated at the WRK III treatment plant in Andijk, the Netherlands. It was compared with the full-scale WRK III treatment line comprising of flocculation, lamella sedimentation and filtration, during the cyanobacteria blooms (primarily *M. aeruginosa*) occurring in the raw water reservoir. Analytical techniques for the assessment of process efficiency included turbidity, residual coagulant (Fe), TOC/DOC, UV absorbance at 254 nm, and residual Mn. They were supported by more advanced methodologies including bromate and MFI measurements, particle count (in the size range of d_p>2.75 μm and in a limited number of cases d_p>0.3 μm) and computer image analysis (d_p>1.9 μm). Coupled with the high resolution SEM (scanning electron microscopy), the latter provided valuable information for assessment of process mechanisms and kinetics.

DAF proved to be a viable and efficient alternative to sedimentation for the treatment of heavily algae (*M. aeruginosa*) laden water. It consistently resulted in equal or better algae removal efficiency than the sedimentation process (71% versus 87% for jar test conditions, and 96.3% versus 95.9% for pilot plant conditions), requiring at the same time an up to 50% lower coagulant dose than the sedimentation (3 mg Fe(III)/L versus 10 mg Fe(III)/L in the jar test conditions, and 7-12 g Fe(III)/m^3 versus 20-24 g Fe(III)/m^3 + 0.2-0.5 g Wisprofloc-P/m^3 in the pilot plant conditions and the full scale sedimentation), which is significant from an environmental point of view. Also, the produced sludge was of high solids content. Other positive aspects included the relatively short flocculation time (however, not shorter than 15 min, compared to >30 min for sedimentation) and the low flocculation energy input requirement (G=10 s^{-1} compared to G=30 s^{-1} for sedimentation), as well as the high process loading rate (leading to 5-6 times overall lower space requirements). Furthermore, the DAF-filtration treatment scheme resulted consistently in high (>2 log) algae removal efficiency. The relatively low coagulant demand for DAF, however, produced less efficient sweep coagulation conditions than in the case of the sedimentation, resulting in organo-Fe complexes formation and consequently higher DAF effluent Fe and turbidity residuals. This was overcome by coagulation at pH<IEP (iso-electrical point), suggesting the potential of using cationic polyelectrolyte coagulant aids. These improved the particle (algae) removal efficiency significantly in case of the model water experiments (20-40%), compared to a minor rise of only 1-2% in pilot plant experiments. This is owed to the promotion of particle (algae) adsorption coagulation based on charge neutralisation. Furthermore, the observed charge neutralisation phenomena resulted in improved particle-bubble attachment and hence more efficient DAF. Non-ionic and anionic polyelectrolytes (and the bridging mechanism) were found to be less efficient, partly because of the lower particle concentration (lower coagulant dose) and the negative charge of the air bubbles.

The particle (algae) removal efficiency the pilot plant achieved on reservoir water was by 20-30% higher than that in the jar tests with model water. The NOM (natural organic matter) concentration and composition was found to be the likely cause of the differences, especially as it is present in the form of organic matter meshes and fibrilar structures. Although the complexing of metal coagulants by NOM imparts additional coagulant demand, these structures may eventually promote flocculation, serving as sites for easier floc growth and embedment, as well as providing a gel-like structure that is easier to be removed by the rising air bubbles in DAF.

Relatively low doses of ozone (0.2-0.5 mg O_3/mg TOC, or 0.6-1.5 mg O_3 /L) as an algae conditioner under bench-scale conditions significantly (by 30-40%) increased DAF efficiency towards 2 log removal. Combined with cationic polyelectrolytes, the DAF process efficiency rose further typically by another 5%. Although not yet confirmed by pilot plant results with reservoir water, the application of the lower dose range of 0.2 mg O_3 /mg C at pH 7 resulted in bromate levels within the range of the 10 μg/L MAC prescribed by the EU. The further improvement of organic matter removal by down-stream GAC (granular activated carbon) filtration would enhance interest in this treatment, especially if ozone is already used for other purposes within the plant (e.g. disinfection), and providing the MAC value remains within the currently prescribed range.

The use of $KMnO_4$ as an algae conditioner also tended to increase DAF process efficiency towards 2 log particle removal, resulting in equal or better performance than the full-scale conventional sedimentation, on occasions even better than the full-scale sedimentation + filtration in terms of particle (d_p>2.75 μm) removal. However, the resultant higher effluent Fe (organo-Fe complexes) and Mn (mostly MnO_2) concentration raised the particle count in the colloidal size range (<0.5 μm) and the turbidity. This situation improved when cationic polymers were introduced as coagulant aids, significantly improving the DAF process efficiency (typically by 5%) on the account of reduced Fe and Mn residuals below their respective MAC values. This option, however, did not lower the accompanying filtrate MFI value to the desired 5 s/L^2 value. The lowest MFI value achieved by the non-optimised filtration (within the DAF + filtration scheme) was in the range of 20 s/L^2. This suggests the need for optimisation of the filtration step to fully utilise the conditioning benefits of $KMnO_4$.

The improved algae removal in the case of conditioning by an oxidant is probably caused by a number of process mechanisms. Motile algal cells are immobilised by the oxidation of the outer cell layer, and their metabolic processes are disrupted. This is accompanied with the creation of stress conditions in the algal environment, resulting in EOM (extra-cellular organic matter) excretion, partial algae lysis and IOM (intra-cellular organic matter) leakage. The EOM and IOM act as natural coagulant aids, resulting in spontaneous microflocculation even before coagulant addition. Oxidation and removal of the organic coating on the particles, as well as oxidation of NOM, result in more favourable coagulation conditions through the production of in-situ coagulant, which otherwise would have been complexed and unavailable for coagulation. The resultant MnO_2 in the case of $KMnO_4$ conditioning increases the particle concentration and promotes sweep coagulation.

The particle size frequency and volume distributions obtained under different process conditions and after different process stages served as the basis for an empirical calculation of the relative and absolute floc density, an input variable in the single collector collision efficiency DAF reaction zone model. Other input variables such as the mean particle (floc) and bubble size, as well as their number and volume concentrations, were also directly measured or calculated. The calculated values for the particle-bubble attachment efficiency α_{pb} support the previous assertion regarding the positive impact of cationic polyelectrolytes, and conditioners. Results suggest that under optimal DAF process conditions almost every second particle (floc)-bubble collision resulted in their attachment. The role of the recirculation ratio (5-10%) and the saturator pressure (500-700 kPa), although significantly affecting the bubble size distribution, proved less important for DAF process efficiency. Thus, critical attention should be paid to the coagulation/flocculation process as the major determinant of DAF efficiency. Although difficult to model, the impact of NOM concentration and characteristics on the particle-bubble attachment efficiency is of great significance. Thus, the simplified modelling approach that particle-bubble agglomeration is solely the result of their collision is not fully justified.

DAF proved to be an efficient, robust and flexible water treatment process, and a highly appropriate alternative to sedimentation for the treatment of algae laden water. The lower civil engineering costs and chemicals consumption often will compensate for the higher energy and maintenance costs. Its high algae removal efficiency makes it particularly attractive as pre-treatment before direct filtration, or membrane filtration. Where it is already applied, it's efficiency can be significantly enhanced temporarily (e.g. during short or long term algae blooms) by the application of e.g. cationic polyelectrolytes, oxidants, or a combination of the two. The application of $KMnO_4$ seems particularly attractive, since no hazardous by-products are produced. Similarly existing sedimentation units can be retrofitted with (enhanced) DAF in case of increased algae problems. Nevertheless, the final choice of the treatment technology or mode should be made after pilot plant investigations and cost-benefit analysis.

Chapter 1

THE ALGAE PROBLEM IN THE NETHERLANDS FROM A WATER TREATMENT PERSPECTIVE

-This chapter is based on A. Vlaški, A.N. van Breemen and G.J. Alaerts (1996), The algae problem in the Netherlands from a water treatment perspective, *J. Water SRT-Aqua*, 45, 4, pp.184-194, and The algae problem in the Netherlands, problem analysis and treatment strategies in five water treatment companies, (1994) Working Paper EE-1, IHE, Delft.

ABSTRACT : Surface water treatment in the Netherlands receives increased attention requiring better characterisation and evaluation of the 'algae problem'. Investigations and analysis carried out at five drinking water production locations in the Netherlands utilising surface water (amounting to 60% of the overall Dutch surface water production) show that the most significant problem are the cyanobacteria. Seasonal blooms of *Microcystis aeruginosa* and *Oscillatoria agardhii* force short- and long-term process modifications, affecting water production cost. Assessment of possible solutions suggests that, apart from the application of appropriate water quality management measures, new insights on more efficient (pre)treatment processes are critical. In treatment, the agglomeration (coagulation/flocculation) phase is considered the essential step in the removal of the algae. It should be optimised in relation to the pre-treatment preceding it (e.g. application of chemical oxidant/s, and/or microstraining), as well as to the applied solid-liquid separation treatment process. Dissolved air flotation (DAF) emerges as a feasible option given the natural tendency of algae to float. The existing discrepancies between literature and practice call for further research on process mechanisms and suggest that considerable gains can still be made in process optimisation. This research should notably increase insight into the influence of algae morphology and physiology on treatment efficiency.

1.1 INTRODUCTION

1.1.1 Algae in water supply - a growing concern

VEWIN (the Association of Dutch Water Supply Companies) forecasts the drinking water production in the Netherlands to amount to 1360 Mm^3/year by the year 2000, of which 16 % (220 Mm^3/year) will originate from treated surface water, another 16 % (220 Mm^3 /year) from dune-infiltrated water, and the remainder (930 Mm^3/year) from groundwater [1]. Although this projection for all the sources shows a substantial amount of the infiltrated water is of surface water origin, surface water is not recognized by VEWIN as the preferred alternative. Very often it contradicts one or more of the four basic principles that have to be met for any water source : (i) good water quality, (ii) constant water quality, (iii) low treatment cost and (iv) security from calamities. Nevertheless, surface water is an inevitable choice due to increased water consumption and increasing number of restrictions on groundwater abstraction. The main surface water sources in the Netherlands are the rivers Rhine and Meuse. These two rivers receive a substantial industrial, municipal and agricultural pollution load, resulting in water treatment problems such as high turbidity, colour, organic micropollutants content and high nutrient loads. Eutrophication is recognized as the cause of numerous problems, both in surface water quality and in drinking water treatment. It causes plankton blooms in stagnant and slowly flowing water bodies; their removal emerges as a priority problem in surface water treatment.

Algae cause a wide range of technical and health problems in the treatment and in the distribution step [2]:

i) problems related to algae laden water treatment:

- interference with coagulation and flocculation;
- increased disinfectant demand of the water;
- production of haloform precursors;
- filter clogging and increased use of backwash water; and

- filter penetration;

ii) problems related to algae presence in the distribution system:

- taste and odour;
- precipitation of algal mucopolysacharides at low pH;
- corrosion of iron mains and discolouration of water;
- shielding of pathogenic bacteria from disinfection;
- aftergrowth of bacteria and higher organisms in the mains;
- infestation of zooplankton; and
- release of toxins.

Nevertheless, water supply experts disagree about the required degree and approach to algae removal in water treatment. Lacking a proper water quality standard, one has to rely on the philosophy that the more algae are removed during common treatment, the better, accepting however that removal efficiency for different types of algae varies widely. The diversity of algal species with respect to their size, form, surface characteristics, production of extra-cellular organic matter (EOM), motility, etc., makes it almost impossible to propose a simple solution that fits water quality management and treatment concerns. Therefore, the 'best available technology' for treatment tends to set the standards.

This chapter intends to address the issue of raw water quality for five Dutch surface water resources, contributing to approximately 60% of the total surface water origin treated water in the Netherlands. Statistical analysis of the water quality parameters contributing to the eutrophication phenomenon will be performed. This will be related to the resultant impact on the algal population, quantity- and species-wise, defining the most significant species from the point of view of imparted treatment problems. The treatment problems, as well as the adopted short and long term reservoir management and treatment strategies will also be discussed. Finally, an overview of existing conventional treatment technologies, as well as more advanced options for their enhanced operation, will be discussed in the context of most efficient algae removal in drinking water treatment.

1.1.2 Eutrophication and algae

Long term changes in the nutrient supply to water bodies cause shifts in the balance and spectra of resources to which phytoplankton is known to respond quantity- and quality-wise. Such changes occur naturally due to a variety of causes. The large increases in the amount of nutrients discharged into water bodies during the past decades, especially in the industrialized countries, and the subsequent changes in the ecology of these water bodies, have called attention to the problem of nutrient enrichment, or eutrophication. Generally, the adjective 'eutrophic' (='well feeding') is used to describe biological aquatic systems with a high input of otherwise growth limiting nutrients, and which therefore support a high level of organic productivity [3].

Eutrophic conditions occur naturally (by leaching of soils and vegetation decomposition), but are nowadays primarily due to human activities (domestic sewage, agricultural run-off, etc.) [4]. The anthropogenic eutrophication, should be tackled at the source of the nutrients discharge, rather than by the symptoms. In the Netherlands, as well as in some other West European countries, measures like tertiary sewage treatment and a ban on detergent phosphate have drastically reduced nutrient inputs. However, the nutrient rich under-water sediments, the contained fertilizer use in agriculture, and the inevitable leaching of N and P from wastewater, will keep the trophic levels high.

Eutrophication manifests itself as increased water fertility which results in higher primary productivity, i.e. algal growth. Under extreme conditions this can lead to temporary excessive algal biomass and ecosystem instability. As a result of increased eutrophication, the diversity of algae, especially with respect to the planktonic forms, is reduced in favour of large concentrations of a limited taxonomic range of cyanobacteria, chlorophytes and diatoms. The trophic level of the water body thus defines the range and concentration of species present within it [3].

A common annual cycle for eutrophic impounded water in northern temperate regions like the Netherlands consists of a spring population of diatoms which may include *Asterionella formosa, Fragilaria crotonensis, Tabellaria flocculosa, Melosira* spp., *Cyclotella* spp., *Stephanodiscus astraea*, and/or *S. hantzschii*. During early summer, diatoms may be replaced by small species of cryptophyceae or chlorophyceae, and at a later stage by dinoflagellates such as *Ceratium hirundinella*, chrysophytes such as *Dynobrion spp.*, *Synura* spp., or chlorophyte genera such as species of *Chlorella, Scenedesmus, Dictyosphaerium, Ankistrodesmus, Eudorina* and *Pandorina.* However, in late summer and throughout autumn, dominant populations of cyanobacteria (blue-green algae) may occur, including *Anabaena* spp., *Aphanizomenon flos-aquae, Gloeotrichia*, and *Coelosphaerium*; especially *Microcystis aeruginosa* and *Oscillatoria* spp. can be abundant and are able to form surface blooms of aggregated cells under windless conditions [4].

In addition, cyanobacteria tend to dominate a large part of the growing season. Objectionable odours may emanate from water bodies due to decaying biomass, and the water may suffer from offensive tastes and odours, while serious disruption of water treatment processes is experienced. Predation of algae is reduced by colony forming and filamentous cyanobacteria, resulting in the preclusion of effective filter feeding zooplankton (algae size related). This further influences the plankton abundance and composition. In addition, toxins produced by cyanobacteria, such as microcystin, have been responsible for occasional outbreaks of wildlife and livestock illness and death, and has been related to human illnesses [4].

Principles of freshwater ecology phytoplankton form the basis for appropriate design and management strategies of lakes and reservoirs [3]. On the other hand, knowledge of the freshwater ecology does not guarantee the accurate prediction of the plankton responses to particular changes in their environment. Furthermore, the primary problem of water treatment engineers in practice is the problem of biomass 'peaks', composed of different dominant species, regardless of the average level of biomass. Therefore, more knowledge is needed on the behaviour and characteristics of individual organisms. This knowledge should preferably overcome the indiscriminative algae related lumped material parameters approach (chlorophyll, organic carbon, turbidity, etc.). Species differ in their ecological, morphological and physiological characteristics that allow them to survive and grow, and that can strongly influence water treatment.

1.1.3 Growth strategies and characteristics of cyanobacteria

The spatial and temporal distributions of an organism within an ecological system are the result of the evolutionary strategies the organism adopted for growth and survival [5]. Strategies can be defined as sets of similar morphological, physiological, reproductive or behavioural traits that have evolved among species or populations and that are better suited to particular environmental conditions than that of others. Accordingly, different organisms that adopted similar strategies, are likely to resemble in their ecological behaviour.

Based on the logistic growth equation phytoplankton communities have commonly been classified into two main categories : r (characterised by rapid growth and colonisation of a nutrient rich environment) and K (characterised by the ability to out compete other organisms in a nutrient limited environment) [3]. However, other approaches exist that are seemingly more suitable for algae and which classify organisms into three categories marked C, S or R [5]:

- C/competitors : exploitation of environments saturated with light and nutrients, through the investment in rapid growth and reproduction, and to do so before other species;
- S/stress tolerant : operation under conditions of severe depletion of externally supplied essential nutrients; and
- R/ruderals : tolerate frequent or continuous turbulent transport through the light gradient.

There is a considerable overlap of morphometric and physiological characteristics (e.g. unit volume, surface area/volume, maximum linear dimension, photosynthetic efficiency, projected area, maximum growth rate, cellular phosphorus uptake, temperature sensitivity, threshold dose of exposure to saturating light intensity, motility, minimum sinking rate and susceptibility to grazing) of these classes, but the growth and recruitment of a particular strategist will be preferentially promoted in function of specific environmental circumstances. These species are then more likely to build up the largest fraction of sustainable biomass (i.e. become dominant) for as long as the same circumstances persist.

Environmental variables that are supposed to influence the selection among the primary strategists (C, S or R) may be grouped as physical factors (e.g. temperature, stability of the water column, absolute mixed depth, turbidity of mixed layer), chemical factors (notably nutrient concentration) and biotic factors (grazers filtering rate). These may be referred to as externally imposed (allogenic) perturbations. On the other hand, self-imposed (autogenic) perturbations play a substantial role in the definition and selection among the strategists, like euphotic depth (depth of light penetration), reduction due to high biomass concentration, nutrients concentration decline, or increased rate of zooplankton grazing.

Extreme eutrophication expressed in a range of unusual physio-chemical conditions has been proven advantageous to the physiologically and ecologically adaptive cyanobacteria. Ideally, physical and to a lesser extent chemical stability must accompany such extremes. The more a species is versatile in adapting itself to a wider set of survival associated factors, the more abundant, and the longer lasting is its population during the yearly cycle [5]. The cyanobacteria have proven to be highly versatile and adaptable, able to survive and to dominate in conditions highly unfavourable for other species. This versatility makes cyanobacteria able to survive and dominate in water bodies during protracted periods of the annual plankton cycle. In some relatively large and shallow lakes, e.g. the Hartbeespoort Dam and the Vaalkop Dam in South Africa, cyanobacteria compose the most significant part of the population throughout the year; this situation probably reflects a final eutrophication stage of a lake, considering the comparatively constant climatic and nutrient conditions [6].

Carbon, nitrogen and phosphorus are the essential nutrients for cyanobacteria. The nutrient uptake pattern influences the population structure in the water body. It has been noticed that high phosphorus levels generally result in cyanobacterial dominance, though blooms also occurred in low or undetectable phosphate concentrations [7]. The latter is probably caused by fast and almost complete uptake of phosphate by blooming species, leaving the water with low phosphate concentrations for a period of time. A large diversity in response towards phosphorus deficiency has been found, with little or no growth rate dependence on external (i.e. soluble reactive phosphorus, or total dissolved) phosphorus. The rate of phosphate uptake is optimal between pH 7.5 and 8.5 and declines sharply below neutral pH; it is larger in light than in dark conditions.

Most cyanobacteria are N_2 fixing organisms [7]. Fixation of dissolved atmospheric N_2 fixation is carried out by unicellular organisms, by filamentous, heterocystic forms (such as *Anabaena* and *Nostoc*) and by filamentous strains lacking heterocysts (heterocysts and akinetes emerge in nutrient exhausted surroundings, akinetes having a preservatory role and being able to germinate in better nutrient conditions).

The role of carbon in the nutrition and life cycle of cyanobacteria is still not sufficiently clarified, but together with nitrogen it is readily available from the atmosphere, unlike phosphorus. The deficiency in nitrogen and phosphorus is known to lower the respiration rate, and causes accumulation of carbohydrates during the light period of the day, preparing cells for survival in the dark [7].

Cyanobacteria depend on efficient photosynthesis. Like other taxa, they show phototactic/phobic and chemotactic/phobic responses, indicating their affinity to light and limnological conditions. Cyanobacteria, however, are versatile and adaptive, which allows them to out compete other algae in unfavourable environments with regard to nutrients and light. They can adapt to the light-dark regime in a way that gives them the possibility to maintain a higher photosynthetic activity during short photoperiods, so that they may survive for prolonged periods in the dark. They can also adapt to irradiance intensity, whereas a planktonic species from a temperate lake may be killed by prolonged exposure to irradiance exceeding a certain high intensity. A characteristic adaptive mechanism adjusts their chlorophyll content to the light's wave length, to maximize light-efficiency [7].

Cyanobacteria also show motility within the water column. Two types of movement are exerted by cyanobacteria when searching favourable living conditions [7]: phototactic gliding of cyanobacteria attached to surfaces (of suspended sand, clay particles, etc.), and light regulated buoyancy. The rate of displacement, and the frequency of path reversals are related to e.g. the vertical movement in a water body, and are influenced by temperature, viscosity and chemical factors. On the other hand, factors that change rapidly or a concentration gradient, like light intensity, wind and chemical factors, influence the direction of the movement.

Finally, circumstantial evidence exists that some cyanobacteria secrete organic substances of which some are inhibitory or toxic to other organisms, including fish, waterfowl domestic livestock, and humans [3]. An assessment from almost 300 sites all over the world, in the period 1981-1989, shows that the likelihood of an individual bloom being toxic is 45-75 %. Usually, *Microcystis, Anabaena, Oscillatoria* and *Aphanizomenon*, and less often *Gomphosphaeria, Coelosphaerium, Nodularia, Nostoc* and *Cylindrospermum*, were involved [8]. Research conducted on toxic algae blooms in the UK in 1989 indicated associated death of dogs and sheep, and illness in people. The toxicity was attributed to microcystin poisoning, originating from *M. aeruginosa*. Such blooms and associated toxicity were also recorded in Norway, Denmark, Sweden, USSR, Finland [8], Australia [9] and Brazil [10]. In Finland, cattle poisoning attributed to cyanobacterial blooms took place at three sites in 1985 and 1986, and similar blooms occurred in drinking water sources as well. The isolation of strains of cyanobacteria confirmed the presence of hepatotoxic and neurotoxic strains of *Anabaena*, as well as hepatotoxic strains of *Microcystis* and *Oscillatoria* [11]. The National Rivers Authority in UK reported in 1990 about the 1989 cattle kill and human infestations. 68% of the 78 water bodies tested for toxicity were found positive, which was related mostly to the presence of *M. aeruginosa*, though other cyanobacteria were present as well, notably *Aphanizomenon flos-aquae, Anabaena*, and *Oscillatoria* [6]. In 1990 the incidence of *Anabaena* spp. bloom in a recreational lake in the Netherlands was found to have caused summer flu in humans. Toxicity analysis of water containing *Microcystis* spp. blooms and originating from the Andels Meuse and Braassemer Lake in the Netherlands showed the presence of hepatotoxic microcystin toxin, This led to the conclusion that consumption of these toxins via drinking water is possible [12]. Further research was advocated on the influence of environmental factors on the characteristics and levels of cyanobacterial toxicity, as well as on the stability and persistence of the toxins involved.

No consensus exist in literature on the question whether cyanobacterial toxins are relevant from the point of view of water treatment and distribution. However, it is suggested that the public health aspect should receive more attention. Distinction is to be made between physical contact with surface water containing cyanobacterial toxins, consumption of such water, and consumption of water that contained cyanobacteria before treatment. Similarly, it still is uncertain whether the toxic substances, rarely identified, are secreted by viable algae, or whether they are produced during putrescence of algal matter [3]. Significant levels of toxins found during water treatment may be provoked by cell lysis due to for example oxidants application.

Controversy also exists regarding the nature of the human infestations caused by ingestion of cyanobacterial toxins. Some species of cyanobacteria have been identified as causing gastroenteritis, however, only when consumed in large numbers [13]. High cyanobacteria concentrations in the raw water supply have been found to cause an epidemic of pyrogenic reaction among kidney patients consuming such water [14]. Lysis of cyanobacteria, when an algicide or oxidant disinfectant is applied, may result in harmful concentrations of toxins in domestic water supplies causing liver injury, gastroenteritis, or hepatoenteritis [15]. Others on the other hand, state that conclusive evidence exists that cyanobacterial toxins are a health hazard in drinking water [4].

However, not all cyanobacteria species are toxic, and differences exist between strains of the same species. The recognition of different environmental conditions (nutrient composition, light availability, presence of competitor algae species, etc.) and the governing mechanisms which may result in differing toxicity of the same strain at different locations, has not been addressed so far and deserves further research attention.

In conclusion, it appears that the excretion of toxins is a widely recognized problem, and that high cyanobacteria concentrations warrant much caution. It is therefore recommended that a careful approach is adopted to the application of certain chemicals such as oxidants (e.g. ozone or $KMnO_4$) as particle (algae) conditioners due to the potential danger of toxins leakage from ruptured algae cells. The fate of toxins along the treatment line, especially the final product toxin levels deserves research attention. The issue of long term toxins consumption and health implications has not been addressed so far and also attracts attention. The recommended approach is to face and assess the potential toxicological problems regardless of the high cost and long duration of such analysis.

1.1.4 *Microcystis* and *Oscillatoria* spp.

Whilst cyanobacteria species are generally recognized as K-strategists, species differ with regard to their R or S character. The most widely spread cyanobacteria are *Microcystis* spp., which are characteristic for eutrophied lakes, and a typical S-species, able to dominate periods of late summer (high temperatures), and severe nutrient depletion. Their physiological and morphological characteristics make them a particular nuisance in water treatment. On the other hand, in the Netherlands concern grows about the incidence of the *Oscillatoria* spp. in some water resources. *Oscillatoria* spp. are known to possess an effective light-capturing mechanism across a wide band of the visible spectrum, as well as a buoyancy regulating mechanism. These make for a pronounced R-species, allowing the species to inhabit constantly mixed lake environments [5]. There is obvious differentiation between these two cyanobacterial genera, with respect to their shape, size, extra cellular organic matter (EOM) characteristics, nutrient uptake, buoyancy regulating mechanism, etc. On the other hand, they are both identified as cause of numerous problems in water treatment [4, 6, 8, 12, 16, 17, 18]. Therefore, they are of particular further interest to this study.

Microcystis spp. are encountered worldwide. They are spherical in shape, of 3-7 μm in size, and can occur in irregularly shaped colonies of often more thousands cells, as well as single cells. There are indications that the single cell form is an adaptation of this species developing at later stages of their life cycle (i.e. later during the season), due to the dissociation of colonies in order to provide better light availability for all cells. These species favour warmer water (>14 °C) and stable, relatively low nutrient availability, combined with stable hydraulic conditions [3, 16].

Oscillatoria spp. are characterized by an oscillating movement and lack of definite sheath. Their cells exceed 100 μm in length and are 3-5 μm in diameter; trichomes are solitary or intermingled, forming non-parallel bundle-like structures. Generally, they inhabit temperate lakes of two distinct types : (i) mildly eutrophied, large, deep, and usually alpine lakes, and (ii) shallow, enriched, unstratified basins, where they form stable maximum concentrations in the summer metalimnion, provided that this is located within the euphotic zone. They possess an extremely efficient light capturing mechanism, functional across a wide band of the visible spectrum, favouring low irradiance and achieving maximum growth rates at relatively low nutrient availability [3].

Cyanobacteria in general, and especially *Microcystis* spp., are also able to survive in conditions unfavourable for other species. For example, *Microcystis* spp. can overcome O_2 toxicity. With the high light irradiance of late summer, and high O_2 concentrations and low CO_2 concentrations, O_2^- (superoxide) is produced that damages cytochromes, pigments and other segments of the algal photosynthetic system. To prevent the toxicity of O_2^- the enzyme superoxide dimutase (SOD) is manufactured in algal cells. However, prolonged high light intensity triggers a drop in SOD level and die-off of algae. This phenomenon has been observed with most algal species, except for *Microcystis* spp. which are able to maintain high levels of SOD and survive such conditions [7].

Characteristic for *Microcystis* spp. and *Oscillatoria* spp. is the existence of gas vacuoles [7]. Up to 10,000 of these organelles are present in a single cell. Generally they have the shape of a hollow cylinder of 70 nm diameter with a conical cap at each end. They are composed of protein, the wall being permeable to dissolved gases in surrounding liquid, but unpermeable to liquid (the pressure inside being usually atmospheric). Nutrients as well as Cl^- and K^+ uptake are known to play a major role in rise or fall of turgor (difference in hydrostatic pressure between the inside and outside of the cell) and consequent buoyancy. This enables the cells to attain an as good as possible position in a water body (mostly in the vertical sense) with regard to light and nutrients [7].

Other characteristics of the genera establishing their dominance are : allelopathy and light attenuation [7]. Allelopathy involves chemical inhibition of other organisms by the excretion of inhibitory organic compounds. Attenuation of light occurs as the cyanobacteria can increase their biomass very fast, thus increasing turbidity and limiting the growth of competitors. Cyanobacteria are vertically motile during diurnal or higher order nutrient fluctuations, favouring them as inhabitants of water bodies during the late summer and autumn. In addition *Oscillatoria* spp. are able to maintain approximately neutral buoyancy, and cannot be ingested by grazing zooplankton because of their length of >100 μm, ensuring low loss rates [3].

However, characteristic differences exist between the two genera with respect to nutrient affinity and uptake. *Oscillatoria* spp. have been reported to fix nitrogen [3]. *Microcystis* spp. in contrast, have been reported as non-nitrogen-fixing species [7]. Low N/P ratios can be expected to favour N_2 fixing *Oscillatoria* spp., whilst the opposite would favour the growth of *Microcystis* spp.

The toxins related to cyanobacteria blooms may be grouped into three families: neurotoxins, lipopolysacharide endotoxins, and hepatotoxic peptides. Microcystin belongs to the hepatotoxic family and is a potent and specific in vitro inhibitor of phosphatases, with an action similar to that of the tumour promoting toxin of shellfish poisoning, okadaic acid. This particular toxin, consists of polypeptide containing D-serine, L-orithine and some protein L-amino acids (circular peptide) [8, 16].

1.2 PROBLEM ASSESSMENT

Analysis of operational data of five major surface water treatment plants in the Netherlands has been performed, of which a representative cross-section is given in Table 1.1 [19]. These all produce water for different purposes (drinking water, industrial water and dune infiltration water), and face growing problems caused by algae. They are interested in efficient short- and long-term treatment strategies and process modifications to address the algae problem. The enterprises are North Holland Water Supply Companies (PWN), Rhine-Kennemerland Water Transport Company (WRK), Dune Water Supply Company South Holland (DZH), Amsterdam Water Supply (GWA) and Energy and Water Supply Company Rhineland (EWR). At a later stage of the research after the further presented analyses were completed, two more enterprises with similar interests were involved, namely Water Supply Association Overijssel - Regio Oost (WMO, formerly Water Supply Company Oost-Twente - WOT), and Groningen Water Supply Company (GWG) (Fig.1.1).

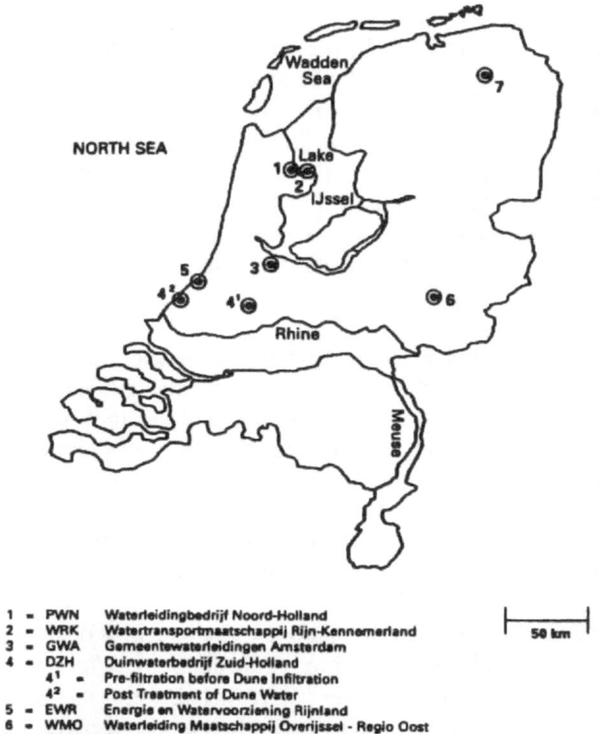

Fig. 1.1 Locations of the water treatment plants of the seven water treatment enterprises.

Although the origin of all concerned surface water is in essence river Rhine or Meuse water, the water quality in the five sources largely varies, because abstraction is almost never practised directly from the rivers. GWA treats water from the Loenderveen Lake, which is regularly replenished with Bethune polder drainage water (in the future to be mixed with Amsterdam-Rhine canal water at a ratio of 1:1). PWN and WRK treat water from the IJssel Lake, which is fed with water from river Rhine via one of its branches, the IJssel river. DZH practices dune infiltration of water abstracted from the Andels Meuse (a dammed arm of river Meuse, practically serving as a reservoir), and EWR of Rhineland polder drainage water. The purpose of their treatment also varies (drinking water, industrial water, water for dune infiltration, etc.) and so does the water treatment technology applied. Table 1.5 summarises the selected water quality management and treatment technology.

Table 1.1 Year-average minimum and maximum water quality of surface water resources, based on average monthly measurements by the water enterprises.

Water Quality Parameter	IJssel Lake (Reservoir) PWN Andijk 1984-89	IJssel Lake (Reservoir) WRK Andijk 1986-92	Loenderveen Lake GWA Amsterdam 1986-92	Andels Meuse DZH Den Haag 1986-92	Polder Water EWR Katwijk 1989-92
Temperature °C	1.2-20.7	1.4-20.3	3.6-22.3	1.8-22.4	1.5-21.6
pH	7.9-9.45	7.7-8.9	7.5-8.4	7.7-8.6	7.6-8.5
PO_4-o mgP/L	N.M.	0.0077-0.095	0-0.014	0.006-0.063	0.124-1.04
PO_4-t mgP/L	0.06-0.76	0.05-0.25	0.007-0.04	0.04-0.2	0.39-1.15
NO_3+NO_2 mgN/L	N.M.	0.57-4.52	0.42-1.44	1.74-5.5	1.1-6.6
NH_4 mgN/L	0.04-0.79	0.04-0.9	0.12-0.89	0.08-0.55	0.17-2.78
SiO_2 mgSi/L	N.M.	0.05-2.8	3.7-13.1	0.25-5.6	1.04-11
Chl *a* µg/L	8.6-198	6.5-101	1.3-8.4	1.6-37	1.5-51
Chl t µg/L	18-238	13-132	2.7-10.3	2.6-58	N.M
Algal Vol. mm^3/L	1.1-411	0.4-97	N.M.	N.M.	N.M.

Chl *a* :Chlorophyll-*a* — PO_4-o : ortho phosphate — N.M.: not measured
Chl t :Chlorophyll-total — PO_4-t : total phosphate

During the collection and processing of the data, it appeared that the degree of attention paid to algae, as demonstrated by the frequency and diversity of algae related measurements, varies between the enterprises, and is proportional to the number of problems experienced in the past. The attention is determined, for example, by whether the end product is drinking, industrial or infiltration water. This inconsistency in the data set is compounded by the absence of a proper algae related standard for product water, and by the absence of a proper measuring technique that is able to include all types of algae in proportion to their actual concentrations.

Statistical analysis of the data related to comparison of the five water resources was performed. The aim of the two way analysis of variance (ANOVA) was to test the similarity between the (year-average) raw water quality of the five water resource localities, and to check whether the raw water quality of each of the water resources has changed within the analysed time period (see Table 1.1). The results show : (i) significant differences between the raw water quality at the five localities (raw water resources), and (ii) no significant changes of the water quality of each of the localities during the analysed periods. Based on a multiple range test, Table 1.2 reflects the similarities for the analysed water quality parameters at different localities.

Table 1.2 Similarity of raw water quality between the five locations. a, b, c, d denote typical water quality categories, based on the two way ANOVA.

Local	°t	pH	PO_4-o	PO_4-t	NO_{2+3}	NH_4	SiO_2	Chl-*a*	Chl-t
1	ab	c	N.M.	c	N.M.	a	N.M.	c	c
2	a	a	a	a	a	a	a	a	a
3	a	b	a	b	b	a	a	b	b
4	bc	a	b	d	b	b	c	ab	N.M.
5	c	ab	a	b	b	a	b	a	a

1 - PWN reservoir 3 - DZH Andels Meuse 5 -GWA Loenderveen Lake
2 - WRK III reservoir 4 - EWR polder water

N.M.: no measurements available for the locality
Chl *a* : Chlorophyll-*a*
Chl t : Chlorophyll-total
NO_{2+3} : nitrites +nitrates

Multiple correlation analysis of the variables (analysed raw water quality parameters) for all the localities (raw water resources) shows high correlation coefficients between chlorophyll-tot on one hand and pH, ortho-PO_4, total PO_4, NO_2 +NO_3 , and SiO_2 on the other, and low correlation coefficients between chlorophyll-tot and °t and NH_4 (Table 1.3). The lack of data for SiO_2 for the polder water may explain to some extent the high negative correlation coefficient obtained.

Similar correlations were obtained for chlorophyll-*a*. Based on the maximum correlation coefficients, simple and multiple linear regression analyses were carried out, showing highest r^2values for simple linear regression between log(chlorophyll-tot) and log(PO_4-t) [r^2=0.87], and for multiple linear regression between log(chlorophyll-*a*), log(PO_4-t) and pH [r^2=0.80]. Fig. 1.2 represents a plot of the observed versus the predicted log(chlorophyll-*a*) values in function of the phosphate concentration and pH. The transformation equation predicting the (chlorophyll-*a*) value on a year-average maximum basis is : pr.log(chlorophyll-*a*)=0.52 log(PO_4-t) + 0.6(pH) - 3.2. The highest r^2 values for PO_4-tot suggest that phosphorus is the key element which controls algae growth in surface water impoundments, consistent with the conclusions of Vollenweider [20] and also elaborated by many authors like Sakamoto [21], Dillon and Rigler [22], and Tillman et al. [23], for water impoundments under a range of conditions.

Table 1.3 Correlation coefficients (*r*) and significance levels (*P*) for correlation of log (chlorophyll-tot) with log values for different variables (except for pH which is a log value itself). Number of data n=15.

Coef.	°t	pH	PO_4 - o	PO_4 - t	NO_{2+3}	NH_4	SiO_2
r	0.11	0.76	0.96	0.93	0.81	-0.09	-0.8
P	0.7	0.0011	0.0000	0.0000	0.0002	0.75	0.0003

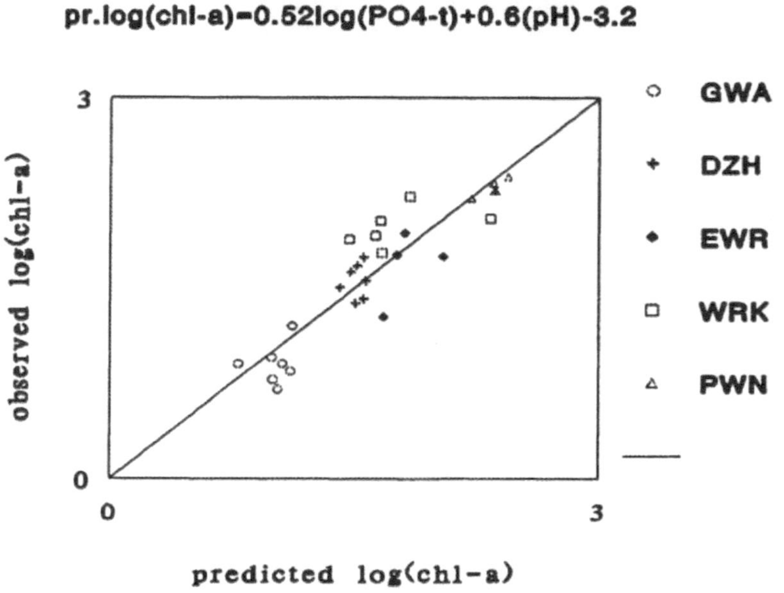

Fig. 1.2 Observed versus predicted log(chlorophyll-*a*) values, r^2=0.8.

The reservoirs' chlorophyll and phosphorus data were superimposed on a graph by Harris [24] (based on results for a combination of lakes analysed by Janus and Vollenweider, and a number of oligotrophic Canadian lakes); the results fall in the 95% confidence limit of the regression line (Fig. 1.3).

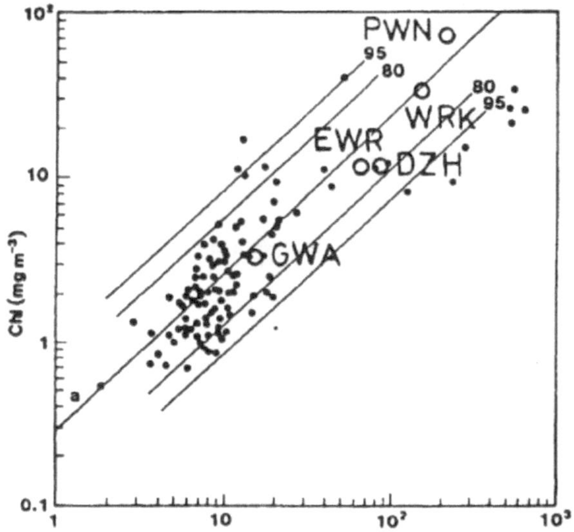

Fig. 1.3 Annual mean chlorophyll vs. P concentrations for the five impoundments, superimposed on a graph from Harris [23].

The close relation between chlorophyll-*a* and phosphorus, emerging from the multiple regression analysis, suggests causality and is again in accordance with Vollenweiders results. However, the causal relation between chlorophyll-*a* and pH is more complex. Although high correlation coefficients do not automatically imply the presence of causal relations, the causal relation between chlorophyll-*a* and pH may be supported by the CO_2/pH relation hypothesis. Indeed, the relatively high pH of the five impoundments (pH 7.5-9.45 on year average basis) favours the growth of cyanobacteria at high eutrophication levels [25], as recorded in our case [19]. In addition it may be hypothesized that a feedback mechanism exists between chlorophyll, PO_4 and pH, in the sense that resuspension of phosphorus from reservoir sediments occurs under the recorded high pH conditions, which supports further algae growth [26, 27].

The intensive photosynthetic activity of high biomass concentrations further shifts pH to high values, possibly to the point where C becomes a limiting factor. As cyanobacteria have lower saturation constants K_m for uptake of total carbon and CO_2, this condition combined with the high pH promotes cyanobacteria dominance [25].

Summaries of the observations on the most significant species encountered in the five impoundments describing their characteristics and treatability, are presented in Table 1.4. The treatability of the algae species, i.e. the extent of their removal in water treatment, is mainly affected by (i) the treatment technology applied, (ii) optimization of the applied treatment technology, and (iii) characteristics of the algal species involved. Although the choice of the treatment technology rarely is made on the basis of efficient algae removal alone, the occurrence of the periodical algae blooms plays an important role. Under normal circumstances (moderate algae concentrations) algae can be removed more or less efficiently by commonly applied treatment technologies.

However, this technology is not necessarily optimal for the removal of all the algae species concurrently or separately (at other time periods) present in the water. The algae characteristics and properties influence their removal, including size, shape, cell surface characteristics (composition of algal cell wall and resultant colloidal surface charge characteristics), excretion of EOM and motility. Many of these characteristics vary with algae age, as well as within subspecies, limiting the removal predictability. In this light, the terms 'Problematic', 'Good', etc. treatability of Table 1.4 are only indicative. They are used to characterize the general degree of problems related to a particular algal species removal, the effect on effluent water quality, and the water production costs.

Each of the studied water impoundments is characterised by seasonal (yearly) variations. Algal blooms are common during certain periods of the year, but they are proportional to the degree of eutrophication and most acute in the cases of WRK (Figs. 1.4 and 1.5) and PWN treating IJssel Lake water. A pronounced seasonal occurrence of cyanobacteria (especially *M. aeruginosa*) is found in all five water resources, again showing proportionality between concentration and the degree of eutrophication. This was found to be most typical in Loenderveen Lake. While diatoms are characterised as superior competitors under phosphorus limitation (high N/P ratio) and water temperature below 14°C, green algae and cyanobacteria are dominant under nitrogen limitation (low N/P and Si/P ratios) and high temperatures [23].

Table 1.4 **The most significant algae species encountered in the five impoundments, indicating their basic characteristics and treatability.**

Algae species and taxonomy Location/Treatment plant	Figurative representation	Characteristics	Treatability
Melosira spp. (Diatomeae) 1. 4. 5.		Porous, capsule like, cylindrical cells forming filaments; blooms; L=10-40 μm; W=5-20μm.	Problematic
Stephanodiscus spp. (Diatomeae) 1. 2. 4.		Single, porous cells, spines at edges; in hard waters or basic lakes; L=8-20 μm.	Problematic
Asterionella spp. (Diatomeae) 1. 3.	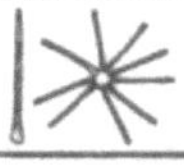	Often very abundant, spoke like arrangement of rectangular frustules; in hard waters; L=40-130 μm; W=2 μm.	Moderately problematic
Nitzschia spp. (Diatomeae) 3.		Commonly solitary frustules, tapered at ends, straight or sigmoid, porous; L=25-110 μm; W=5-10 μm.	Moderately problematic
μ-algae (Diatomeae) 3.		Small capsule like cylindrical cells, very much like *Stephanodiscus hantzscshii*.	Problematic
Microcystis spp. (Cyanophyta) 1. 2. 3. 4. 5.	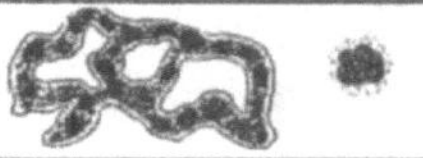	Irregularly shaped colonies of thousands of sphere cells, also single cells; gas vacuoles (N_2);extensive blooms-surface mat; D=3-7μm.	Problematic
Oscillatoria spp. (Cyanophyta) 1. 2. 3. 4.		Cells forming filaments - solitary or intermingled, cells lack definite sheath; blooms;L=2-4 μm, (L_t=100 μm); W=3-5μm.	Problematic
Aphanizomenon spp. (Cyanophyta) 4.		Cells forming filamentous bundles, heterocysts present; blooms; L=5-15 μm, (L_t=150 μm); W=5-6 μm.	Problematic
Anabaena spp. (Cyanophyta) 3.		Cells forming filaments, heterocysts present; upper layers; blooms - surface mats; L=6-8 μm, (L_t=30-35 μm); W=6 μm.	Moderately problematic
Scenedesmus spp. (Chlorophyta) 4.	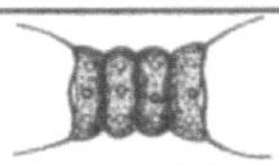	4, 8 or 12 oval or fusiform cells forming colonies, 1 or 2 spines at end cells; L=10-15 μm	Moderately good
Coelastrum spp. (Chlorophyta) 1. 4.	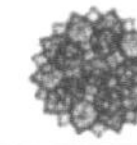	Spherical or polygonal cells forming hollow colonies, by protuberances from mucilaginous sheaths; D=8-12 μm.	Moderately good
Oocystis spp. (Chlorophyta) 1.		Elliptic or lemon shaped, one or more generations of mother cell walls enclose daughter cells; L=14-26 μm; W=10-20 μm.	Good
Cryptophyceae spp. 1. 3. 5.		Solitary cells, rarely colonial, protozoan like, 2 flagella, chloroplast pigments - brown, blue or red.	Moderately problematic
Chrysophyceae spp. 3. 5.		Unicellular or colonial, rarely filamentous, chloroplast pigments - yellow, brown, golden brown, thick wall, 1 or 2 flagella, motile.	Moderately problematic

1.- PWN reservoir **2.- EWR polder water** **3.- GWA Loenderveen Lake;**
4.- WRK III reservoir **5.- Andels Meuse.**
L - cell length **L_t - total length of filament** **W - cell width** **D - cell diameter**

Comment : The morphology of different algal species asserts the need of different algae counting procedures. *Microcystis aeruginosa* is counted as single cells and as colonies. Bundles of filamentous species most often disintegrate during the preparation of the sample; an attempt is made to count all the algae that may form such bundle structures. *Scenedesmus* spp. of 8 or 12 cells are counted as 2 or 3 *Scenedesmus* entities. Similarly other disintegrated colonial species are counted as separate entities.

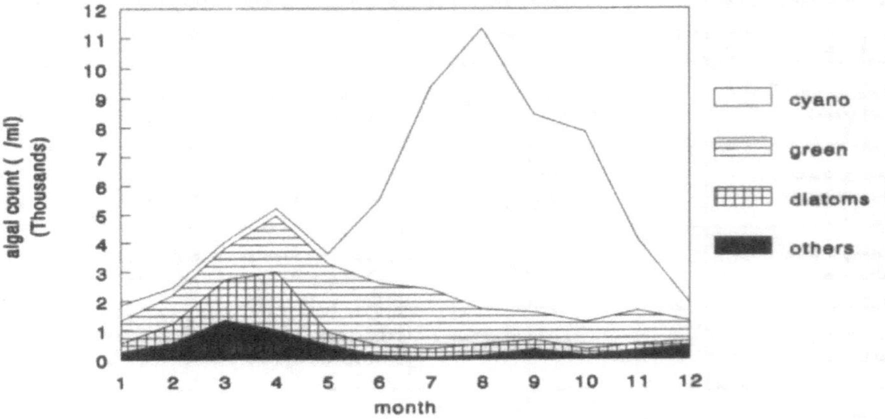

Fig. 1.4 Average monthly algal count for the WRK III reservoir (integrated over the total reservoir depth of 20 m), for different algal taxa in the period 1988-92.

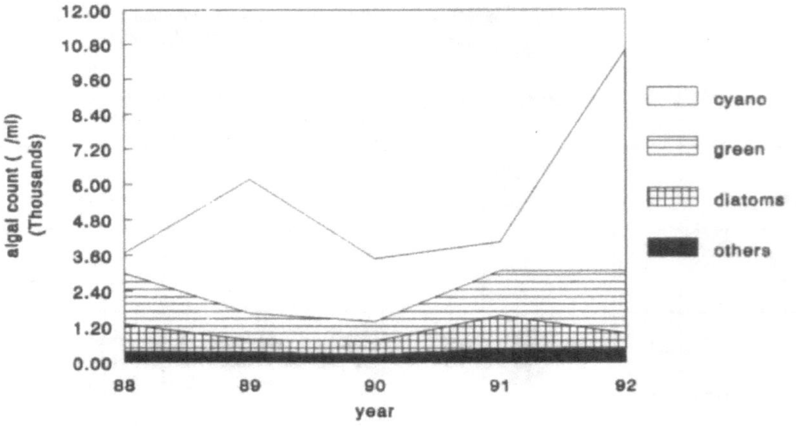

Fig. 1.5 Average yearly algae count for different taxa in the WRK III reservoir for the period 1988-92.

Reservoir management involving chemicals application (affecting nutrients availability, pH, etc.) and water mixing within the reservoir, as well as climatic conditions, influence the phytoplankton composition as for example in the PWN and WRK reservoirs where higher concentrations of *Oscillatoria* spp. were found in the period of 1990-92. This is consistent with other experiences under Dutch climatic circumstances [25]. This suggests that the current raw water quality criteria are insufficient to avoid prolonged cyanobacteria blooms. Cyanobacterial dominance refers to a high biomass concentration, which is correlated (indirectly) to the nutrient level. This is the case in particular with *Oscillatoria* spp., while for *Microcystis* spp. it is the limiting nutrient that most likely will determine the magnitude of its bloom.

1.3 DISCUSSION

1.3.1 Algae and water treatment - problems and remedies

1.3.1.a Reservoir management

To cope with algae efficiently requires a combination of detailed water quality monitoring, reservoir water quality management, and water treatment measures. Appropriate and especially well timed microbiological analysis (i.e. done immediately after sampling and not postponed for weeks, which often is the practice), accompanied with other physio-chemical analysis provides a better tool for prediction of algae concentration and species variations. Although treatment plant operators are found to be increasingly aware of the importance of the prediction capacity, it often is neglected or insufficiently addressed. Delays between sampling and analysis typically result in poor treatment process operation. On the other hand, it is generally very effective to reduce the algal load ahead of the treatment through reservoir water quality management and pretreatment as it appears that no treatment step on its own is able to ensure adequate removal rates. A range of reservoir management techniques are applied in this context with varying degrees of success: phosphorus removal (either by precipitation with coagulant, or by diverting phosphorus contributing flows), artificial destratification by mixing of the water column, use of algicides (mainly copper sulphate), sonic disruption of algal cells, suction dredging, and aeration of the hypolimnetic layer [4]. Reservoir management significantly diminishes the algae load, but the appropriate approach is reservoir and algae specific.

The successful application of better reservoir design is the case presented in Fig. 1.6. The use of the shallow 3 m deep PWN reservoir was temporarily supplemented by the 20 m deep WRK III reservoir, resulting in drastic reduction of the exposure to sun-light and subsequent poor algae growth. Examples of successful pre-treatment include the coagulation with ferrous sulphate and settling of water entering the water reservoir of GWA, the application of ferrous sulphate in Andels Meuse by DZH and in the WRK III reservoir (which also includes periodical mixing of the water column). On the other hand, such measures have also been reported to yield adverse effects. In the Wahnbachtalsperrenverband in Siegburg, Germany, phosphorus reduction in the raw water reservoirs led to a rapid increase in diatom level at the expense of other algae, and resulted in poorer overall treatment performance during periods of their blooms.

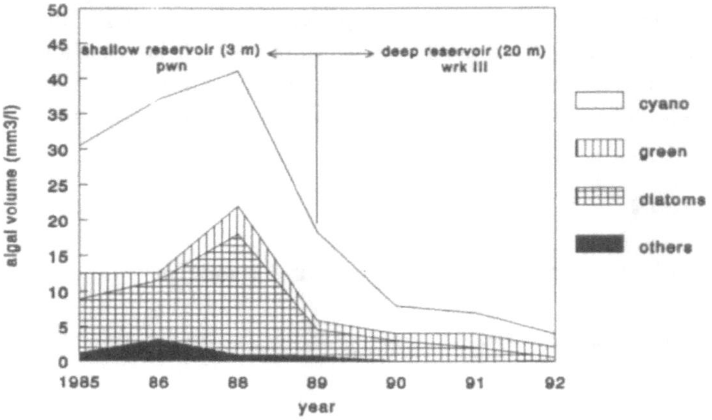

Fig. 1.6 Average algal volume for different taxa in the reservoirs of PWN (3 m deep) and WRK III (20 m deep), (1987 data unavailable).

1.3.1.b Treatment process

Cyanobacteria and in particular *Microcystis* spp. and *Oscillatoria* spp. are known to pass readily through treatment and end up in drinking water in relatively large numbers. A critical feature of the *Microcystis* spp. is their ability to form colonies, as well as to exist in a single cell form, whilst *Oscillatoria* spp. can exist as solitary or intermingled filaments. Small algae pass through filters more readily than larger ones so that when blooms of small algae, such as *Microcystis* spp. occur, their counts in treated water are high especially towards the end of the bloom period when colonies disintegrate naturally, or as a consequence of algicides. Up to 5.3×10^4 cells/mL of the former and 4.5×10^4 cells/mL of the latter species have been recorded by the Severn Trent Water Authority in 1985 and 1986 in the United Kingdom, for filtered water [4]. *Microcystis* spp. cells can rupture and leak intra-cellular substances during turbulent aqueduct transport from the reservoir to the treatment site (rupture rate between 61-72%) [28]. Similar rupture and colony break-up may be expected during initial pumping stages in water treatment plants, representative for Dutch operation circumstances. In other words, water treatment plant design may in itself aggravate the algae related problems.

In general, the treatment approach in the studied plants is complex and aims at the overall removal optimization of different pollutants, including suspended solids, dissolved and colloidal matter, organic micropollutants, heavy metals, algae, etc.; this varies depending on the raw water source as well as the final product user (Table 1.5). Coagulation and flocculation, and rapid sand filtration are applied in all plants, while granular activated carbon filtration is used in three cases (GWA, WRK and PWN). Dissolved air flotation is applied at DZH and EWR, while sedimentation is practised at WRK III and PWN. Slow sand filtration is practised only at the GWA.

Table 1.5 The treatment processes in the five studied treatment plants.

MANAGEMENT &TREATMENT TECHNOLOGY APPLIED	GWA 30Mm³/y Weesperkarspel	WRK III 110 Mm³/y Prinses Juliana	PWN 35 Mm³/y Andijk	DZH 70 Mm³/y Bergambacht + Scheveningen	EWR 15 Mm³/y Lindenbergh
pre-coagulation	###	###		###	
pre-settling	###			###	
reservoir	###	###	###	###	
reservoir mixing		###			
pre-filtration	###			###	
straining 35μm			###		
sieves 200μm		###			
pre-chlorination		#P#	###		
transport Cl_2				#P#	
ozone	###				
softening	###				
coagulation/ flocculation	ooo	###	###	###	###
sedimentation	ooo	###	###		
flotation (DAF)				###	###
rapid sand filters	ooo	###	###	###	###
activated carbon	#o#	###	###		
slow sand filters	###				
straining 30μm			###		
post-chlorination	#P#				
ClO_2 disinfection			###		

- PRESENTLY IN OPERATION ooo - PREVIOUSLY IN OPERATION
#P# - PERIODICALLY APPLIED #o# - REPLACED WITH (IN OPERATION NOW)

Algal blooms and typically those of cyanobacteria, pose the most serious treatment problems. These include : (i) interference with the coagulation and flocculation process, (ii) increased disinfectant demand, (iii) trihalomethane production, (iv) frequent filter blocking, and (v) increased use of backwash water (Table 1.6). This translates into higher production cost. Process modifications to address these problems can also be expensive or have side effects. As no algae standards or objectives exist, water companies tend to deal with the problem on the basis of 'as much as possible' removal, within the technological context and production cost bracket that is considered commonly acceptable. This approach is usually indiscriminatory and poorly focused. Local guide levels exist such as those at the Wahnbachtalsperrenverband in Siegburg, Germany [29], (0.1 μg/L chlorophyll-*a* in treated water).

Table 1.6 Present and anticipated situation in the five studied Dutch water enterprises. Most significant alga species encountered, related treatment and distribution problems, and remedies (illustrations from [30], except for GWA photo material of μ-algae deposits on GAC filter).

Company (% prod. of total surf.water in Nl)	Raw water source, treatment plant and process	Characteristic algal species in water resources	Present situation with respect to raw water source and treatment	Anticipated development
PWN (8 %)	IJssel Lake-Andijk -Straining 35 μm -Cl_2- 6 g/m^3 -$FeClSO_4$- 16 g/m^3 -$Ca(OH)_2$ -30.5 g/m^3 -Polyelect. -0.5 g/m^3 -SED, v=3.5 m/h -RSF, v=10 m/h -GAC, v=13.5 m/h -Straining 35 μm -ClO_2	*Microcystis* spp. (Cyanophyta) *Oscillatoria* spp. *Melosira* spp. (Diatomeae) *Stephanodiscus* spp. *Asterionella* spp. *Coelastrum* spp.(Chlorophyta) *Oocystis* spp. *Cryptophyceae* spp. 	Serious current operational difficulties: ■ 100% coagulant and coagulant aid dose increase (up to 40mg/l Fe^{3+} and 1.0 mg/l polyelectrolyte) ■ trihalomethane formation ■ increased backwash water quantity ■ algae passing through treatment	■ In long term only partial improvement of present eutrophication of raw water resource expected ■ Partial improvement of situation due to deepening of reservoir before treatment
WRK (25 %)	IJssel Lake-Princess Juliana, Andijk -Screens 200 μm -$Fe_2(SO_4)_3$ - 20 g/m^3 -Polyelect.-0.2 g/m^3 -$Ca(OH)_2$ -SED (lamella) -RSF, v=10m/h -GAC (partial)	*Microcystis* spp. (Cyanophyta) *Oscillatoria* spp. *Aphanizomenon* spp. *Melosira* spp. (Diatomeae) *Stephanodiscus* spp. *Coelastrum* spp.(Chlorophyta) *Scenedesmus* spp. 	Serious current operational difficulties: ■ 100% coagulant and coagulant aid dose increase (up to 40 g/m^3 Fe^{3+} and 1.0 g/m^3 polyelectrolyte) ■ trihalomethanes formation ■ increased backwash water quantity ■ algae passing through treatment ■ increased MFI and AOC values and possible infiltration wells clogging	■ In long term only partial improvement of present eutrophication of raw water resource expected ■ Partial improvement of situation due to mixing and coagulant application in reservoir
GWA (7 %)	Loenderveen Lake-Weesperkarspel -Ozone - 2-5 g/m^3 -Softening -$FeCl_3$ (1992) -Polyelect.-(1992) -SED(lamella, 1992) -PAC (1992) -RSF (1992) -GAC -SSF	*Microcystis* spp. (Cyanophyta) *Oscillatoria* spp. *Anabaena* spp. *Nitzschia* spp. (Diatomeae) *Asterionella* spp. μ - algae *Cryptophyceae* spp. *Chrysophyceae* spp. 	Operational difficulties: ■ surface clogging of GAC filters by phyto- and zooplankton cake layer	■ New surface water source of inferior quality (contributing to eutrophication) to be used to double production capacity ■ Reduction of algae load over GAC filters due to: -direct filtration mode of filtration plant at intake (1-3 g Fe^{3+}/m^3) -reduced ozone doses at beginning of process
DZH (16 %)	Andels Meuse-Bergambacht + Scheveningen -RSF -PAC- 2-4 g/m^3 or -$Al_2(SO_4)_3$ -4-6 g/m^3 -DAF, t=7-14 min -RSF, v=10 m/h	*Microcystis* spp. (Cyanophyta) *Melosira* spp. (Diatomeae) *Cryptophyceae* spp. *Chrysophyceae* spp. 	Operational difficulties: ■ algae and cellular products passing through treatment ■ increased AOC values, resulting in reduced pipe flow due to increased k-values of pipes	■Unchanged situation as long as application of phosphate precipitation in Andels Meuse is allowed ■New direct filtration plant for algae removal in Brakel under investigation
EWR (3 %)	Rhineland polder water- Lindenbergh -$FeClSO_4$ - 8 g/m^3 -DAF, t=15 min -RSF, v=5 m/h	*Microcystis* spp. (Cyanophyta) *Oscillatoria* spp. *Stephanodiscus* spp. (Diatomeae) 	Product water quality problems: ■ algae passing through treatment in large quantities, especially during periodical blooms	■ Complete change of present raw water source with one of better quality, possibly Andels Meuse

DAF - dissolved air flotation; GAC - granular activated carbon PAC - powder activated carbon

Nevertheless, their implementation on a wider scale and at different circumstances is not always feasible. It has been argued that the discussed chlorophyll-*a* guide level is too stringent in view of the very low assimilable organic carbon (AOC) value associated with it [31]. Namely, the proposed 0.1 μg/L chlorophyll-*a* guideline value should imply 3 μg/L AOC, which is lower than the already ambitious recommended 10 μg/L AOC required for biologically stable water (proposed by van der Kooij [32]).

Research in the past two decades was mostly directed at optimising conventional treatment processes, commonly centred around sedimentation and filtration. Agglomeration (coagulation/flocculation) has received the largest attention, as it is the prerequisite for any efficient particles (and algae) removal regardless of the subsequent down-stream removal process (sedimentation, flotation or filtration). Efficient water conditioning with the purpose of improving the coagulation/flocculation process was approached in a similar manner. However, it was soon realised that it is typically impossible to achieve adequate algae removal when they are present in concentrations that are typical for some eutrophic reservoirs (periodically up to 10^5 - 10^6 cells/mL). Obtained removal rates strongly depend on prevailing plankton type and range from 50-99.9%. However, even a rate of 99.9% removal can still be inadequate for raw water of such poor quality [29].

1.3.2 Dissolved air flotation (DAF) - a viable solution option

Because of the natural tendency of algae to float dissolved air flotation (DAF) attracts attention as a (pre)treatment option prior to final filtration. Because DAF requires a preceding particle agglomeration (coagulation/flocculation) step, it performs equally well as common flocculation/sedimentation with respect to DOC, UV, true colour and trihalomethane precursor removal [33, 34]. On the other hand DAF tends to be more efficient than sedimentation for removal of turbidity and particulates formed by coagulation [35]. Experience from newly built flotation facilities for drinking water treatment (e.g. Birmingham Frankley, UK; Millwood New Castle, USA; Pietarsaari, Finland), and from adapted existing plants, where e.g. sedimentation was replaced by flotation followed by rapid sand filtration document this [36, 37]. It is generally confirmed that algal removal efficiency can be substantially enhanced by flocculation and flotation upstream of the final filtration. The same applies for algae conditioning, i.e. bringing the algae into a state which results in their more efficient down-stream process removal, by application of oxidants, e.g. ozone [38], or $KMnO_4$ [39].

Janssens [40] has proposed a tentative diagram to be used for process selection based on raw water turbidity and algal content (Fig.1.7). If all other quality requirements are fulfilled regarding DOC, hardness, colour, etc., flotation is the best available (pre)treatment technology for treatment of raw water with chlorophyll-*a* values of typically >10 μg/L. This is the case in all of the five studied companies for shorter or longer periods of the year (Table 1.1). In addition, under the Dutch circumstances the occurrence of *M. aeruginosa* (Table 1.4) favours use of DAF, because this species is characterised by buoyancy [3], which renders them more susceptible to removal by the rising air bubbles in DAF.

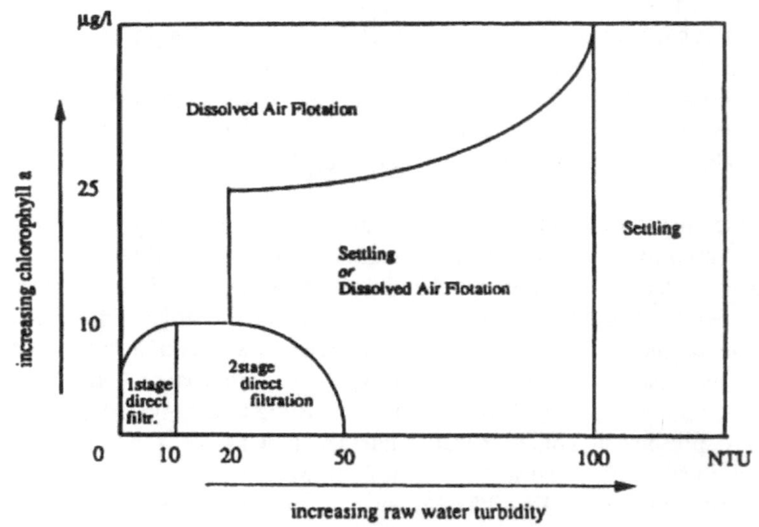

Fig. 1.7 Typical selection diagram for a water treatment process based on turbidity and chlorophyll-*a* concentration (from Janssens [40]).

Although this diagram can be very useful, it does not consider a range of factors which may be of substantial impact on the choice of process technology. Such factors include algae morphology and physiology (different species exhibit different response to treatment) [41], the source and character of turbidity, temperature, colloidal and particulate matter concentration, and possibly other factors as well. Furthermore, the choice of technology is largely a matter of cost-benefit analysis, suggesting that this diagram can be considered only as indicative.

A typical layout of a DAF treatment plant is presented in Fig. 1.8. It consists of a tank in which two zones are distinguished : (i) the contact zone, and (ii) the separation zone.

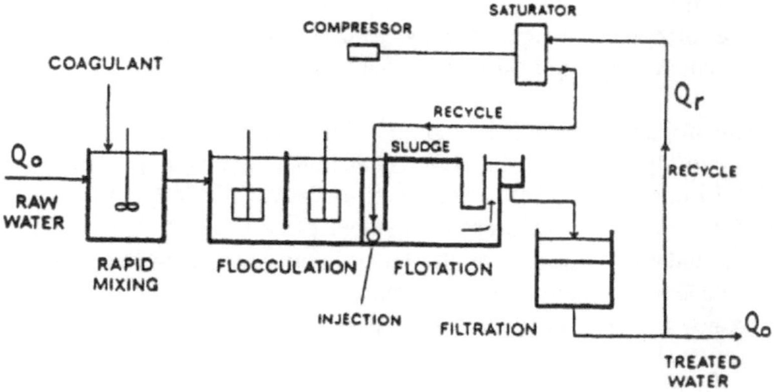

Fig. 1.8 Schematic presentation of a DAF treatment plant [42].

In DAF, air is introduced into the contact zone via a pressurised (approximately 500 kPa) recycled stream of treated water (5-10% of total flow). The change of pressure from e.g. 500 kPa to atmospheric results in the production of numerous microscopic (10-100 μm) air bubbles which attach to the particles (flocs). The commonly large ratio between air bubble and particle concentration of 10^2-10^3, encountered in normal surface water treatment, is suggested responsible for the high DAF efficiency [42]. The particle-bubble agglomerates rise to the surface in the separation zone, where they are easily skimmed off as a light foam.

The following advantages of DAF are proposed [43]:

-small surface needed for unit and equipment;
-low civil engineering cost requirements;
-typical reduction of 10-20% in chemicals consumption as compared to conventional sedimentation;
-high dry solids content of the sludge (4-6%), compared to sedimentation (2-4%);
-lower water losses (1%, compared to 3% for sedimentation) due to production of sludge with a higher solids content;
-less affected by flow variations;
-suitable for intermittent use, reaching steady quality after approximately 60 min;
-operates in cold climates with minimum flotation time less than 10 min, and less sensitive to temperature variations than sedimentation;
-high algae removal efficiency.

Disadvantages include higher operational costs and the requirement to have skilled operating personnel. Nevertheless, for treating water with high algal count and relatively low turbidity DAF appears to be the most cost-effective technology. Water with high turbidity, on the other hand, is also treatable to a certain limit and with appropriate process modifications.

1.3.3 Agglomeration (coagulation/flocculation) of algae

The single collector collision efficiency DAF kinetic model proposed by Malley and Edzwald [45] recognises two sets of variables that define process efficiency :

-pre-treatment variables ahead of the flotation tank, which influence the particle-bubble attachment efficiency, the particle concentration and the particle-bubble collision efficiency;

-design and operation variables within the flotation tank, which influence the bubble size and concentration.

The agglomeration (coagulation/flocculation) step is a critical part of the DAF; it therefore deserves to receive priority in comprehensive research approach. Such comprehensive optimisation should take into account the down-stream filtration. Neither conventional sedimentation, nor advanced DAF are able to remove on their own the number of algae encountered in eutrophied surface water. Model calculations suggest that the particle destabilization degree is more important than the floc size [42]. pH and coagulant dose that produce floc particles with little or no charge, generally result in favourable flotation conditions. Particles (algae) must be destabilized for good particle attachment to bubbles, as algae stability is caused mainly by : (i) the electrostatic repulsive force due to the negative surface charge (at pH 2.5-11.5), and (ii) the steric hindrance due to the water layer associated with hydrophilic surface and the adsorbed EOM at the cell surface [42]. Bernhardt and Clasen [45] conclude that algal cells can be considered to behave as inert particles and that the classical coagulation theory therefore is applicable; algae could be destabilized, aggregated and thus brought into filterable form, by adsorption coagulation or sweep coagulation. This can be achieved by charge neutralization by means of adsorption of positively charged organic or inorganic polyelectrolytes, or by metal hydrolysates when sweep coagulation is applicable. Agglomeration conditions that produce particles with little or no charge generally result in efficient DAF [42]. The selection of inorganic or organic coagulants as sole coagulants or coagulant aids, pH, and other coagulation variables therefore need further study.

The kinetic variables (G and t values) for efficient mixing and coagulation, as well as flocculation (i.e. creating particle collision opportunities) similarly need to be optimised. Unlike mixing and coagulation which take place almost instantly after coagulant addition (floc formation starts within 10^{-2} s up to a few seconds), the floc formation requires lower mixing energy input (low G) during longer time periods (5-20 min), enabling particles to collide. In practice, floc formation and floc break-up occur simultaneously, making application of mechanistic models complicated. However, floc size and strength determine overall efficiency in the floc removal step. While efficient sedimentation requires larger, heavier and easily settleable flocs, the floc size of 10 μm to few tens of micrometer has been shown to be optimal for flotation [42, 44]; similar to filtration, flotation efficiency worsens for particle size of 1 μm and is one to two orders of magnitude lower for particles < 1 μm. Floc growth spurs initially by the presence of small but numerous primary particles, and thereafter it is determined by the collision of 'micro-aggregates'. In case of adsorption coagulation, as floc size increases, the number of polyelectrolyte bridges per particle decreases proportionally to floc size and the floc becomes weaker and more susceptible to shear by hydrodynamic forces.

Algal cell aggregates in particular appear to be susceptible to shear and hence turbulence. A shear -free transport of flocculated algae to further treatment is required. A critical location in DAF is the injection of the recirculated water through the injection nozzles or needle valves. Depending on the configuration of the DAF unit, flocs may be locally subjected to high shear. This suggests that longer flocculation times to produce larger flocs are questionable, if flocs break up anyway near the nozzles. Some authors [35] conclude that the flocculation time therefore does not affect flotation performance (based on turbidity measurements); good results can be obtained without flocculation or slow mixing, with flocs formation limited to the rapid mixing period of about 2 minutes. Others [46] suggest minimum flocculation time of 5-6 minutes. On the other hand, flocculation time e.g. at the Scheveningen flotation plant of DZH is in the range of 20-40 min, while in the Lindenbergh flotation plant of EWR it is 25 min.

Optimal agglomeration finally depends very much on the dominant algal species. Cyanobacteria appear in different sizes, shapes and forms, and differ in their EOM composition, which is known to influence the agglomeration process to a great extent. It can be expected that different species and seasonal conditions require different sets of optimal agglomeration conditions for subsequent flotation. This calls for further study of the influence of morphological and physiological characteristics of alga species on the agglomeration and down-stream removal efficiency.

Particle (floc) characterisation such as automated particle count or computer image analysis becomes an essential tool in the assessment of the coagulation/flocculation and DAF process assessment.

1.3.4 Conditioning of algae

Recent advances in DAF operation pertain to the decrease of the flocculation and flotation time, and other process improvements related to the critical agglomeration (coagulation/flocculation) step. In addition, it was found that oxidants may under certain conditions (depending on raw water quality, coagulant dosage, location of dosing, etc.) improve coagulation and flocculation and hence down-stream floc removal efficiency. The positive experiences, especially with ozone [38, 47, 48] and $KMnO_4$ [39] in direct filtration, warrant further research on these oxidants prior to aggregation and DAF.

Although their use is not yet widespread, the application of oxidants as flocculation aid is becoming a viable option. In this context, the term conditioning seems appropriate to express the effect of oxidants on the algae structure and the resulting flocculation benefits. Namely, in addition to disinfection, secondary benefits arise from their effect on the surface chemistry of colloidal and suspended particles. As a consequence ozone can cause microflocculation (spontaneous flocculation), and hence coagulant dose reduction, improved water quality (measured as turbidity), and reduction in filter bed head-loss [47]. The mechanism of the induced flocculation is not yet fully understood; several mechanisms are proposed by Langlais et al. [49]: (i) reactions with algae, (ii) reduction of particle stability due to loss of adsorbed organic matter from natural particles, (iii) polymerization of natural organic matter (NOM), and (iv) break-up of metal-organic complexes yielding in situ iron production of metal coagulant.

The commonly recommended oxidant in DAF has been ozone, whether as a classical conditioner or for ozoflotation. Ozone reacts with the algal outer cells layer (EOM) and the algal cells themselves; the extent of oxidation depends on the algal species, making its effect seasonally dependent [38, 47]. Similarly, overdosing can break-up colonies, and damage cell structure. As one consequence, DOC and AOC levels may increase, as well as the toxins concentration [50, 51]. The optimization of the ozone dose for efficient aggregation and flotation, and the understanding of the process kinetics require further study. As an alternative $KMnO_4$ is not commonly applied as a disinfectant and little is known about its optimal use. Its efficiency as a conditioner is related to its oxidative activity and the coagulating effect of the hydrous manganese dioxide floc which is formed when the permanganate is reduced [39, 52].

Physical conditioning commonly consists of straining and its efficiency depends on algal size and shape. Microstraining is widely applied with mesh size usually in the range of 35 μm. Mesh sizes as low as 5 μm are commercially available, their application depends largely on local raw water plankton and suspended matter concentration and composition. The experience in the five studied Dutch enterprises shows that microstrainers can substantially reduce the algal loads on the subsequent treatment step [19]. However, algae blooms often create problems in the operation of the microstrainers resulting in deteriorating effluent quality. Single cells of *Microcystis* spp. and *Oscillatoria* spp., although with distinctly different morphological features, have both been noted to pass through this treatment as in case of the WRK III and PWN treatment plant, causing increased algae counts in the final effluent. The current understanding of the role of this treatment step in the optimisation of the overall treatment scheme can benefit from further pilot studies.

Finally, the use of other physical conditioners such as UV light has not been considered to our knowledge. Its applicability as an efficient disinfectant is yet gaining increasing recognition [53]. UV light is not considered to be an efficient algicide as high dosages (~5000 J/m^2) and long exposure times (several minutes) are probably necessary. Yet, its effect on the viability of algae and hence on aggregation and down-stream flotation, should position it as a potentially interesting research topic.

1.4 CONCLUSIONS

The problems related to the periodical high concentrations of algae, are assessed for five Dutch surface water treatment locations. Cyanobacteria, especially *M. aeruginosa* and *O. agardhii* are recognised as the most significant problem causing algae. Examples of efficient reservoir management including deep reservoir operation (WRK III and PWN since recently), coagulation/flocculation and settling within the reservoir (WRK III and DZH), periodical destratification mixing (WRK III), phosphate removal by coagulation/flocculation and settling prior to the reservoir (GWA), can substantially diminish the algae load on subsequent treatment. Water treatment plant managers and operators should carefully consider such positive experiences and investigate their applicability and feasibility in their circumstances. Nevertheless, high algae concentrations, and especially blooms, can still cause numerous treatment problems. This forces water companies to adapt short-term and long-term process modifications, which raises production costs. This raises the issue of further cost-benefit algae treatment optimisation, as well as the viability and applicability of advanced treatment technologies. Increase of the knowledge basis on these issues should improve treatment process operation and raise the confidence in the application of new more efficient technologies. Existing algae related treatment problems and recent research results suggest that there is still a large operational window for increasing the efficiency of existing treatment capacities, based on the cost-benefit principle.

Considering the natural tendency of algae to float, and the positive experiences elsewhere, dissolved air flotation emerges as an attractive treatment technology, as alternative to conventional sedimentation. It features a high rate, high efficiency, low chemicals consumption, and high versatility and adaptability. On the other hand, discrepancies exist between literature and practice, which suggests that optimisation would benefit from DAF further research. The fact that in the nineties many treatment plants in Western Europe will require an increase in production capacity (in our study GWA, DZH and GWG) due to increased water consumption, or to upgrade the process efficiency due to more stringent water quality regulations, compound this opinion. DAF is also an attractive option to be considered as pre-treatment in combination with emerging membrane technology in case of high algal loads.

Research in the Netherlands should preferably focus on the process optimization of a comprehensive DAF-based treatment scheme for the removal of cyanobacteria species, notably *Microcystis* spp. and *Oscillatoria* spp. An integrated approach is preferred for this purpose, which includes the up-stream agglomeration (coagulation/flocculation) of particles, as well as down-stream filtration. The emphasis is to be placed on the agglomeration phase which is considered critical, whether of particles (algae) in the coagulation/flocculation stage or of particles (algae) and bubbles in the DAF stage. The effects of pre-treatment and conditioning steps on overall process efficiency should be assessed. Positive and negative side effects of the treatment must be assessed, especially with regard to the level of DOC and its AOC fraction in the effluent, because of the possibility of cell disruption by the oxidants. The same applies to the toxin levels in the effluent, which are considered a second priority, but are more difficult to quantify. Finally, the influence of algal morphology and physiology on the process mechanisms need to be investigated.

The evaluation of the process efficiency should be related to the kinetic parameters and rely on more comprehensive particle characterisation. This should include comparison of particle (floc) size frequency and volume distribution for raw water, for water subject to aggregation (coagulation/flocculation), and for treated water. Bubble size characterisation (size distribution) and particle charge related phenomena are also to be considered. This would enable characterisation of process efficiency in a qualitative and quantitative manner, and further clarify important aspects of process kinetics.

REFERENCES

1. VEWIN Tienjarenplan-hoofdrapport, 1989. Ten year plan - main report (published by the Dutch Water Supply Companies' Association).
2. Watson A. M., 1990. *The filtration of algae* - Ph.D. thesis, University College, London.
3. Reynolds C.S., 1984. *The ecology of freshwater phytoplankton*, Cambridge University Press, Cambridge, UK.
4. Hutson R.A., Leadbeater B.S.C. and Sedgwick R.W., 1987. Algal interference with water treatment processes. *Progress in Phycological Research*, **5**, pp. 266-297, (edited by Round/Chapman), Biopress Ltd.
5. Sandgren C.D. (editor), 1988. *Growth and reproductive strategies of freshwater phytoplankton*, Cambridge University Press, Cambridge, UK, pp. 261-316.
6. National Rivers Authority (NRA) report, 1990. Toxic blue green algae, *Water Quality Series*, UK.
7. Carr N.G. and Whitton B.A. (editors), 1982. *The biology of cyanobacteria - Botanical Monographs*, Vol. 19, Blackwell Scientific Publications, UK.
8. Lawton L. and Codd G.A., August 1991. Cyanobacterial (blue-green algal) toxins and their significance in UK and European waters - *J. IWEM.*, pp. 460-465.

9. Hallegraeff G.M., 1991. Massive bloom of toxic blue-green algae in Australian rivers. pp.7.
10. Di Bernardo L., 1995. *Algas e suas influências na qualidade das águas e nas tecnologias de tratamento (Algae and their influence on water quality and treatment technology)*. Ediçâo Patrocinada pela ABES Associaçâo Brassileira de Enhenharia Sanitária e Ambiental, Brasil.
11. Sivonen K., Niemela S.I., Niemi R.M., Lepisto L., Luoma T.H. and Rasanen L.A., 1990. Toxic cyanobacteria (blue-green algae) in Finnish fresh and coastal waters, *Hydrobiologia* **190**, pp. 267-275.
12. Hoekstra A.C., Bol J. and Seinen W., 1991. Blauwwieren en hun toxinen in voor drinkwaterproduktie gebruikt oppervlaktewater (Blue-greens and their toxins in surface water used for drinking water production), *H_2O*, **14**, 91, pp. 387-393.
13. Speedy R.R., Fisher N.B. and McDonald D.B., June 1969. Algal removal in unit processes - *J. AWWA*, **61**, pp. 289-292.
14. Kay P., Sykora J.L. and Burgess R.A., March 1980. Algal concentration as a quality parameter of finished drinking waters in and around Pittsburgh, Pa. - *J. AWWA.*, **72**, pp. 170-176.
15. Falconer I.R., Runnegar M.T.C., Buckley T., Huyn V.L., and Bradshaw P., 1989. Using activated carbon to remove toxicity from drinking water containing cyanobacterial blooms -*J. AWWA*, **81**, pp. 102-105.
16. Codd G.A. and Bell S.G., 1985. Eutrophication and toxic cyanobacteria in freshwaters - *Wat. Poll. Control.*, pp. 225-231.
17. Sugiura N., Inamori Y. and Sudo R., 1990. Degradation of blue-green alga, *Microcystis aeruginosa* by flagellata, *Monas guttula - Environmental Technology*, **11**, pp. 739-746.
18. Vlaški A. and A.N.van Breemen, 1992. A preliminary investigation of the membrane filtration index and characterization of particles, of water produced at WRK III treatment plant, (internal research report), TU-Delft, the Netherlands.
19. Vlaški, A., van Breemen, A.N., Alaerts, G.J., 1994. *Algae and Water Treatment in the Netherlands, Problem Analysis and Treatment Strategies in Five Water Companies.* International Institute for Infrastructure, Hydraulic and Environmental Engineering -Delft (IHE), Working Paper EE - 1, Delft, the Netherlands.
20. Vollenweider, R.A. Scientific fundamentals of the eutrophication of lakes and flowing waters, with particular reference to nitrogen and phosphorus as factors in eutrophication. *OECD, report DAS/CSI/68.27*, Paris, 1971.
21. Sakamoto M., 1966. Primary production by phytoplankton community in some Japanese lakes and its dependence on lake depth. *Arch. Hydrobiol.*, **62**, pp. 1-28.
22. Dillon P.J. and Rigler F.H., 1974. The phosphorus-chlorophyll relationship in lakes. *Limn. and oceanography*, **19**, 5, pp. 767-773.
23. Tillman D., Castling R., Sterner R., Killam S.S. and Johnson FA, 1986. Green, bluegreen and diatom algae: Taxonomic differences in competitive ability for phosphorus, silicon and nitrogen. *Arch. Hydrobiol.*, **106**, 4, pp. 473-485.
24. Harris G.P., 1986. *Phytoplankton ecology : structure, function and fluctuations.* Chapman and Hall Ltd., London ,UK.
25. Schreurs H., 1989. *Cyanobacterial dominance, relations to eutrophication and lake morphology*. Ph.D. thesis University of Utrecht, the Netherlands.
26. Mc Dougal B.K. and Ho G.E., 1991. A study of the eutrophication of North Lake, Western Australia, *Wat. Sci. Techn.*, **23**, pp. 163-173.

27. Seitzinger S.P., 1991. The effect of pH on the release of phosphorus from Potomac estuary sediments: Implications for blue-green algal blooms, *Estuarine Coastal and Shelf Science*, **33**, pp. 409-418.
28. Dickens C.W.S. and Graham P.M., 1995. The rupture of algae during abstraction from a reservoir and the effects on water quality. *J. Water SRT-Aqua*, **44**, 1, pp.29-37.
29. Bernhardt H. and Clasen J., 1991. Flocculation of micro-organisms. *J.Water SRT-Aqua*, **40**, 2, pp. 76-87.
30 Streble H. and Krauter D., 1973. *Das leben im Wassertropfen*, Kosmos Gesellschaft der Naturfreunde Franckh'sche Verlagshandlung, Stuttgart, Germany.
31 Oskam G. and van Genderen J., 1995. Eutrophication and development of algae in surface water - a threat for the future ? *Special subject 8, IWSA - Congress*, Durban, South Africa.
32 Van der Kooij D., 1990. Assimilable organic carbon (AOC) in drinking water. In : *Drinking Water Microbiology*, editor G.A. McFeters, Springer-Verlag, pp. 57-87.
33. Gehr R., Swartz C. and Offringa G., 1993. Removal of trihalomethane precursors from eutrophic water by dissolved air flotation. *Wat. Res.*, **27**, 1, pp. 41-49.
34. Malley J.P., 1990. Removal of organic halide precursors by dissolved air flotation vs. conventional water treatment. *Environmental Technology*, **11**, pp.1161- 1168.
35. Malley J.P. and Edzwald J.K., 1991. Laboratory comparison of DAF with conventional treatment. *J. AWWA*, **83**, pp. 56-61.
36. Roux le J.D., 1988. *The treatment of odorous algae-laden water by dissolved air flotation and powdered activated carbon*, National Institute for Water Research, CSIR, South Africa, 1988, pp. 72-79.
37. Arnold, R.S. and Harvey P., 1994. Recent applications of DAF pilot studies and full scale design, *IAWQ-IWSA-AWWA Joint Specialised Conference on Flotation Processes in Water and Sludge Treatment*, Orlando, USA.
38. Petruševski B., van Breemen A. N., van Aelst A.C. and Alaerts G.J., 1994. Pre-ozonation : key for efficient particle and algae removal in direct filtration, *Proceedings IOA Regional Conference*, Zurich, Switzerland.
39. Petruševski B., van Breemen A.N. and Alaerts G.J., 1995. Effect of permanganate pre-treatment and coagulation with dual coagulants on algae removal in direct filtration, *IAWQ/IWSA Workshop on Removal of Microorganisms From Water and Wastewater*, Amsterdam, the Netherlands.
40. Janssens J.G., 1992. Design concepts and process selection for particle removal in surface water treatment. *Lecture notes - Advanced design concepts for integral drinking water treatment* IHE-Delft, the Netherlands.
41. Petruševski B., Vlaški A., van Breemen A.N. and Alaerts G.J., 1993. Influence of algae species and cultivation conditions on algal removal in direct filtration. *Wat. Sci. Tech.*, **27**, 11, pp.221-220.
42. Edzwald J.K. and Wingler B.J., 1990. Chemical and physical aspects of dissolved-air flotation for the removal of algae. *J.Water SRT- Aqua*, **39**: 24-35.
43. Zabel T. , 1985. The advantages of dissolved-air flotation for water treatment. *J. AWWA*, **77**, pp. 42-46.
44. Malley J.P. and Edzwald J.K., 1991. Concepts for dissolved-air flotation treatment of drinking waters. *J. Water SRT-Aqua*, **40**, 1, pp. 7-17.

45. Bernhardt H. and Clasen J., 1992. Studies on removal of planktonic algae by flocculation and filtration *Water Malaysia '92, 8th ASPAC - IWSA Regional Water Supply Conference & Exhibition*, Technical papers, Volume 2.

46. Janssens J.G., 1990. The application of dissolved air flotation in drinking water production, in particular for removing algae. *DVGW Wasserfachlichen Aussprachetagung* Essen - Federal Republic of Germany.

47. Edzwald J.K. and Paralkar A., 1992. Algae, coagulation, and ozonation. *Chemical Water and wastewater treatment II*. Klute R. and Hahn H.(Eds.), Springer-Verlag, Germany, pp. 263-279.

48. Bourbigot M-M, Martin N., Faivre M., Le Corre K. and Quennell S., 1991. Efficiency of an ozoflotation-filtration process for the treatment of the River Thames at Walton Works. *J.Water SRT-Aqua*, **40**, 2, pp 88-96.

49. Langlais B., Reckhow D.A. and Brink D.R. (eds.), 1991. Ozone in water treatment application and engineering. *AWWARF Cooperative report*, Lewis Publ., Chelsea, Mich.,USA, pp. 190-213.

50. Falconer I.R., Runnegar M.T.C., Buckley T., Huyn V.L., and Bradshaw P., 1989. Using activated carbon to remove toxicity from drinking water containing cyanobacterial blooms. *J. AWWA*, **81**, pp. 102-105.

51. Sukenik A., Teltch B., Wachs A.W., Shelef G., Nir I. and Levanon D., 1987. Effect of oxidants on microalgal flocculation - *Wat. Res.*, **21**, 5, pp. 533-539.

52. Cleasby J.L., Baumann E.R. and Black C.D., 1964. Effectiveness of potassium permanganate for disinfection - *J. AWWA*, **56**, pp. 466-474.

53. Kruithof J.C. and Leer van den R.C., 1992. Practical experiences with UV-disinfection in The Netherlands. *J. Water SRT-Aqua*, **41**, pp. 88-94.

Chapter 2

RESEARCH OBJECTIVES, HYPOTHESES AND METHODOLOGY

2.1 Research scope and relevance

This research was initiated and sponsored by five Dutch water enterprises sharing the common problem of periodical algae blooms in their raw water resources - reservoirs :

-North Holland Water Supply Companies (Waterleidingbedrijf Noord-Holland-PWN),
-Rhine-Kennemerland Water Transport Company (Watertransportmaatschappij Rijn-Kennemerland- WRK),
-Amsterdam Water Supply (Gemeentewaterleiding Amsterdam-GWA),
-Dune Water Supply Company South Holland (Duinwaterbedrijf Zuid-Holland-DZH), and
-Energy and Water Supply Company Rhineland (Energie- en Watervoorziening Rijnland-EWR).

During the research, the last two companies fused while the following two other Dutch enterprises with similar interests joined the research. The research was also joined and supported by Kiwa Research and Consultancy (Kiwa Onderzoek en Advies).

-Water Supply Association Overijssel - Regio Oost (Waterleiding Maatschappij Overijssel - Regio Oost-WMO, formerly Waterleidingbedrijf Oost-Twente-WOT) ,
-Groningen Water Supply Company (Gemeente Waterbedrijf Groningen-GWG).

The research started as a joint venture project between these companies, the Technical University Delft and the International Institute for Infrastructure, Hydraulic and Environmental Engineering, Delft, the latter taking the final responsibility for carrying out the research. The research progress was guided by staff members of the participating companies and academic/scientific institutions, through an Advisory Committee.

The first chapter studied the algae related problems and the treatment practice of the participating companies. This chapter gives an analysis of eutrophication and algae presence in the utilised water resources, it specifies water treatment problems, and it discusses short- or long-term treatment strategies. It also offers a short overview of current experiences from elsewhere with respect to coping with the algae problem in a cost effective manner, which served as a rationale for outlining the experimental part of the research.

Currently 32% of the total water production in the Netherlands originates from surface water bodies, each having a different degree of existing and forecasted eutrophication and algae problems. In all five studied cases periods of algal blooms were common. Some of the companies face the problem of highly eutrophied raw water resources (PWN, WRK), characterised by the tendency of phytoplankton population shift towards cyanobacteria species, which are most often responsible for seasonal blooms and particular treatment problems. Similarly, for the future increase of production capacity the GWA considers using a new surface water source (Amsterdam-Rhine canal water mixed with Bethune polder water at a 1:1 ratio) with higher eutrophication levels than the one currently utilized. Production capacity expansion is also planned by DZH, by increasing the abstraction from the currently utilised eutrophic source (pretreated Andels Meuse water). In case of the EWR, a complete change of the raw water source has occurred (River Meuse), caused partly by eutrophication and algae problems.

Although not a preferred alternative, surface water is expected to receive more attention in the

Netherlands in the future, due to restrictions on ground water use. This will carry the burden of coping with periodical algal blooms. Although water quality management and pollution control have considerably reduced eutrophication on most locations, more efforts are required especially in multinational rivers like the Rhine and the Meuse, and in lowering agricultural and diffuse pollution which are now the main causes of eutrophication. Being complex and costly, these protection programmes are still very difficult to apply in less developed countries, leaving them with growing eutrophication and water treatment problems. Despite all efforts, a moderate but visible degree of eutrophication is to remain with us.

The recent availability of efficient membrane technologies offers new avenues for treatment (PWN, GWA and GWG consider them for their extension programmes). Although successful ultrafiltration experiments without pretreatment (or including microstraining only) have been conducted in the Netherlands, one can predict that the impact of pretreatment, especially in case of high algal concentrations, will increase the productivity and the life time of the membranes. In the short term, existing constraints in conventional treatment call for treatment solutions such as:

- solving problems within the existing treatment technology structure (e.g. adjust coagulant dose or coagulation pH on a short term basis),
- modifying the existing technology structure (e.g. introduce an oxidant as a conditioner),
- retrofitting with more efficient treatment technology (e.g. an option to be considered is to replace existing sedimentation with DAF within the existing construction parameters),
- applying the most cost-efficient treatment technology for newly considered production capacities.

The natural tendency of algae to float, and several positive experiences in the Netherlands and elsewhere, make the dissolved air flotation (DAF) potentially attractive as an alternative to conventional sedimentation. DAF is nowadays practised at DZH, EWR and WMO (formerly WOT), while (lamella) sedimentation is practised by WRK and PWN. Consequently, the comparison of sedimentation and DAF, was identified as a priority research topic for the participating enterprises. The critical importance of the coagulation/flocculation process, as well as of down-stream filtration, calls for a comprehensive approach.

The application of oxidants (ozone and $KMnO_4$) as chemical conditioners is considered as a particularly important factor for improving the DAF process efficiency. The application of ozone and $KMnO_4$ is to be optimised and critically assessed in terms of their by-products. Similarly , the application of polyelectrolytes of different origin (natural or synthetic) and charge (cationic, anionic and non-ionic) will be assessed in the context of enhancing DAF efficiency. An attempt to qualify the involved process kinetic mechanisms will be made. This part of the research will provide deeper insight and knowledge over potentially valuable tools for improved algae removal within existing production capacity as well as for future expansion of existing plants.

To summarize, this research is intended to broaden and improve the understanding of conventional and advanced water treatment technologies, in the context of existing and expected algae, especially cyanobacteria related problems. Our previous analysis in Chapter 1 showed that these are the species which deserve our attention in terms of expected further eutrophication, as well as the amount of existing problems in treatment. It is intended to arm the participating companies with a firmer knowledge basis of their production capacities and provide tools for short and long term algae problem solutions. It is expected to encourage their consideration and application of potentially beneficial chemical conditioning techniques of algae. Finally, it is to potentially contribute to the development of a rationale that would help define and quantify the 'algae problem', and help setting-up a standard in future.

2.2 Research objectives

Based on the previous discussion and chapter-wise supported rationale, the objectives of the research are :

❒ Optimize the process of agglomeration (coagulation/flocculation) of cyanobacteria (in particular *Microcystis* spp.) by metal coagulants (with or without the addition of organic or inorganic coagulant aids), under different coagulation and flocculation modes (G and t values), as a function of efficient water treatment centered on DAF, and to compare this with a treatment centered on sedimentation.

❒ Identify cyanobacteria (*Microcystis* spp.) characteristics that are process relevant , e.g. size, shape (form), cell surface properties, surface charge, EOM presence, etc.

❒ Assess and evaluate the benefits and the draw-backs of polyelectrolytes and oxidants, notably ozone and $KMnO_4$ for improved algae agglomeration (coagulation/flocculation) and down-stream DAF.

❒ Evaluate and contribute to the existing knowledge on DAF (and sedimentation) process kinetics; identify and assess qualitatively the influence of essential process parameters, the effect of coagulation/flocculation conditions on particle size and volume distribution, particle charge, and on resultant particle-bubble interaction. Similarly assess the impact of different process conditions on bubble characteristics and on resultant particle-bubble interaction.

❒ Evaluate and contribute to the existing knowledge on ozone and $KMnO_4$ algae conditioning process kinetics; identify and assess qualitatively the influence of essential process and water quality parameters, notably the effect of oxidant dose, organic matter concentration and composition, particle charge, etc., and their effect on resultant DAF efficiency.

2.3 Research hypotheses

The following hypotheses were developed to guide the experimental work and its design :

❒ Knowledge on the algal biology and ecology is necessary to understand and improve the removal process optimisation. The single celled *M. aeruginosa* is a suitable species representative of Dutch surface water circumstances, which ranks among the most problematic in water treatment. Its size, shape and cell surface characteristics make it an ideal surrogate for simulating undesired algae bloom conditions.

❒ The natural characteristic of algae to float suggests dissolved air flotation (DAF) could be an equally or more efficient particle (algae) removal technology when compared to sedimentation.

❒ The agglomeration (coagulation/flocculation) of particles (algae) is the main prerequisite for their efficient removal regardless of, but influenced by the down-stream solid-liquid separation process.

❒ Oxidants (ozone, $KMnO_4$) can significantly affect agglomeration (coagulation/flocculation) of algae and hence improve or deteriorate down-stream DAF removal efficiency (in terms of particles, turbidity, residual coagulant, etc.). The optimisation of this process is seasonally dependent driven by changes of species and their concentration.

❒ Apart from particle (algae) count, turbidity, residual coagulant, TOC/DOC, UV absorbance and other conventional analyses, the quantitative and especially qualitative characterisation of the effects in different process steps by computer image analysis will be a valuable tool for process kinetics assessment and modelling.

❒ Model water experimental research with laboratory cultured algae spiked into tap or reservoir water must be verified by pilot plant research using reservoir water and algae in their natural circumstances and environment.

2.4 Research methodology

Experimental work will be conducted on two levels :

1. At laboratory bench-scale, with model water, and using a DAF jar test experimental set-up (Fig. 2.1), and

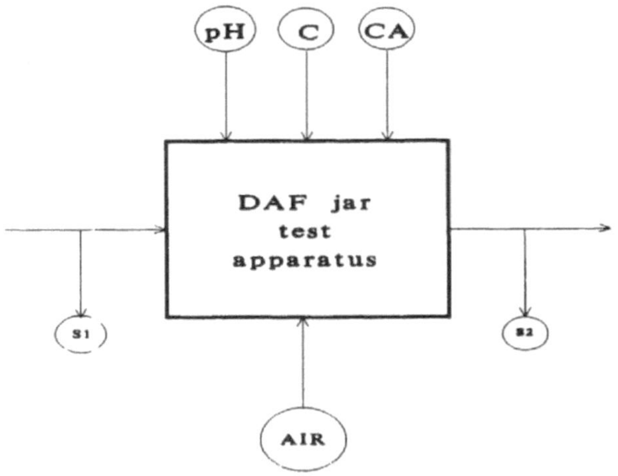

Fig. 2.1 Scheme of the DAF jar test apparatus (6x1.8 L jars). S1 and S2 : sampling points; C : coagulant; CA : coagulant aid; and pH : pH correction (by HCl or NaOH).

2. At pilot plant scale for experimental research with actual reservoir water of one location (Fig. 2.2).

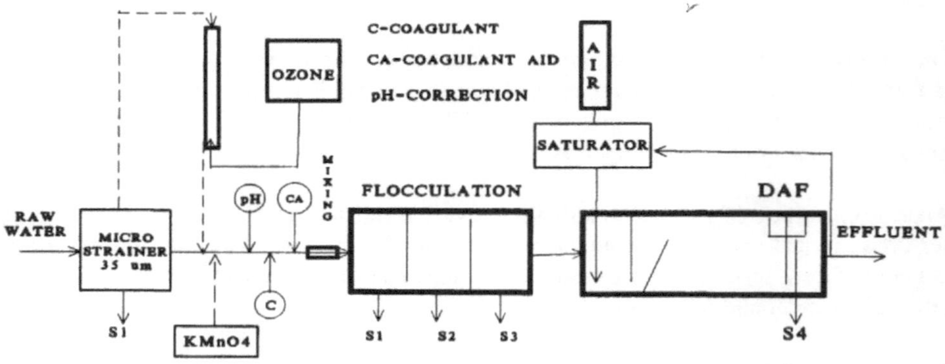

Fig. 2.2 Scheme of the commercial mobile DAF pilot plant (Q_{max}=13 m^3/h).

The model water research will be conducted by spiking semi-continuously, laboratory cultured *M. aeruginosa* to water originating from the Biesbosch reservoirs. Our ability to work with cyanobacteria in natural reservoir water is restricted due to their seasonal occurrence; this defines the period of the year for the pilot plant research from August to October.

The DAF jar test apparatus intended for the laboratory model water research is a standard apparatus equipped with DAF facilities. The pilot plant research is to be conducted with a commercial mobile DAF pilot plant unit with flexible coagulation and flocculation provisions. The research on ozone and $KMnO_4$ application is also to take place at the two levels. For ozone production on both levels a Trailigaz LABO LO, France, ozone generator will be used.

The standard analytical methods that will be applied for the assessment of process efficiency comprise :

- Turbidity (Sigrist L-65, Switzerland),
- Residual coagulant (residual total Fe coagulant measured by atomic absorption spectrometry at 248.3 nm [NEN 6460]),
- Residual manganese (total and dissolved Mn measured by atomic absorption spectrometry at wavelength 279.5 nm for manganese [NEN 6466],
- TOC/DOC,
- UV absorbance at 254 nm, and
- Inverted microscope inspection (M40 Wild Leitz, Switzerland and Nikon Optiphot, Japan).

Measurements and analyses which are not commonly applied in practice, however offering an opportunity for improved process efficiency and kinetics assessment, are also to be applied :

- Particle (algae) count of particle >2.75 μm (HIAC-Royco PC-320, USA, flow-through, light beam scatter-based particle counter),
- Particle count of particles in the size range 0.3-5 μm (Particle Measuring Systems Inc./Liquid Batch Sampler LBS-100 and Microlaser Particle Spectrometer, Boulder, Co., USA),
- Electrophoretic mobility (Tom Lindström AB-Repar apparatus),
- O_3 concentration (iodometric method),
- BrO_3^- concentration (ion-chromatography method),
- SEM - high resolution scanning electron microscopy,
- MFI - modified fouling index (KIWA standard measurements apparatus),
- Computer image analysis (Mini-Magiscan, IAS 25/IV25 Joyce-Loebl Ltd.).

The computer image analysis was used as one of the main tools for qualitative process kinetics assessment. This analytical technique is comprised of a microscope stage guided by a computerised stage controller. It enables computerised registration, measurement and storage of a wide range of viewed object parameters such as its length, width, breadth, surface area, circularity, etc. The image analyses were performed on fresh samples, as well as on photographs of samples. For this purpose, a specially devised flat photo cell was connected at different locations of the jar test apparatus or the pilot plant unit, and samples were photographed or taken before and after flocculation, and after particle removal. Special care was taken to avoid disruption of flocculated material during the passage of the sample from the jar-test unit to the photo cell, by allowing low flow velocities adjusted through a system of valves. In case of DAF, the released air bubbles were also photographed and analysed. Professional high resolution black-and-white film AGFA 25 or Kodak TMAX 100 was used for the purpose, while forced development was applied for film processing. 30-100 fields, or more than 500 particles of each film shot were processed and analysed, using a ccd-camera (604*288 pixels) mounted to a Nikon Optiphot (Japan) microscope (at 40x magnification). The minimum detectable size under these circumstances was 1.9 μm. The software package Genias25 developed by Joyce-Loebl (UK) was used for data processing.

The computer image analysis results are presented in the form of particle size frequency and volume distributions. The frequency and volume distributions represent the number and the volume of recorded particle size range. For this purpose the recorded particles and/or floc material were approximated to spheres and presented in size ranges of 10 μm, from 0-200 μm. Particles >200 μm were also recorded (by image analysis and visually), however, they represented only a small percentage of the overall particle concentration which was effectively removed in treatment. The adopted size limit of 200 μm was found more relevant for the assessment of the overall DAF efficiency, since it included a more significant portion of the overall particle (floc) concentration, of size which is more problematic for DAF removal than the larger particles (flocs). The particle size of 50 μm, which is roughly the size limit of visually observable aggregates, was tentatively adopted as the size dividing the small from the large particle size range.

Statistical analysis of data was performed with the software package Statistica. One or two way analysis of variance (ANOVA), followed by multiple comparison and regression analysis among treatments were performed when appropriate. The same applies for the Mann-Whitney U-test.

Chapter 3

OPTIMISATION OF COAGULATION CONDITIONS FOR THE REMOVAL OF *Microcystis aeruginosa* BY DISSOLVED AIR FLOTATION OR SEDIMENTATION

-Parts of this chapter were published by A. Vlaški, A.N. van Breemen and G.J. Alaerts in (1996) *J. Water SRT-Aqua*, 45, 5, pp.253-261.

ABSTRACT : Analysis of surface water treatment practice in the Netherlands shows that cyanobacteria, especially *Microcystis aeruginosa* and *Oscillatoria agardhii*, are the main problem causing algae. Conventional (sedimentation) and advanced (dissolved air flotation) treatment of algae laden (model) water were studied. The agglomeration (coagulation/flocculation) phase was found to determine the process efficiency and hence to be the prerequisite for process improvement. Both processes were assessed in terms of their removal efficiency for the cyanobacterium *M. aeruginosa*. Relevant process parameters were studied, including the influence of coagulant ($FeCl_3$) dose, coagulation pH, flocculation time, energy input (G value), single-stage versus tapered flocculation, application of a cationic polyelectrolyte (Superfloc C-573) as coagulant aid, and surface loading. The process efficiency was assessed as a function of the agglomeration and the resultant particle (floc) size distributions. The results were evaluated in function of highlighting the occurring process mechanisms.

3.1 INTRODUCTION

The new interest in surface water treatment in the Netherlands has called for study of the specific problems caused by algae. An investigation carried out at five drinking water production plants in the Netherlands utilising surface water (relating to 60% of the total Dutch drinking water production from surface water), showed that the main problem causing algae are the cyanobacteria, in particular *Microcystis aeruginosa* and *Oscillatoria agardhii* [1] (Chapter 1). Their seasonal blooms impose short- and long-term treatment modifications with significant additional costs. Apart from developing water quality management measures, the understanding of efficient treatment technologies needs to be expanded.

In treatment, the agglomeration, i.e. coagulation/flocculation phase is likely the most critical. It can be influenced by conditioning of the algae (application of a chemical oxidant), and/or physical conditioning of the water (microstraining), as well as by optimising the subsequent solid-liquid separation process [2, 3, 4]. In practice, for the latter conventional (sedimentation) and more advanced (dissolved air flotation [DAF]) processes are used, followed by final filtration. It has been recommended to evaluate the flotation-filtration and sedimentation-filtration treatment options as integrated processes [5]. By consequence this pertains equally to the agglomeration (coagulation/flocculation) step [6].

The low density of algae (0.99-1.2 g/cm^3) and their phototactic/phobic and chemotactic/phobic responses tend to keep them in a continuously floating state. Optimised coagulation with respect to coagulant dose and pH is required both for DAF and sedimentation in order to destabilise the algae. The stability of algae in suspension is ascribed to the repulsive electrostatic interactions caused by surface charges that are negative for pH 2.5 - 11.5 [7], and to steric effects of water bound to cell surface and of adsorbed macromolecules or EOM [8]. DAF without the use of coagulants has been found to be inefficient [5, 9]; apart from the mentioned steric hindrance, the slightly negative surface charge of air bubbles contributes to this [10, 11]. Thus, as in DAF particle destabilisation is more important than large floc size [12], coagulation conditions that produce flocs or particles of little or no charge are required.

According to classical coagulation theory [6] efficient particle destabilisation and high process efficiency is to be expected at pH below the iso-electric point (IEP=6-7 for iron hydroxo complexes), however, other reports have suggested that optimal algal removal (in direct filtration) occurred at pH of 8-8.8 [2]. This, however, does not necessarily imply that more efficient coagulation takes place at the higher pH. Namely, in case of direct filtration the time interval between coagulant addition and the filtration step is relatively short (in the range of a few minutes), suggesting that higher pH (>7.5) is needed in order to assure faster and complete coagulant (iron) hydrolysis. Thus, the formation of difficult to remove, colloidal iron-organic complexes is partially avoided. Furthermore, in cases of higher Ca^{2+} concentrations, preferential complexation of Ca over Fe with organic matter occurs at pH>8. Both factors render more coagulant available for more efficient destabilisation of colloidal and particulate matter present, including algae. In the same study [2] the application of cationic polyelectrolyte enhanced process efficiency and reduced coagulant dose. Other experiences with the use of polyelectrolytes as coagulant aids in the context of DAF were controversial with respect to obtained benefits [5, 13].

Furthermore, small (10-30 μm) and light flocs should be produced by the coagulation/flocculation process [10]. Due to the inverse relation between flocculation mixing intensity and time on one hand, and floc size on the other [14], flocculation G values of 30-80 s^{-1} have been proposed, while flocculation times as short as 5 to 6 minutes were found to suffice [4, 5]. With respect to DAF, one can assume that the rising air bubbles act as a rising filter bed which collects particles on its way to the water surface, in analogy to the DAF kinetic model proposed by Malley and Edzwald [12]. In this context, it seems feasible to investigate the possible existence of an optimum flocculation Gt range for DAF (optimum direct filtration conditions occur within a flocculation Gt range of 15,000-40,000 [15]). On the other hand, coagulation conditions that produce flocs with good settling characteristics are essential for efficient sedimentation. The settling characteristics of a flocculent suspension depend on floc size, floc density, water temperature [16] and in some cases initial particle number concentration [17]. Good settling is generally achieved at high(er) coagulant doses, low flocculation energy input (G=10-40 s^{-1}) and long(er) flocculation times (30-60 min). Although tapered flocculation seems more appropriate for sedimentation (providing longer flocculation times and favourable floc structure due to diminishing energy input) it is also applied in DAF, as in the case of the DZH (Dune Water Supply Company South Holland) and EWR (Energy and Water Supply Company Rhineland) treatment plants.

Malley and Edzwald [18], comparing sedimentation and DAF performance on laboratory scale in terms of particle removal, showed that as temperature decreases, the performance of DAF over that of sedimentation increases. They also found comparable performance with flocculation periods of 5 minutes for DAF, and of 20 minutes for sedimentation. These findings were based on experiments with model water prepared by spiking water with aquatic fulvic acid and montmorillonite clay, as well as with natural water. Compared to earlier research, the removal efficiencies are modest, especially in the case of DAF. Janssens [5] reported 65 - 92 % removal of naturally occurring cyanobacteria including *M. aeruginosa* by DAF, but lower efficiency was observed for other algae including the cyanobacterium *M. pulvera* (small size of 2-3.4 μm and without gas vacuoles); it was not mentioned whether the higher removal efficiency was achieved when using coagulant aid, which might have enhanced the efficiency of the tested alum, PACl (polyaluminium chloride) and ferric chloride coagulants. Edzwald and Wingler [19] reported a DAF removal efficiency of 96.8 and 99.8% for laboratory cultured *Chlorella* (Chlorophyceae) and *Cyclotella* (Bacillariophyceae) respectively, spiked into natural reservoir water; the initial concentrations used in these experiments were 10 and 5 times higher than in our experiments, while alum and PACl (polyaluminium chloride) were tested as coagulants at a low pH of 6.5 for the first and 5.5 for the second. The high algae concentration asserted a positive impact on the coagulation/flocculation and DAF efficiency. Furthermore, the low coagulation pH conditions are not representative for treatment practice and result in adsorption coagulation, which may have positively affected the efficiency. Zabel [20] found that similar initial concentrations of *M. aeruginosa* (10^5 cells/mL) resulted in 98% removal in full-scale DAF operation, while under the same conditions upflow floc blanket sedimentation efficiency was 76%. These results were achieved by using aluminum sulfate as coagulant at a relatively low pH of 6.5-7.2. In the case of chlorinated ferrous sulfate coagulant at pH of 8.3-8.7, the removal efficiencies (of other algae species) were lower; worst results were obtained for *Stephanodiscus* : 40% by sedimentation and 83% by DAF.

These and other data from literature and practice strongly suggest that the removal efficiency of algae varies for both processes, depending on raw water characteristics, alga species, their morphology and physiology, and treatment process parameters, notably coagulation pH, type and dose of coagulant, flocculation energy input and time and process loading rates. In treatment, it can be surmised that the algal removal efficiency relies most significantly on good particle agglomeration (coagulation and flocculation). Therefore, when working with laboratory cultured algae, the potential interference of the complexing agents used in the preparation of the algae growth medium on the coagulation/flocculation step (binding of coagulant making it inaccessible for floc growth) must be taken into account [21]. Furthermore, the occasional particle count of acidified effluent samples showed that a substantial part (15-20% for the higher coagulant doses) of the particle count can be attributed to particulate residual coagulant. The definition of particle removal efficiency and the comparison of results among different studies, become even more complicated.

In this chapter the optimisation of the agglomeration phase for both alternative processes is discussed and assessed. Model water experiments were used for determining removal efficiency for *M. aeruginosa*. The efficiency of the two processes is compared in the context of cost-effective *M. aeruginosa* removal, while the results are evaluated in the light of existing theoretical and operational knowledge related to the two processes. Relevant process parameters were studied : coagulant ($FeCl_3$) dose, pH, flocculation time and energy input (G value), single stage versus tapered flocculation, application of cationic polyelectrolyte (Superfloc C-573) as coagulant aid, and surface loading. This is done in order to clarify some of the inconsistencies met in literature and practice, e.g. the applicability of very short flocculation times for DAF, the effect of tapered verus single stage flocculation, determination of the optimal enegy input (G value), the impact of polyelectrolytes on DAF efficiency, etc. The underlying process mechanisms and kinetic aspects are also discussed, while a broader discussion on the mechanistic theory of DAF is given in Chapter 7.

3.2 EXPERIMENTAL PROCEDURES

3.2.1 Experimental design

The experimental set-up consisted of a modified KIWA - standard jar test apparatus with incorporated DAF facilities that could be connected to small-scale filters (diameter 60 mm, height 400 mm) with sand as filtration medium (commercial fraction 0.80-1.25 mm, bed thickness 200 mm) (Fig. 3.1). Initially, both sedimentation and DAF were optimised without considering further treatment by rapid filtration; at a later stage (specifically mentioned in Results and Discussion) rapid filtration was also included at a constant filtration rate of 10 m/h, similar to the five full scale filtration installations studied. The filtration itself was not optimised due to time constraints and it was applied in a limited number of experiments in combination with previously optimised up-stream DAF and sedimentation conditions.

Fig. 3.1 The modified jar test apparatus with DAF facilities, including a small scale filter unit.

For both, DAF and sedimentation, the experimental procedure included coagulation at a G value of 1,000 s^{-1} for 30 s, temperature of 20°C at a pH of 8, while the flocculation G (G_f=10, 30, 50, 70, 100 and 120 s^{-1}) and flocculation time (t_f=5, 10, 15, 25, 30 and 35 min) were varied in a range which aimed at including low, medium and high values respectively. The adopted G values were converted into the appropriate number of revolutions of the stainless steel flocculation paddle (bxh=7.5x2.5 cm) by using a previously established calibration line (log-log scale). Three-stage tapered flocculation for DAF was tested at diminishing energy input (G_f= 82, 49, 23 s^{-1}) and equal flocculation time (t =3x8 min), same as in the Scheveningen DAF plant of DZH. The tested coagulant ($FeCl_3 \cdot 6H_2O$) doses were in the range of 0 - 15 mg Fe(III)/L. The cationic polyelectrolyte Superfloc C-573 was used as a coagulant aid in the range of 0 - 1.5 mg C-573/L for DAF and 0 - 3 mg C-573/L for sedimentation, combined with the previously optimised Fe coagulant dose. The polyelectrolyte dose range choice was based on providing roughly 1/10th of the determined optimal coagulant dose. The sedimentation time range covered the low, medium and high range of values (t_s=10, 20, 30, 45, 60 and 90 min), while for DAF typical recirculation ratios (R=5, 8 and 10%) and saturator pressures (P=500, 600 and 700 kPa) were tested [10]. After releasing the recycled flow into the jar test unit, slow mixing at a G<10 s^{-1} was applied for 30 s, followed by a period of 4 min flotation (without any mixing), after which sampling took place.

3.2.2 Algal suspension

The investigated cyanobacterium were semi-continuously laboratory cultured, single cells of *M. aeruginosa*. The algae were provided from two sources : the Microbiology Laboratory, University of Amsterdam, the Netherlands, and the Institute of Freshwater Ecology, Windermere Laboratory, United Kingdom. The algae were cultured on Jaworski growth medium in 5 L vessels, at 20 °C and 1,000 lux light intensity, according to directives and recipes [22] (Appendix 3.1). The algae were spiked into water originating from Biesbosch storage reservoirs. This reservoir system supplies water to the city of Rotterdam, partly to the city of Dordrecht, the provinces of South-West Netherlands and North-West Brabant. The water is originally eutrophic river Meuse water, but efficient water quality management during the six months of impoundment results in the production of almost oligotrophic conditions with generally low algae counts and average chlorophyll concentrations of 5.6 μg/L [23]. The initial algal concentration in our model water was standardised by particle (algae) count (HIAC) as ≈ 10,000 cells/mL. The rationale behind this standardisation was to simulate an algae concentration which is considered problematic and representative for water treatment. This concentration roughly corresponds to the average monthly cyanobacteria concentration in one of the most seriously algae affected water resources, namely the WRK III water treatment plant reservoir (Fig. 1.4). Colonies of *M. aeruginosa* are the prevalent cyanobacterium species during such blooming episodes. The colonial form of this species is relatively easily removed by e.g. the initial WRK III microstraining, while the single cell form which is a product of natural disintegration of colonies (by the end of their life cycle), or mechanically induced disintegration due to e.g. high shear related to the initial pumping stage, regularly penetrates treatment and is encountered in treatment plant effluents [1]. Thus, the simulated situation is even more serious and difficult with respect to treatment, since we are dealing with the same concentration of the single cells form of *M. aeruginosa*.

Providing algae continuously at constant and comparable quality from the culture was a difficult task. Spiking identical volumes of algae suspension from the two algae cultures resulted in the desired particle (algae) count, however, the corresponding turbidity varied from 1.5 to 2.5 FTU in the case of the Amsterdam algae and from 3 to 4.1 FTU in the case of the Windermere algae. In both cases the spiked algae were in the stationary growth phase which was reached approximately 14 days after the culture start-up. Prior to spiking, the appropriate volume of algae culture was strained through a 35 μm mesh, in order to discard large lumps - agglomerates of algae debris and cellular organic matter, which was increasingly formed at later growth stages of the cultured algae. Thus, a simulation of the single cell form of *M. aeruginosa* was accomplished. The model water temperature was kept constant at 20°C in order to simulate conditions during late summer and early autumn blooms.

3.2.3 Analysis methods

In the context of providing efficient down-stream filtration and longer filter runs the performance goal for sedimentation and DAF was set for turbidity <1 FTU, residual coagulant <1 mg Fe/L, and minimal particle (algae) count (HIAC); DOC removal was to be quantified.

Previous investigations [2] have shown that the use of turbidity as a lumped parameter for the evaluation of process efficiency is often inadequate, especially in the lower turbidity range (<0.5 FTU), e.g. in the case of filtration product water. Namely, the low filtrate turbidity generally cannot be correlated with the accompanying low particle count. Particle count measurements are essential in such cases in order to better identify the mechanisms and phenomena underlying coagulation conditions. Thus, process efficiency evaluation was based on measurements of turbidity (Sigrist L-65, Switzerland) and particle (algae) count measurements (HIAC-Royco PC-320, USA), while inverted microscope count (M40-Wild Leitz, Switzerland) and computer image analysis were performed periodically. Residual coagulant as Fe_{total}, DOC and electrophoretic mobility (EM) (Tom Lindstrom AB-Repar) were also measured. The sampling locations are given in Fig. 2.1 (Chapter 2). Samples were taken for raw water and after full treatment, except for EM, inverted microscope analysis, and computer image analysis, when additional samples of flocculated water were analysed. Flocculated water samples contain algae and Fe hydrous oxides flocs. A standard sample volume of 0.5 L was abstracted after DAF from the taps at the bottom of all the jars simultaneously, within a standardised time of one minute. For sedimentation the same procedure was performed, using a vacuum pipe system which abstracted water from the upper half of the jars simultaneously. In both cases attention was paid not to disturb the floated/settled matter which would affect the measurements. Glass bottles were used for sampling. Most measurements were conducted on spot, except for Fe_{total} and DOC. These were done on preserved samples kept at a temperature of 4 °C, with one-two days delay. Fe samples were preserved in plastic bottles by adding 0.5 M HCl to reach a pH<2 and dissolve all iron present. DOC samples were kept in glass bottles to avoid the impact of plastic on measurements.

To ensure reproducibility of the experimental results, the performance of the jar test apparatus was periodically (weekly) tested under identical experimental conditions in all six jars. The results which are further discussed are representative of multiple experimental analysis. The impact of variations of initial experimental conditions on the results, i.e. initial algae count and turbidity was also considered and statistically analysed. Statistical analysis of data was performed with the Statistica software package. Two way ANOVAs followed by multiple comparison among treatments were performed when appropriate. The same applies for the Mann-Whitney U-Test.

The computer image analysis was performed with a Mini-Magiscan (IAS 25/IV25 Joyce-Loebl Ltd., UK) image analysis system (Fig.3.2), producing information about particle size frequency and particle volume distributions.

Fig. 3.2 The Mini Magiscan image analysis system (IAS 25/IV25 Joyce-Loebl Ltd., UK) used for particle analysis.

These analyses were performed on fresh samples, as well as on photographed samples using previous experience from a similar research [24]. For the purpose of the latter, a specially devised flat photo cell was connected to the jar test apparatus. Representative samples of raw, flocculated, and treated (DAF, sedimentation and filtration) water were passed through the photo cell and photographed. The same applies for (algae free) water into which the DAF recycled stream was introduced and released air bubbles were photographed. Professional high resolution black-and-white film AGFA 25 or Kodak TMAX 100 was used, while forced development was applied for film processing. Hundred fields of each picture were processed and analysed, using a ccd-camera (604*288 pixels) mounted to a Nikon Optiphot microscope (at 40x magnification). For verifying the representativity of the samples, multiple shots of the same sample were analysed occasionally. The software package Genias25 developed by Joyce-Loebl Ltd., (UK) was used for data processing.

Results from computer image analysis are presented in the form of particle size frequency or volume distributions. These depict the recorded frequency (absolute or relative percentual value) or volume of suspended material (detection limit of 1.9 μm), in a particular particle size range, after different treatment steps. For this purpose the recorded particles (flocs) material were approximated to spheres. For verifying the reproducibility of the applied analytical technique, multiple shots of the same sample were analysed regularly. Particle size distributions departed strongly from normality. Particle size frequency distributions from different sets of experiments performed under identical conditions were compared by the application of the chi-square test. Testing the hypothesis that the obtained particle size distributions were identical, resulted in a probability level of $P=0.999$, suggesting well comparable particle size distributions and good results reproducibility.

3.3 RESULTS AND DISCUSSION

3.3.1 Turbidity and particle (algal) count

For turbidity in the range of our interest (with sedimentation and DAF effluent turbidity mostly >0.5 FTU) we found a linear correlation between turbidity and particle count (r^2=0.99) for each of the cultured algae strains, however, the regression lines had different slopes (Fig. 3.3).

This suggests different morphology of the two strains. Particle count/turbidity ratio differences were also observed with different algae growth stages, as well as under slight changes of growth conditions, e.g. light intensity. The phenomenon of different algae morphology features of the same species under different growth conditions has been noted in nature also, e.g. in the case of the diatom *Stephanodiscus hantzschii*, a species known to develop bristle like appendages in autumn and winter, which hampers its removal [21].

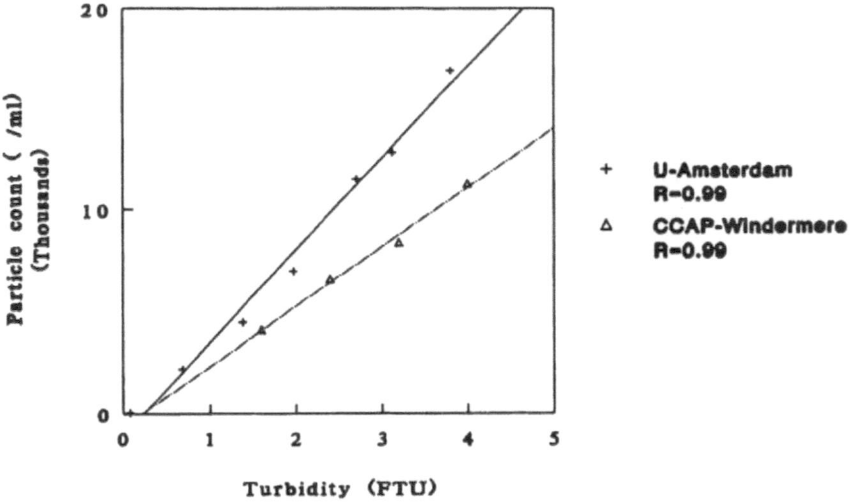

Fig. 3.3 Particle count versus turbidity for tap water spiked with cultured *M. aeruginosa*.

3.3.2 Optimisation of agglomeration

a. Coagulation pH

At pH 8 organic particle agglomeration is likely to be solely achieved through sweep coagulation because of the higher solubility of the metal hydrolysate and the dominance of negative metal hydroxide compounds [25]; here this was confirmed by the negative surface charge of all floc material, also under conditions of optimal removal. Figs. 3.4 and 3.5 describe particle size frequency distributions after flocculation and after DAF at pH 6 and pH 8. After flocculation, a more pronounced reduction of the smaller size particles (<50 μm), and a larger share of formation of larger size particles (>50 μm) frequency were observed at pH 6 as compared to pH 8.

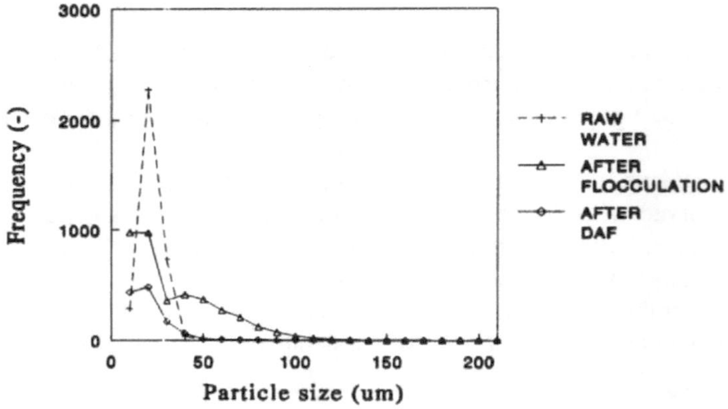

Fig. 3.4 pH 6, 10 mg Fe(III)/L, particle size frequency distribution before and after DAF. Experimental conditions : R=5%, P=500 kPa.

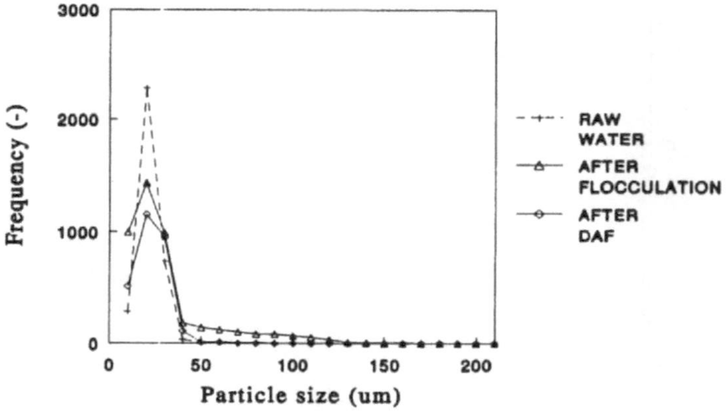

Fig. 3.5 pH 8, 10 mg Fe(III)/L, particle size frequency distribution before and after DAF. Experimental conditions : R=5%, P=500 kPa.

In both cases the applied coagulant dose produced flocs larger than 50 μm which were efficiently removed. However, it is evident that at the lower pH conditions adsorption coagulation and charge neutralisation substantially contributed to the flocculation beside the simultaneously occurring and dominant sweep coagulation which is coagulant dose dependent.

This resulted in more efficient flocculation and almost doubled particle removal efficiency at the lower pH, suggesting that adsorption coagulation plays an important role in algal flocculation. Electrophoretic mobility measurements confirmed that algae charge neutralisation occurred under these conditions, which is also in accordance with findings of Bernhardt and Clasen [6, 26]. However, the practical implications of these findings are mostly indicative. Namely, pH control for the sole purpose of coagulation enhancement is in practice not always feasible. The application of polyelectrolytes, in this case preferably cationic, could be a suitable alternative to achieve particle charge neutralisation.

b. Flocculation conditions

Results here confirmed that DAF required considerably shorter flocculation times than sedimentation (t_f=10 compared to 30 minutes), however, a lower flocculation G value of 10 s^{-1} performed equally well or better in DAF than higher G values. The same low G value was applied by Edzwald and co-workers resulting in better particle removal by DAF compared to sedimentation [8], and in excellent algae removal [19].

Here, results did not allow determination of an optimal Gt region for DAF. This was due to the fact that a flocculation G value of 10 s^{-1} and of 70 s^{-1} consistently resulted in similar removal efficiencies, better than the other tested G values, thus suggesting other specific factors that influence DAF efficiency. Similar to what Janssens [5] reported, tapered flocculation which is generally considered favourable for sedimentation did not significantly affect DAF efficiency.

Beside chemical factors, floc size played an important role in process efficiency in both, sedimentation and DAF. Image size analyses showed an inverse relation between flocculation energy input (G value) and floc size (see also Chapter 6). A low G value of 10 s^{-1} produced a floc volume distribution which led to the highest DAF removal efficiency (Fig. 3.6), as verified by (HIAC) particle count, turbidity and residual Fe measurements. Comparison of the particle (floc) volume distributions after flocculation at G=10 s^{-1} (Fig. 3.6) and G=23 s^{-1} (Fig. 3.7), suggests that the increase of energy input resulted in denser floc structures (see also Chapter 6).

DAF efficiently removed the particles in the larger particle size region (>50 μm) both at G=10 (Fig. 3.6) and 70 s^{-1} (data not presented). Therefore, the common recommendation that DAF requires high G values in order to produce small (10-30 μm), strong, shear resistant flocs, is not supported by these results. Sedimentation on the other hand, was found to be favoured here with floc size distributions obtained by G values of 10-30 s^{-1}, yielding slightly better results at G=30 s^{-1} (data not presented). A Mann-Whitney U-Test, for this G value, testing significant differences in efficiency between flocculation times t_f of 20 and 30 min, showed no significant differences for particle (algae) count, but significant differences for turbidity. The initial (particle count and turbidity) conditions for the analysed sets of experiments did not differ significantly suggesting that the results can be reliably compared. Finally, it is suggested that the floc size/density ratio plays an important role in the DAF and sedimentation processes. This aspect will be discussed in more details in Chapter 6.

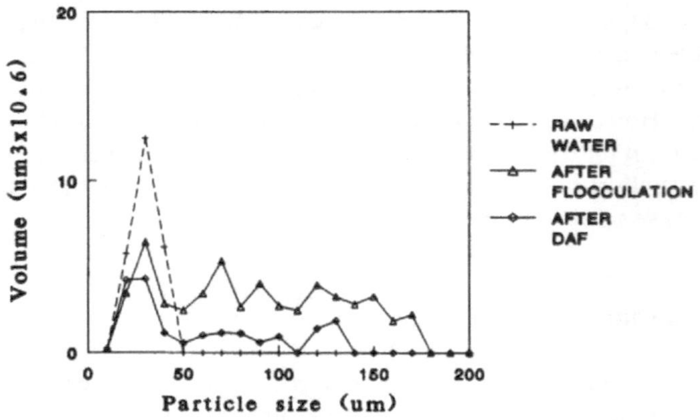

Fig. 3.6 $G=10\ s^{-1}$, particle volume distribution before and after DAF. Experimental conditions: 3 mg Fe(III)/L, $G_f=10\ s^{-1}$, $t_f=8$ min, R=8%, P= 600 kPa.

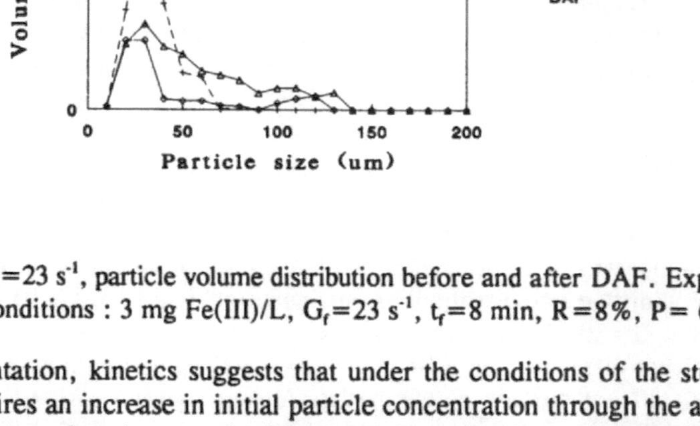

Fig. 3.7 $G=23\ s^{-1}$, particle volume distribution before and after DAF. Experimental conditions : 3 mg Fe(III)/L, $G_f=23\ s^{-1}$, $t_f=8$ min, R=8%, P= 600 kPa.

In case of sedimentation, kinetics suggests that under the conditions of the study efficient sedimentation requires an increase in initial particle concentration through the application of higher coagulant dosages than those required for DAF. Low G values are expected to promote formation of floc size volume distributions with a higher frequency in the larger particle size range, as well as flocs with lower density. The low G values also produced floc size distributions favouring DAF, of which the performance is not solely dependent on large floc sizes. Mean floc and bubble size dimensions calculated by image analysis enabled calculation of the coefficient of particle-bubble attachment efficiency α_{pb} [8, 28], which will be discussed in more detail in Chapter 7.

c. Polyelectrolyte as coagulant aid

The high DAF efficiency achieved at pH below the iso-electric point (IEP) for iron hydroxo complexes (IEP=6-7) (see earlier), emphasizes the role of adsorption-coagulation and charge neutralisation for particle removal. This suggests potential process efficiency benefits in case of applying cationic polyelectrolytes as coagulant aids.

Investigations here showed that low cationic polyelectrolyte (Superfloc C-573) doses of 0.5 and 1.0 mg C-573/L, in combination with the previously established optimal primary coagulant doses (3 mg Fe(III)/L for DAF and 10 mg Fe(III)/L for sedimentation), significantly raised particle (algae) removal efficiency : for DAF up to 94.5% and for sedimentation up to 98.9%. This was accompanied by charge neutralisation of the algae at a coagulant aid dose somewhat higher than the established optimum, as verified by electrophoretic mobility measurement. Particle volume distributions of flocculated water with and without the addition of coagulant aid (Fig. 3.8) showed a distinct difference: a clear improvement was observed in agglomeration performance in the case with polyelectrolyte addition, expressed in a more pronounced larger particle size (>50 μm) contribution.

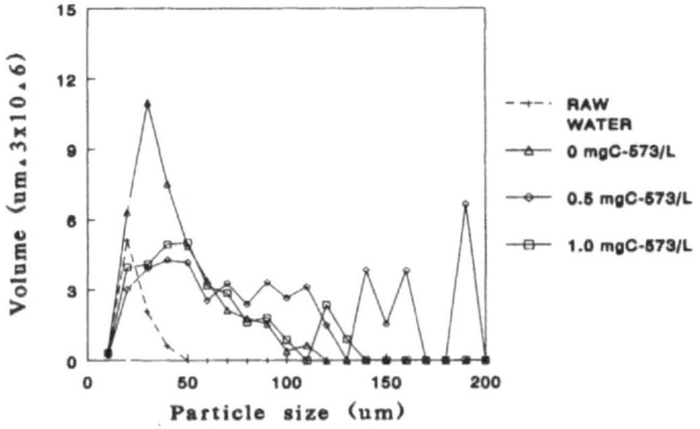

Fig. 3.8 10 mg Fe(III)/L and Superfloc C-573 as coagulant aid, particle volume distribution of flocculated water with and without polyelectrolyte. Experimental conditions: pH 8, G_f=30 s^{-1}, t_f=30 min.

Although the floc volume distribution for the case of 0.5 mg Superfloc C-573 was more pronounced in the large particle size range compared to that for 1.0 mg Superfloc C-573/L, the latter produced better results after sedimentation. It is suggested that the better sedimentation efficiency is related to a denser floc structure resulting from the higher polyelectrolyte dose. The smaller, 0.5 mg Superfloc C-573/L dose, favoured efficient DAF, which improved the algae flocculation and resulted in bigger and lighter floc material than was produced by the 1.0 mg C-573/L dose.

Though the filtration was not included in this particular set of experiments, it can be that the low DAF and sedimentation effluent algae concentration would enable longer and more economical filter operation. This may possibly be disturbed to a limited extent by the residual polyelectrolyte which may account for filter blockage and filter backwash problems.

3.3.3 Sedimentation versus DAF

The goals effluent turbidity (<1 FTU) and residual coagulant (<1 mg Fe/L) were achieved by both processes. The DOC removal was around 25%, relying on coagulation efficiency, but not on the solid-liquid separation process, as was also reported by Malley and Edzwald [18]. However, in the case of initial model water turbidity of 3-3.5 FTU (Windermere algae), the 1 FTU turbidity goal was achieved only when cationic polyelectrolyte was used as coagulant aid, or when longer and not very feasible flocculation times were applied (>30 min). Figs. 3.9 and 3.10 show that in order to attain maximum particle (algae) removal efficiency, sedimentation required approximately three times higher coagulant dose (10 mg Fe(III)/L vs. 3 mg Fe(III)/L).

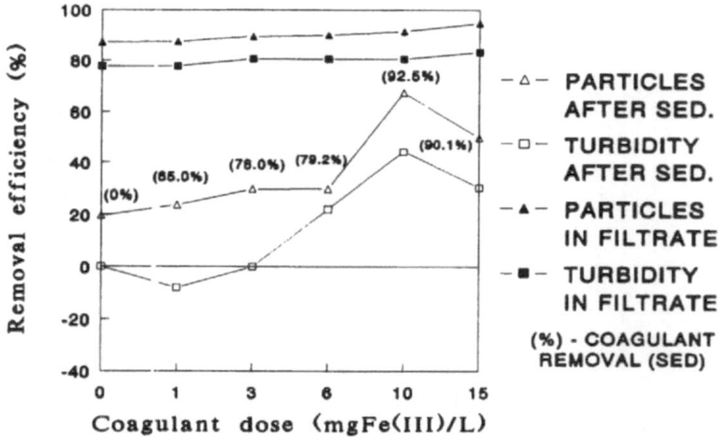

Fig. 3.9 Removal efficiency after sedimentation (SED) and after sedimentation-filtration. Initial conditions : see Section 3.2. Experimental conditions : pH 8, G_f=30 s^{-1}, t_f=30 min, t_s=60 min, v=10 m/h.

Consequently, the sedimentation process would also result in higher sludge production, that in addition has worse dewatering characteristics [20]. Similarly, the sedimentation alternative required a three times longer flocculation time (t_f=30 min vs. 10 min) and higher flocculation energy input (G_f=30 s^{-1} vs. 10 s^{-1}), as well as a significantly lower process loading rate (t_s=60 min vs. standardly applied 10-30 min) in order to achieve optimal process efficiency. The flocculation G values of 10 s^{-1} and 70 s^{-1} (the latter being in the range generally considered as favourable for DAF) consistently resulted in similar DAF efficiency.

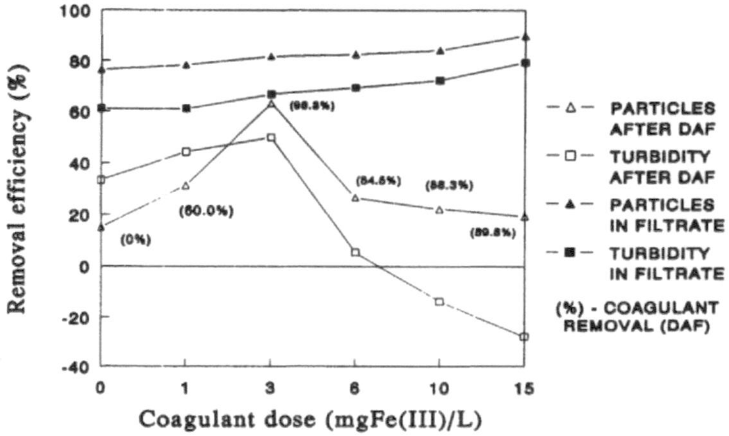

Fig. 3.10 Removal efficiency after DAF, and after DAF-filtration. Initial conditions: see Section 3.2. Experimental conditions : pH 8, G_f=70 s^{-1}, t_f=10 min, R=8%, P=600 kPa, v=10 m/h.

The coagulant doses for which maximum particle and turbidity removal was achieved in the DAF and the sedimentation were 3 and 10 mg Fe(III)/L, respectively. Although the residual coagulant concentration was higher for higher coagulant doses, the percentual coagulant removal efficiency after DAF and after sedimentation generally increased with the coagulant dose. The respective residual coagulant concentration carried over from DAF or sedimentation to the filtration step was 0.1-1.5 mg Fe/L, and thus lower than dosages commonly applied in direct filtration.

Figs. 3.11 and 3.12 represent particle volume distributions created by coagulant doses of 6 and 15 mg Fe(III)/L. Flocculation with 6 mg Fe(III)/L resulted in flocs of 10-50 μm and 50-100 μm, while the 15 mg Fe(III)/L dose resulted in larger amounts of flocs in both sizes. The particle volume distribution after the sedimentation step shows that the two coagulant doses resulted in a relatively similar floc size carried over to the filtration, however, a better filtration efficiency was observed for the 15 mg Fe(III)/L dose, as determined by particle distribution measurements and also corroborated by HIAC particle count. This can be ascribed to the higher residual iron concentration after the sedimentation at this coagulant dose, and hence an enhanced "direct" filtration effect. The model water in these experiments was relatively free of organic matter (TOC≈2.5 mg C/L) which otherwise might have complexed significant quantities of the hydrolysed iron coagulant species and render it unavailable for the direct filtration effect.

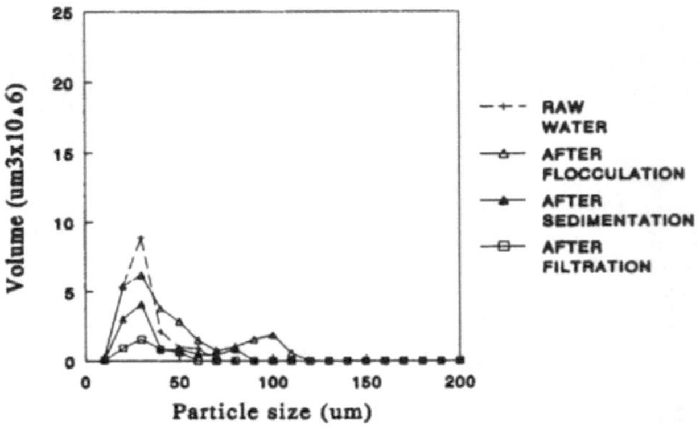

Fig. 3.11 Sedimentation-filtration, 6 mg Fe(III)/L, particle size volume distribution. Initial conditions: see Section 3.2. Experimental conditions : pH 8, G_f=30 s^{-1}, t_f=30 min, t_s=60 min, v=10 m/h.

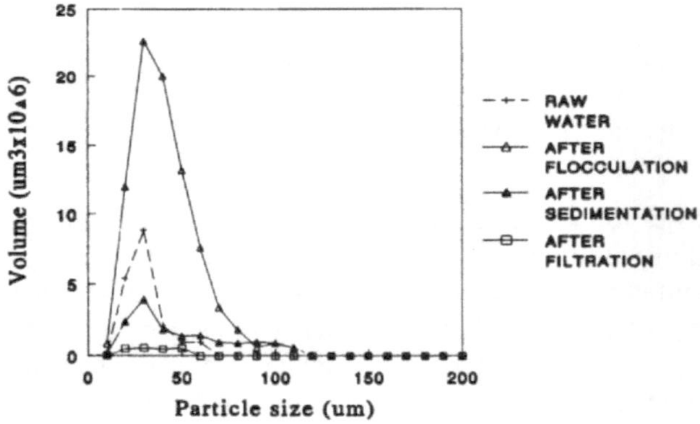

Fig. 3.12 Sedimentation-filtration, 15 mg Fe(III)/L, particle size volume distributions. Initial conditions : see Section 3.2. Experimental conditions : pH 8, G_f=30 s^{-1}, t_f=30 min, t_s=60 min, v=10 m/h.

These results may be considered only as qualitative as the small diameter of the filters may cause wall effects; similarly, the absence of filter ripening (due to the small water quantities filtered), may lead to underestimation of the filters performance and efficiency.

From the point of view of final product water quality - treatment plant effluent, the achieved algae removal efficiency by both treatment processes is to be considered inadequate. At optimal conditions in this study, sedimentation performed somewhat better than DAF in terms of algal removal efficiency (maximum removal of 87% for sedimentation versus 71% for DAF). Achieving more than 99% removal efficiency in our study would suggest a residual of < 100 cells/mL, a concentration which was often close to the demineralised water particle count. Although no standards exist for algae in drinking water, a residual algae concentration of 100 - 500 cells/mL (achieved by DAF and sedimentation under optimal process conditions with the use of cationic polyelectrolyte as coagulant aid) is relatively high. The down-stream filtration is the final barrier to retain colloidal and particulate matter.

Optimisation of the filtration step is expected to further raise the particle removal efficiency from 94.7% (sedimentation-filtration) and 90% (DAF-filtration). The generally higher recorded removal efficiency of the filtration following the sedimentation (Fig. 3.9 and 3.10), is attributed to the larger size of the residual floc material after sedimentation. This is the consequence of the lower flocculation G value of 30 s^{-1} as compared to the 70 s^{-1} applied in the DAF-filtration. The fact that removal efficiency is critically dependent on coagulant dose in the sedimentation-filtration case (Fig. 3.9) also suggests that it is the upper filter layer that has functioned in filtration, as reported also elsewhere [28]. On the other hand, the slightly steeper slope of the efficiency versus coagulant dose line in the case of DAF-filtration (Fig. 3.10), suggests that apart from the flocculation G value and concomitant floc size, other factors influenced the filtration efficiency.

3.3.4 Statistical analysis

A Mann-Whitney U-Test comparing the particular combinations of optimal DAF and sedimentation conditions (Fig. 3.13) resulted in insignificantly different particle removal efficiency (*P*=0.087) and significantly different turbidity removal efficiency (*P*=0.017).

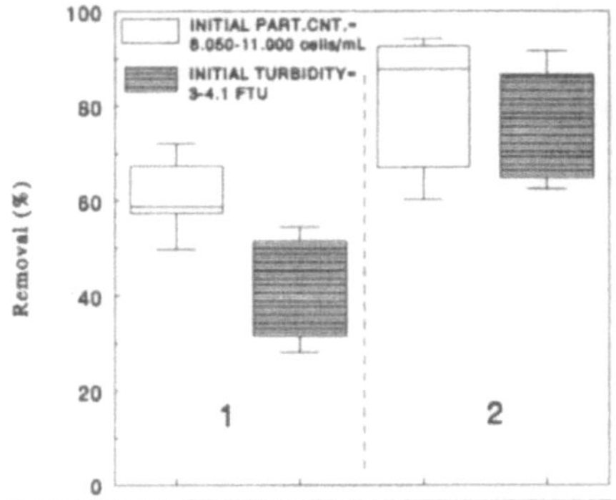

CONDITIONS	Fe(III) (mg/L)	G (s-1)	t (min)	R/P (%/kPa) or t sed (min)
1. (DAF)	3	10	10	8/500
2. (SED)	10	30	30	60

Fig 3.13 Particle (algae) and turbidity removal efficiency for optimised DAF and sedimentation (SED). Number of data sets analysed n=10. Notation : Fe(III)-coagulant dose; G-flocculation energy input; t-flocculation time; R/P-recirculation ratio/saturator pressure; t_{sed}-sedimentation time.

A Mann-Whitney U-Test comparing DAF (G_f=10 s^{-1}) and sedimentation (G_f=30 s^{-1}) on a broader scale (different flocculation time values being pooled within each of these two treatments) yielded no significant differences both, for particle (algae) and turbidity removal efficiency (P>0.05). The flocculation G values of 10 and 70 s^{-1} were tested by two way ANOVA for a flocculation time of 10 and 20 minutes (Fig. 3.14), and are presented in the Table 3.1.

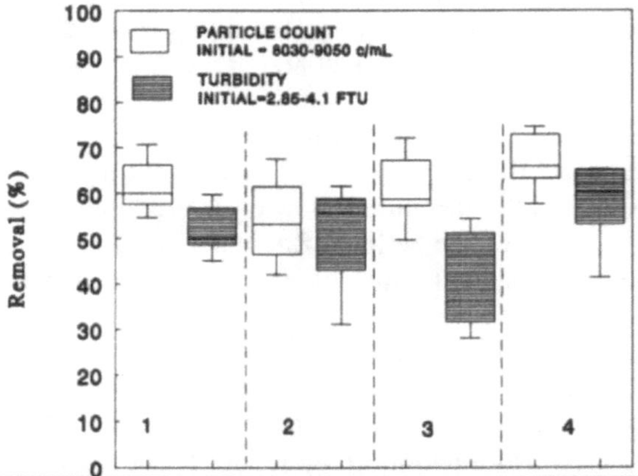

CONDITIONS	Fe(III) (mg/l)	G (s-1)	t (min)	R/P (%/kPa)
1 (DAF)	3	70	10	8/500
2 (DAF)	3	70	20	8/500
3 (DAF)	3	10	10	8/500
4 (DAF)	3	10	20	8/500

Fig. 3.14 Two way ANOVA of DAF particle (algae) and turbidity removal efficiency for different flocculation G and t values (R-recirculation ratio, P-saturator pressure). Number of analysed data sets n=24.

Table 3.1 Results from the ANOVA, quantifying the degree of correlation between flocculation energy input (G) and time (t_f), and their effect on particle and turbidity removal efficiency (n=24).

Parameter	Turbidity		Particle count	
Factor	P	Significance	P	Significance
G	0.66	N.S.	0.061	N.S.
t_f	0.087	N.S.	0.78	N.S.
G x t_f	0.049	*	0.049	*

P: significance level N.S.: not significant *: significant (P<0.05)

It can be concluded that the effects of G and t_f on the particle and turbidity removal efficiency, independent of each other, were not significant. The only significant effect on efficiency was asserted through the interaction of the flocculation G and t_f. This implies that for the lower flocculation G value increase of t_f increases the efficiency and for a higher flocculation G value increase of t_f decreases the efficiency.

Multiple comparison between different t_f within each G value, and between different G within each t_f (CER=0.05, EER=0.013) shows that the only significant difference exists between DAF at G=10 s^{-1} and t_f=20 min, and G=70 s^{-1} and t_f=20 min. Table 3.2 sums up the results of the multiple comparison analysis.

Table 3.2 Multiple comparison test between different treatment (varied G and t_f) and the effect on particle and turbidity removal efficiency (n=24).

Treatment	Particle count	Turbidity
G=10 s^{-1}, t_f=10 min	ab	a
G=10 s^{-1}, t_f=20 min	b	b
G=70 s^{-1}, t_f=10 min	ab	ab
G=70 s^{-1}, t_f=20 min	a	ab

Note : a, b and ab represent similarities between removal efficiency under different treatment conditions.

The ANOVA of the initial particle count for the four sets of experimental conditions showed that the initial particle count didn't vary significantly, suggesting that the particle removal results could be compared. Initial turbidity, however, differed significantly for different t_f, which may have affected the results from the experiments and made the above observations not fully correct. This implies that the results obtained from this set of experiments can not lead to fully reliable conclusions. Similar results were obtained if the initial particle (algae) count and turbidity were introduced as covariates in the ANOVA.

3.4 CONCLUSIONS

A correlation was found between particle count and turbidity (r^2=0.99), however of different slopes for the two different algal strains. Consequently, turbidity may be a useful parameter for prosess efficiency assessment, however less accurate than particle count.

Highest particle and turbidity removal was obtained at pH below the iso-electrical point (pH 6), related to particle (algae) charge neutralisation phenomena, and improved flocculation. The coagulation pH had a pronounced effect on the particle size distribution after flocculation; the concentration of smaller particles (<50 μm) was more significantly reduced on account of the formation of a a more significant concentration of larger particles (>50 μm) at pH 6 versus pH 8. Furthermore, the algae charge neutralisation which occurred at coagulation pH 6 contributed to the increased DAF efficiency due to improved particle (floc)-bubble attachment, the air bubbles being negatively charged.

Tapered flocculation did not improve DAF efficiency, while the possible existence of an unambiguous optimal Gt range for DAF, although suggested by statistical analysis, could not be confirmed. Sedimentation required an increase of initial particle concentration through the addition of high coagulant dosages in the range of 10 mg Fe(III)/L, and low flocculation G values in the range of 10-30 s^{-1}. This resulted in floc volume distributions with high frequency both in the smaller (<50 μm) and larger (>50 μm) particle size ranges. A low flocculation G value of 10 s^{-1} resulted in similar and better algae removal efficiency by DAF than the value of G=70 s^{-1}, producing a floc size distribution with larger and lighter flocs, favouring efficient DAF. The interaction between the flocculation G and t values was found as the only significant factor affecting the DAF process efficiency, each of them separately having no statistically significant effects on removal efficiency. Thus, apart from the coagulation related charge effects, the particle (floc) size and density which are determined by the flocculation process, play a highly important role in determining the down-stream process efficiency. Under optimal process conditions for DAF and sedimentation no statistically significant efficiency difference was observed for the removal of particles (algae), while the achieved turbidity removal was significantly better by sedimentation than by DAF.

Application of relatively low doses of cationic polyelectrolyte (Superfloc C-573) as coagulant aid (0.5 for DAF and 1.0 mg Superfloc C-573/L for sedimentation) in combination with previously optimised coagulant doses and flocculation conditions, increased process efficiency of both processes (without down-stream filtration) up to 99%. Together with an improved removal efficiency achieved by coagulation at low pH (pH 6 vs. pH 8), this suggests adsorption coagulation and charge neutralisation are the main coagulation mechanisms that support the dosage dependant sweep coagulation and improve the overall particle (algae) removal.

Results from bench scale experiments with model water and laboratory cultured *M. aeruginosa* showed that sedimentation or DAF alone is insufficient to cope with moderate to high initial concentrations of this species. Maximum particle (algae) removal efficiency achieved under optimal process conditions was 71% for DAF (3 mg Fe(III)/L, G=10 or 70 s^{-1}, t=10 min), and 87% for sedimentation (10 mg Fe(III)/L, G=30 s^{-1}, t=30 min). An integrated approach including the up-stream agglomeration (coagulation/flocculation) step, as well as the down-stream filtration step is essential. Adding a non-optimised filtration step in the experimental set-up increased removal efficiency in both cases, for DAF up to 90% and for sedimentation up to 95%. The filtration efficiency in this case (small scale filters connected to the jar test apparatus) was generally influenced by the flocculation G value and the resultant floc size/density ratio. It is also speculated that given the relatively low model water TOC of ≈2.5 mg/L, the residual coagulant (after DAF or sedimentation) is utilised in the filter in a "direct filtration" mode. Thus, the relatively low residual coagulant concentrations (from the point of view of direct filtration) result in relatively low filtration efficiencies. The filtration efficiency itself increased with the residual coagulant increase. Additionally, this situation may be negatively influenced by the (small) scale of the filters - filter wall effects and the non-ripened state of the filter, a situation which will be improved if a fully optimised filtration step is applied.

REFERENCES

1. Vlaški A., van Breemen A.N. and Alaerts G.J., 1994. *Algae and water treatment in the Netherlands, problem analysis and treatment strategies in five water companies.* IHE Working Paper EE-1, IHE, Delft, the Netherlands.

2. Petruševski B., van Breemen, A. N. and Alaerts, G.J., 1994. Optimisation of coagulation conditions for in-line direct filtration. *Workshop on Optimal Dosing of Coagulants and Flocculants*, Mûlheim an der Ruhr, Germany.
3. Edzwald J.K., 1993. Coagulation in drinking water treatment : particles, organics and coagulants. *Wat. Sci.Tech.*, **27**, 11.
4. Janssens J.G., 1992. Improved flocculation and dissolved air flotation for algae removal. *Emerging Technologies V Conference*, Vienna, Austria.
5. Janssens J.G., 1990. The application of dissolved air flotation in drinking water production, in particular for removing algae. *DVGW Wasserfachlichen Aussprachetagung Essen*, Germany.
6. AWWA Coagulation Committee, 1989. Committee report: coagulation as an integrated water treatment process - committee report, *J. AWWA*, **81**, pp. 72-78.
7. Ives K.J., 1956. Electrokinetic phenomena of planktonic algae. *Proc. Soc. Water Trtn. And Exam.*, **5**, pp.41-58.
8. Edzwald J.K., 1993. Algae, bubbles, coagulants and dissolved air flotation - *Wat. Sci. Tech.*, **27**, 10, pp. 67-81.
9. Vlaški A., van Breemen A.N. and Alaerts G.J., 1997.Evaluation and verification of the existing dissolved air flotation (DAF) kinetic model (article in preparation), IHE-Delft, the Netherlands.
10. Edzwald J.K., 1994. Principles and application of dissolved air flotation - *Joint IAWQ-IWSA Specialised Conference on Flotation Processes in Water and Sludge Treatment*, Orlando, USA.
11. Malley J.P., 1994. The use of selective and direct DAF for removal of particulate contaminants in drinking water treatment. *IAWQ-IWSA Joint Specialised Conference on Flotation Processes in Water and Sludge Treatment*, Orlando, USA.
12. Malley J.P. and Edzwald J.K., 1991. Concepts for dissolved-air flotation treatment of drinking waters - *J. Water SRT-Aqua*, **40**, 1, pp. 7-17.
13. Puffelen van J., 1993. Flotation - State of the art - Cursus moderne drinkwaterzuiveringstechnieken (Course Modern Water Treatment Technologies), Technical University-Delft, Delft, the Netherlands.
14. Boller M.A. and Kavanaugh M.C., 1995. Particle characteristics and headloss increase in granular media filtration. *Wat.Res.*, **29**, 4, pp.1139-1149.
15. Bernhardt H. and Clasen J., 1992. Studies on removal of planktonic algae by flocculation and filtration - Technical papers, Volume 2, *Water Malaysia '92, 8th ASPAC - IWSA Regional Water Supply Conference*, Kuala Lumpur, Malaysia.
16. Bhargava D.S. and Rajagopal K., 1992. An integrated expression for settling velocity of particles in water - *Wat. Res.*, **26**, 7, pp. 1005-1008.
17. Patry G.G. and Takacs I., 1992. Settling of flocculent suspensions in secondary clarifiers - *Wat. Res.*, **26**, 4, pp. 473-479.
18. Malley J.P. and Edzwald J.K., 1991. Laboratory comparison of DAF with conventional treatment - *J. AWWA*, **83**, pp. 56-61.
19. Edzwald J.K. and Wingler B.J., 1990. Chemical and physical aspects of dissolved-air flotation for the removal of algae - *J. Water SRT- Aqua*, **39**, pp. 24-35.
20. Zabel T., 1985. The advantages of dissolved-air flotation for water treatment - *J. AWWA*, **77**, pp. 42-46.
21. Petruševski B., Vlaski A., van Breemen A. N. and Alaerts G.J. 1993. Influence of

algal species and cultivation conditions on algal removal in direct filtration - *Wat. Sci. Tech.*, **27**, 11, pp. 211-220.
22. Tompkins J., De Ville M.M., Day J.G. and Turner M.F. (eds.), 1995. *Culture collection of algae and protozoa, Catalogue of strains* - Titus Wilson & Son Ltd, Kendal, United Kingdom .
23. Vlaški A., 1991. Aggregation of algae and direct filtration - IHE/TU-Delft, M.Sc. thesis, Delft, the Netherlands.
24. De Rijk S.E., van der Graaf J.H.J.M. and den Blanken J.G., 1994. Bubble size in flotation thickening - *Wat. Res.*, **28**, 2, pp. 465-473.
25. Alaerts G.J. and Van Haute A., 1982. Stability of colloid types and optimal dosing in water flocculation - *Physicochemical Methods for Water and Wastewater Treatment*, Pawlowski L. (ed.), Elsevier Publ., Amsterdam, the Netherlands, pp. 13-29.
26. Bernhardt H. and Clasen J., 1991. Flocculation of micro-organisms - *J. Water SRT-Aqua*, **40**, 2, pp. 76-87.
27. Schers G.J. and van Dijk J.C., 1992. Flotatie, de theorie en de praktijk (Flotation, the theory and practice), *H_2O*, 11, pp. 282-290.
28. Mackie R.I. and Bai R., 1993. The role of particle size distribution in the performance and modelling of filtration - *Wat. Sci. Tech.*, **27**, 10, pp. 19-34.

APPENDIX 3.1 Composition of the algae cultivating medium (Jaworski) used for semi-continuous culturing of *M. aeruginosa*).

ELEMENTS :	CONCENTRATION (μM) :	TRACE ELEMENTS :	CONCENTRATION (μM) :
Ca	84.67	B	40.0
N	1038.00	Mn	7.00
K	91.10	Fe	5.60
P	191.70	Mo	0.81
Mg	203.00		
S	203.00		
C	200.00		
Na	1213.00		

VITAMINS :	CONCENTRATION (mg/L) :	COMPLEXING AGENTS:	CONCENTRATION (μM) :
B-12	0.04	EDTA Fe Na	5.60
B-2	0.04	EDTA Na	6.04
Biotin	0.04		

CULTURING PROCEDURE : The nutrients were added into 1 L of demineralised water at pH 7, followed by autoclave for a period of 20 min at 120°C. The vitamins solution which had been previously sterilised by filtration at pH 7, was added to the medium afterwards. The incubation of algae (80 mL of *Microcystis aeruginosa* suspension at the end of the logarithmic growth phase) was done in 5 L glass vessels (Duran 50, Schott Mainz, Jena, Germany) at 20°C. The algae suspension was aerated with compressed air passed through a sterile 0.2 μm filter, while continuous mixing was applied at the bottom of the vessel with a magnetic stirrer. The light cycle was continuous at light intensity of ± 1,000 lux. For this purpose a vertically placed T.L. lamp Osram-Cool White L18 W20 was used, placed at a distance of 0.2 m from the walls of the vessel. The stationary algae growth phase was reached after an incubation period of 14-17 days, after which 2 L of the algae suspension were sampled in sterile conditions and used in experiments. The abstracted 2 L of algae suspension were substituted with fresh medium and incubated under the same conditions for 7-10 days, after which algae were sampled for experiments again. This completed the culturing cycle, after which a new culture was started-up.

Chapter 4

POLYELECTROLYTE ENHANCED COAGULATION AND ALGAE CONDITIONING BY OZONE OR $KMnO_4$ IN THE CONTEXT OF EFFICIENT DISSOLVED AIR FLOTATION

ABSTRACT : The research investigated ways to improve particle (algae) coagulation and down-stream dissolved air flotation (DAF) removal, by particle (algae) conditioning. The application of organic and synthetic polyelectrolytes, of cationic, non-ionic and anionic charge nature was investigated. The polyelectrolytes were applied as sole coagulant or as coagulant aid with Fe(III) primary coagulant. Bench-scale experiments on model water (tap water spiked with laboratory cultured algae) confirm that cationic polyelectrolytes inducing charge neutralisation and the adsorption-coagulation mechanism are superior to anionic and non-ionic polyelectrolytes associated with the bridging mechanism. The application of oxidants was investigated for the same purpose. Ozone and $KMnO_4$ improved the particle (alga) removal. Ozone doses as low as 0.2 mg O_3/mg C (0.48 mg O_3/L) increased the DAF process efficiency. Similarly, 0.7 mg $KMnO_4$/L resulted already in improved DAF process efficiency, however with lower reproducibility and the drawback of high Mn residual, mainly caused by the colloidal MnO_2 with an accompanying turbidity increase. Chemical modifications of the alga cell layer, and the production of in-situ coagulant aid caused by extra-cellular organic matter (EOM) release and in some cases intra-cellular organic matter (IOM) leakage, are considered to be the major process mechanisms responsible for the efficiency improvement. Cationic polyelectrolytes had a positive impact on DAF process efficiency in a combination of ozone or $KMnO_4$ with inorganic coagulant, as well as in combination of ozone or $KMnO_4$ with the polyelectrolyte as sole coagulant. The dual coagulant system lowered the overall coagulant demand and reduced the coagulant and Mn residual concentrations. The polyelectrolyte coagulant alone proved to be more sensitive to the ozone dose, but avoided any metal coagulant residue.

4.1 INTRODUCTION

4.1.1 Algae laden water treatment

The variety and the severity of problems related to the treatment of algae laden water originates from the wide diversity of algal species, as well as from their seasonal composition and concentration variations. The differences in their morphological and physiological characteristics include size, shape (form), cell surface characteristics (composition and charge), mobility, extra-cellular organic matter (EOM) production and excretion, growth stage, ability to quickly respond to nutrient composition and concentration changes. This can result in 'blooms', excessive concentrations of certain algal species, and periodically results in lower removal efficiency of water treatment. It has been attempted to describe algae removal by approaching them as inert particles and applying the classical coagulation theory whereby they can be destabilised, aggregated and thus brought in filterable form [1]. However, the great diversity of algae accounts for the very different responses to their removal. This calls for a more complex and integrated approach, requiring the establishment of combinations of conventional and advanced treatment unit operations that are suitable for particular circumstances. Membrane technology suggests it can cope efficiently with the problem of high algae concentrations. However, little specific experience is available. Although membranes would in principle be able to remove even the small μ-algae (in the size range of 1 μm), the fact that the very high algae concentrations encountered during blooms generally clog rapid filters and microstrainers, suggests similar problems may occur with membranes.

Conventional water treatment processes for algae laden water usually comprises sedimentation or dissolved air flotation, followed by filtration. The selection of either of the two first processes depends on evaluation of cost and performance, carefully assessed on a per case basis [2]. A bench scale study on model water (with laboratory cultured *Microcystis aeruginosa)* showed a two to three times lower coagulant demand for the optimal algae removal by DAF (3-5 mg Fe(III)/L coagulant) compared to sedimentation (10 mg Fe(III)/L coagulant) [3]. This means lower sludge production and higher solids content. The removal efficiency of dissolved, colloidal and particulate matter in sedimentation, flotation or filtration is primarily defined by the agglomeration (coagulation and flocculation) step. Particle coagulation and flocculation may also prove to be of considerable importance in membrane filtration technology. Depending on pores size, coagulated and flocculated matter may foul the membrane surface, rather than allowing penetration and clogging of membrane pores by suspended and colloidal matter. This suggests easier membrane cleaning [4]. Clearly, 'optimal' coagulation conditions are defined by the down-stream treatment process.

Optimisation of the flocculation of model water (*M. aeruginosa*) via DAF jar tests, resulted in relatively poor algae removal efficiencies in the range of 50-70 % (Chapter 3) [3]. Control of pH (pH$<$IEP) and application of cationic polyelectrolytes increased the removal efficiency to the 99% range, confirming the importance of coagulation conditions. Based on the same rationale, algae conditioning by oxidants such as ozone and $KMnO_4$ should be considered.

4.1.2 Coagulation with metal salts

The stability of mineral and organic particles is attributed to electrostatic charge interactions, hydrophilic effects, or to steric interactions from adsorbed macromolecules [5]. Coagulation with metal salts causes destabilisation of colloidal suspensions by changing the colloids' surface charge, or by physically destabilising them through encapsulation in large precipitating metal-hydroxo flocs. It is a conditioning process during which the concentration, composition, particle size and particle volume distribution are modified for more efficient down-stream solid-liquid separation by sedimentation, dissolved air flotation, or filtration. The natural organic matter (NOM) which consists of a mixture of various hydrophobic (negatively charged humic and fulvic acids due to carboxylic and phenolic groups comprises about 45% of DOC in rivers) and hydrophilic compounds, rather than the particles, controls the coagulant dose and selection [5]. Coagulation with metal coagulants can also facilitate the conversion of dissolved natural organic matter or molecules (NOM) into insoluble metal-organic products, which are presumably easier to remove physically. The metal-organic reaction products are formed via NOM-coagulant co-precipitation reactions, sorption of NOM onto precipitated coagulant, or combination of these two mechanisms [6]. However, even at very high coagulant doses, a portion of 5-60% of the total NOM is not affected by coagulation [7], presumably due to lack of functional groups capable of coordinating with the hydroxide surface [8].

Iron and aluminum salts are the most common coagulants due to their price and availability, as well as their high efficiency [9]. In the Netherlands, iron coagulants are preferred because of health and environmental concerns related to aluminium. The destabilisation of colloidal and particulate matter and the dominant NOM-coagulant interaction mechanisms are affected by a range of factors including the water quality (ions concentration and composition, temperature related viscosity, nature and concentration of colloidal and particulate matter, NOM origin, etc.), coagulant type and dose, and coagulation pH [9, 10].

When introduced into water, iron and aluminum salts undergo a series of very fast hydrolytic reactions. In the case of iron, these reactions result in the formation of the initial monomeric hydrolysis products of $Fe(OH)^{2+}$, $Fe(OH)_2^+$ and $Fe(OH)_3$. The continuing process of nucleation or polymerisation of the monomeric $Fe(OH)_3$ into amorphous polymeric $Fe_n(OH)_{3n}$ results into precipitation and enmeshment of colloidal and suspended matter within these precipitates [10]. The coagulation pH defines the concentration and composition of the metal hydrolysis products and their solubility. The precipitation of metal hydroxide species with enmeshment of impurities is referred to as sweep coagulation, as opposed to adsorption coagulation resulting from surface charge neutralisation followed by particle agglomeration. Although sweep coagulation often is the dominant coagulation mechanism under regular water treatment circumstances, adsorption coagulation is more efficient if one considers the effect per amount of added coagulant. However, it requires adjustment of the coagulation pH below, but close to the iso-electrical point (IEP), i.e. the point of zero average surface charge of the iron hydrolysis products. This enhances flocculation of the negatively charged colloids by the positively charged iron hydrolysis products [3, 11, 12]. The coagulation pH affects the resulting particle (floc) size distribution after flocculation and hence significantly affects the down-stream removal efficiency (Chapter 3) [13]. The IEP for iron-hydroxo complexes is most frequently reported to be in the acidic pH range (between 6 and 7) [9, 14], and strongly influenced by the ionic content and composition of the water (notably Ca^{2+}, HCO_3^-, SO_4^{2-}, HPO_4^{2-}) [12, 14, 15]. These ions can shift the IEP to lower or higher values or even prevent charge reversal. The colloid concentration usually affects the coagulant dose in an inversely proportional manner when sweep coagulation dominates [16], but a stoichiometric relation tends to exist with spherically shaped algae such as the cyanobacterium *Synechocystis miniscula* (mean diameter of 3 μm, similar to the lower size range of single cell *M. aeruginosa*) in adsorption-coagulation conditions [11]. Drinking water treatment for floc removal with sedimentation or DAF occurs with sweep coagulation given the slightly alkaline pH values under natural circumstances.

4.1.3 Coagulation with polyelectrolytes

Polyelectrolytes or polymers are sometimes applied as single coagulants, but more often as coagulant aids in conjunction with either alum or ferric chloride [8]. Together with metal coagulants they can facilitate conversion of soluble NOM molecules into insoluble NOM-coagulant reaction products, either through precipitation of NOM or sorption of NOM molecules onto the surface of suspended particles. When low molecular weight (MW) cationic polymers are added to a solution, precipitation is the only likely mechanism of soluble NOM removal [6].

The chemical structure of natural or synthetic polyelectrolyte molecules is characterised by repeating (monomer) chemical units in a branched or linear manner. Their molecular weight is usually 10^4 - 10^7, and they are water soluble due to hydration of their functional groups. The polymers carry an overall cationic or anionic charge, or they may be non-ionic [10]. The charge, and adsorption characteristics define the particle destabilisation capacity, the most relevant for drinking water treatment being the bridging and the adsorption coagulation mechanisms which in some cases operate simultaneously [10, 17]. The first mechanism is based on the presumption that the polyelectrolyte molecules are of similar dimension as many of the colloidal particles and operate via adsorption of polyelectrolyte to more than one particle, thus binding the particles together. This may be the case when non-ionic or anionic polyelectrolytes are applied. In the case of anionic polyelectrolytes, the particle adsorption occurs through interaction with metal ions, such as calcium, on the particle's surface. As for the non-ionic polyelectrolytes, their adsorption to negatively charged particles is mostly via hydrogen bonding. The high efficiency of cationic polyelectrolytes is based mainly on their charge being opposite to that of the colloidal and particulate matter. The strong attraction of polymer and colloidal surface would result in a configuration of the polyelectrolyte molecules lying flat on the particle surface, making the bridging mechanism less likely. Similar phenomena can be expected to occur with anionic and non-ionic polyelectrolytes if the ionic strength of the water is very high. In has been suggested [17] that the polyelectrolyte can assume a localised 'patch'- like structure on the particle resulting in a mosaic type of charge distribution, with patches of positive and negative charges co-existing on the particle. When oppositely charged 'patched' surfaces come into contact with each other, this results in strong inter-particle attachment and agglomeration.

Polyelectrolytes can be very efficient as sole coagulants in the treatment of turbid or coloured water containing humic acids as well as various microorganisms [10]. Polyelectrolytes don't need major pH adjustment, prevent the carry over of soluble organo-metal complexes to the solid-liquid separation phase, minimise the carry over of light flocs formed in the case of metal coagulant application, and finally reduce the sludge volume and the amount of soluble anions. These advantages promoted intensive use of synthetic polyelectrolytes, notably in the USA [18]. Disadvantages pertain to their high degree of selectivity for certain types of colloids, high unit cost, poor biodegradability for synthetic polyelectrolytes, and uncertainty regarding carcinogenicity and mutagenicity of their monomers [18]. This is compounded by the absence of an appropriate analytical methodology for their quantification, as well as of data on their exact chemical structure. However, natural polyelectrolytes such as chitosan (cationic) and sodium alginate (anionic) can be as efficient as synthetic ones, without the related toxic hazards [18]. This has resulted in a cautious approach to polyelectrolyte application in many countries including the Netherlands where officially only one non-ionic and one anionic organic (starch based) polyelectrolyte is currently approved.

Therefore, the application of natural polyelectrolytes, as well as lower polyelectrolyte doses seems attractive. Numerous cases of improvement of solid-liquid separation efficiency by small concentration of coagulant aid have been reported [3, 18, 19, 20, 21, 22, 23]. Their application typically leads to larger, but also denser floc material (Chapter 3) [13]. One of the main advantages is the accompanying reduction of prime coagulant dose requirement. The appropriate polyelectrolyte needs to be determined on a case to case base, depending on the raw water quality. Anionic and non-ionic polyelectrolyte coagulant aids have been proven efficient mostly for sedimentation where relatively large mineral coagulant doses induce sweep coagulation; by increasing the overall particle concentration they introduce surface area for the bridging mechanism. However, cationic polyelectrolytes have generally been recognised as more effective. For coagulation of humic substances, bacteria and algae they are considered the only option [21]. In the case of cationic coagulant aid, the sweep coagulation is significantly aided by the polyelectrolyte induced adsorption coagulation. This option is attractive if pH reduction below the IEP is not feasible, for example due to high buffering capacity [22].

Critical in the use of polyelectrolytes is the mixing intensity in the coagulation unit. Velocity gradients (G values) of 400-1,000 s^{-1} have been recommended. Larger values possibly give rise to floc break-up, while lower values provide for insufficient contact opportunities. The polyelectrolyte dose may also affect the optimal velocity gradient value, and is generally inversely proportional [10].

The application of cationic polyelectrolytes in the DAF saturation unit has been shown to cause coating of air bubbles with these polyelectrolytes. This may result in increased particle removal efficiency, based on more intensive attraction and efficient adsorption between negatively charged particles and coated positively charged air bubbles [23].

4.1.4 Conditioning with oxidants

Numerous experiences at both, laboratory and operational scale [22, 24, 25, 26, 27, 28, 29], suggest that oxidants conditioning enhances coagulation. These effects accompany the primary disinfectant role of most of the oxidants including chlorine, chlorine dioxide, ozone and potassium permanganate. They are caused by chemical interactions between the oxidant and the dissolved, colloidal and particulate matter (including algae). However, hazardous oxidation by-products are associated with certain oxidants, such as trihalomethanes (THM), haloacetic acids (HAA) and haloacetonitriles (HAN) with chlorine, chlorite with chlorine dioxide, and bromate with ozone [30]. The formation of these by-products is influenced by a range of factors; careful setting of, for example ozone dose or ozonation pH, allow to maintain bromate levels below the guideline values [31, 32, 33]. The application of advanced oxidation processes where combinations of oxidants is considered, may result in improved process efficiency at reduced health hazard risks [34].

The process kinetics of coagulation enhancement by oxidant conditioning depends on the oxidant, its dose, and the raw water quality characteristics. In case of ozone conditioning a series of chemical reactions takes place with a net positive or negative effect on the coagulation process. The process mechanisms of ozone induced coagulation improvement include [35]:

- reactions with NOM leading to a decrease of their molecular weight and increase in their functional group content, and resulting increase of sorption of these compounds on metal-hydroxide flocs;
- increased complexation of calcium with NOM;
- reduced particle stability, due to ozone reacting with the NOM coating the colloids;
- break-up of organo-metal complexes, yielding coagulant in-situ;
- reactions of ozone induced meta-stable ozonides, peroxides and free radicals with NOM,

after the initial ozone has been depleted;
- immobilisation of zoo- and phytoplankton (especially motile algae), and disruption of their metabolic processes;
- reactions with functional groups on alga cell surface reducing cell surface charge, as well as production of surface active substances which behave as biopolymers and natural coagulant aids.

As for $KMnO_4$ conditioning, coagulation is facilitated by similar mechanisms as far as reactions with the NOM and algae are concerned, but the effect includes the additional benefits caused by the formation of the colloidal hydrous MnO_2. The MnO_2 can improve coagulation efficiency by increasing the overall particle concentration and aiding the sweep coagulation, by acting as an adsorption agent for organic and inorganic matter, and by adsorbing on algal cells and assisting their further coagulation by metal coagulants [36].

THM formation discredited chlorine based oxidants in the seventies. Ozone gained considerable attention in the eighties, especially in the USA and France. However, the establishment of a relation between ozone and the formation of brominated organohalogens or bromate in bromide containing water at potentially hazardous levels, reflected negatively on this technique [32, 33, 37, 38, 39]. The formation of bromate is significantly influenced by the ozone dose and ozonation pH which may prove to be decisive for its feasibility [30, 32, 33, 34, 39]. Furthermore, the application of ozone raises the assimilable organic carbon (AOC) in the treated water, and if no additional retainment measure for AOC is present before the distribution network, this may result in bacterial after-growth [39, 40]. If such a barrier, e.g. in the form of granular activated carbon (GAC) filtration exists, ozone will improve the biodegradability of the organic matter and thus may prove beneficial for GAC filter operation [39, 41].

A non-hazardous alternative to ozone is $KMnO_4$ [22, 30, 42]. This very strong oxidant has extensively been used, mainly for taste and odour control, and for Fe(II) and Mn(II) removal. It has gained increased attention lately because of its disinfecting and algicidal effects, its effect on the trihalomethanes formation potential, and as a particle (algae) conditioner before coagulation. However, overdosing of permanganate results in a characteristic, objectionable pink colour of the treated water, but this also allows to optimise its application. Apart from the high cost, its drawback is the potentially higher residual Mn in the effluent.

Application of ozone or $KMnO_4$ in combination with polyelectrolytes is proposed to be a viable option for further increasing removal efficiency. It has been shown that dual coagulant systems (Fe or Al salts in combination with primarily cationic polyelectrolytes) in combination with ozone or $KMnO_4$ perform better than the combination of ozone or $KMnO_4$ and coagulant alone [36]. If one defines a critical coagulant concentration (CCC) as the concentration of each of the considered polyelectrolyte and metal coagulant necessary to reach a 0 mV particle charge when used separately, their joined impact is more efficient and not according to the additive principle (linear additivity model) [6]. The impact on coagulant demand and solid-liquid separation process efficiency depends on the characteristics of the polyelectrolyte and its ratio to the Fe or alum salt, under particular raw water circumstances (particle concentration, ionic composition, pH, etc.) [6]. Ozone or $KMnO_4$ conditioning can have an impact on the raw water quality (e.g. ozone can modify organic matter and influence the DOC composition via EOM and IOM leakage, while $KMnO_4$ additionally results in MnO_2 production), and thus affect the outcome of the combined polyelectrolyte-metal coagulant treatment in many ways. Efficiency improvement may be caused by i.e. a decrease of the negative particle (algae) charge in the case of ozone conditioning [3, 22, 36, 43], or adsorption of the negatively charged colloidal MnO_2 in the case of $KMnO_4$ conditioning [22, 36].

4.1.5 Ozone conditioning

Ozone exerts a strong disinfectant activity (inactivation of viruses and other pathogenic microorganisms) [39], and is effective in colour removal, taste and odour improvement, reduction of phenolic compounds and UV_{254} absorbance, whilst the THM formation potential can be reduced [22, 36].

The ozone dose is very critical. It is primarily determined by the NOM content. Coagulation and down-stream process efficiency are known to improve very significantly at ozone doses of 0.3-0.4 mg O_3/mg C (0.5-1.5 mg O_3/L), while higher doses can lead to deteriorated effluent quality [35]. High ozone doses may lead to algal cell break-up and release of IOM comprised of polysaccharides, proteins and nucleic acids into the water, which may in some cases inhibit the flocculation process. The critical ozone dose differs for alga species [25, 43, 44], depending on the outer algal cell wall composition (hence on age) as well as on the raw water concentration and composition of the NOM. IOM leakage may also benefit the flocculation process, since it can serve as a natural coagulant aid.

Raw water quality strongly influences the ozone conditioning. Ozone conditioning of the same water (Seine River, France) at different locations and times resulted in variable, even conflicting results [35]. These differences are mainly the result of variations in NOM concentration and composition. A minimal concentration of organic matter is needed as a prerequisite for the occurrence of the beneficial effect of ozone on coagulation [45]. Seasonal algae blooms and their EOM seem to play a crucial role. Consequently, different species may cause different effects since they respond differently to ozone treatment; for example, some species are considered the cause of the formation of a thick organic foam layer in water treatment plants, due to ozone induced release of surface-active biopolymers [26]. The effect of ozone on particles coagulation is closely related to its reactions with their NOM coating [29]. The surface properties of particles are altered in the sense of changes in the functional group content and an increase of the carboxyl content of NOM. These reactions provoke the formation of organic radicals that may lead to polymerisation reactions. Thus, by changing the surface properties of the particles the reactions with metal coagulants proceed in a different manner, aided by the formed polymers [45]. Ozone conditioning generally shifts the particle size distribution towards larger particles and changes the organic matter phase from dissolved to colloidal, with a positive down-stream flocculation effect.

Ozone application generally decreases the molecular weight (MW) and increases the acidic functional group content and the hydrophilicity of the NOM [6]. This leads to a net decrease of the tendency of NOM molecules to be removed via precipitation reactions with the metal hydroxide coagulant. On the other hand, the impact of ozone on NOM removal via sorption reactions is more complex; while the increased acidic functional group content would tend to increase sorption, the decreased MW has the opposite effect. Conceptually, the overall affinity of metal-hydroxide surfaces for ionised organic adsorbates may be divided into two components. One component accounts for the non-electrostatic forces between the ion and the surface and depends on their quality. The other component accounts for the electrostatic portion of the affinity, i.e. the repulsion/attraction between the ion and the surface, which is a function of the adsorbate charge and the surface potential (consequently sensitive to pH, ionic strength and the presence of other adsorbates). Thus, the overall coagulation effect will be a net effect of the precipitation and sorption NOM-coagulant reactions. Organic matter (TOC) removal in the case of ozonation is a combined effect of oxidation, stripping and coagulation; It has been found that it improved with increased ozone dose at very high and at low coagulant doses, and it decreased at moderate coagulant doses. Furthermore, the application of ozone in combination with polyelectrolyte coagulants may result in a masked effect of 'reduced coagulant demand', which is a result of the transformation of the NOM to lower MW, and the creation of more hydrophilic species, thus decreasing the required polyelectrolyte coagulant dose for NOM precipitation, as the amount of precipitable NOM decreases. The complexity of the ozone induced NOM changes and the variations in NOM concentration and composition at different locations and time explain the often inconsistent data on TOC removal in the literature [6].

Similarly, calcium hardness influences ozone induced changes of the coagulation process. The coagulation of NOM coated particles improves in the presence of divalent cations such as calcium [26, 43, 46]. The effects of Ca^{2+} include adsorption and charge neutralisation through calcium complexation with both, surface functional groups on the metal oxide and with the NOM coating of particles, double layer compression by calcium, and calcium bridging between negatively charged humic matter or solid surfaces. Fig. 4.1 describes the relationship between the influence of the raw water characteristics (hardness and TOC) and the ozone induced coagulating effect, expressed via the particle collision efficiency coefficient α. Accordingly, optimal ozone induced coagulation takes place at hardness to TOC ratios larger than 25mg $CaCO_3$/mgC and ozone doses in the range of 0.4-0.8mg O_3/mgC [47].

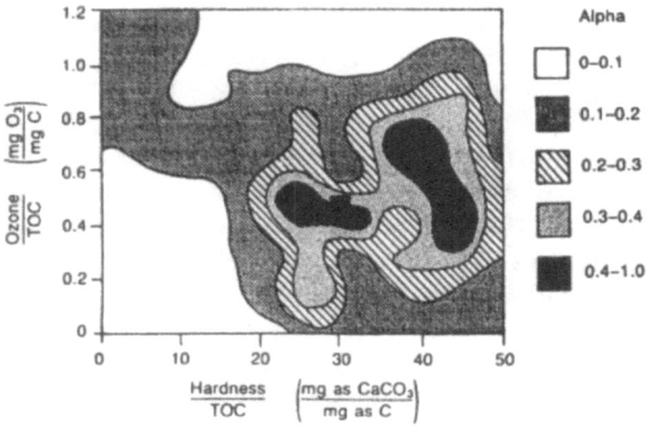

Fig. 4.1 Effect of raw water quality (hardness, TOC) and ozone dose on collision efficiency α [47].

The time lag between ozone application and application of coagulant is also important for coagulation and flocculation efficiency. It has been speculated [36] that the ozone induced flocculation improvement is a reversible process affected by the time lag between ozone and coagulant addition. The shorter it is (in the range of 1-2 min), the better the results from the down-stream (direct) filtration were. This implies that there is no need of large contact units if ozone is applied for conditioning purposes.

It is often difficult to distinguish all occurring phenomena, however, a comprehensive overview of mechanisms involved in the ozone induced algae coagulation improvement includes [36] :

- splitting of EOM molecules;
- alteration and removal of EOM and NOM adsorbed on particles;
- reaction with adsorbed organic coating on algal cell surface, responsible for enhanced algae stability;
- reaction with and modification of the algal cell wall structure resulting in reduction of their stability;
- release of algae IOM (Intracellular Organic Matter) which (further modified by ozone) may act as a natural coagulant aid biopolymer and enhance coagulation.

4.1.6 $KMnO_4$ conditioning

$KMnO_4$ conditioning as an option for coagulation improvement has been considered only recently. The application of this strong oxidant dates back to the beginning of this century, mainly to control noxious taste and odour causing compounds [48, 49, 50]. Ever since it has been used also for Mn(II) and Fe(II) oxidation and conversion from dissolved to colloidal form in ground and surface water treatment [48, 51], as an algicide for phyto- and zooplankton growth control in reservoirs [52, 53, 54], and as an oxidant to replace chlorine for THM avoidance [55, 56, 57]. Recent studies on the applicability of permanganate in the context of improved particle coagulation and flocculation have proven its beneficial effect, showing that it may behave as a coagulant aid and substantially improve direct filtration efficiency in particular [22, 42]. Finally, permanganate has also been used for disinfection purposes and for hydrogen sulphide removal [58].

$KMnO_4$ oxidizes chemical contaminants and biomass [50]. The standard electric potential for the oxidation reaction of permanganate indicates that it can oxidize most inorganic compounds, as well as virtually any organic compound [59]. Its oxidation rate is influenced mainly by pH, temperature and the presence of certain catalytic ions [48]. Under normal pH conditions of 7-8.5, the insoluble hydrous MnO_2 (of 0.3-0.4 μm size [59]) is the end product of permanganate reduction, the reaction rate being higher for pH greater than 7. $KMnO_4$ owes its overall efficiency to the combined oxidative activity and the adsorption characteristics of its oxidation by-product, i.e. the MnO_2. This applies for particles (algae) conditioning as well. The mechanisms involved in the $KMnO_4$ induced coagulation and flocculation improvement are similar to those for ozone application, but in addition include the activity of the MnO_2 :

- increase of flocculation efficiency due to the increase of overall particle concentration aiding sweep coagulation;
- adsorption of/on inorganic and organic matter (including algae), as well as adsorption of cations (and anions) of inorganic and organic origin.

Depending on the formation conditions, a more correct notation of manganese dioxide would be $MnO_{(1+x)}$, where x varies from 0.1-0.95 [60]. The initial (colloidal) size of its precipitate is 0.3-0.4 μm, its surface atoms assuming a well ordered crystal structure with exchangeable hydrogen and hydroxide ions bound on the surface. The hydrous MnO_2 is amphoteric, however, under the pH conditions of water treatment it is of negative charge [60]. Its colloidal size and resulting large surface area (about 300 m^2/g) promote it as a very fast and efficient sorption agent, especially for multivalent metal cations and cationic organics. This does not apply for non-ionic and anionic organics [46, 60]. The principal mechanisms underlying the metal and organic ion sorption are ion exchange sorption and direct ion attraction [60]. The ion exchange properties of MnO_2 are related to acidic functional groups at its surface, resulting in a weak acid behaviour and a strong dependency on pH (increase of pH resulting in increased negative charge and cation exchange capacity).

The increased residual Mn concentration may limit $KMnO_4$ application to a certain degree. The residual is of characteristic pink colour. MnO_2 may impart a dark brown to yellowish colour, dependent on pH. Unlike the permanganate it can presumably be removed by filtration, however, shortening the filter runs. Its size ranks it into a particle category which is difficult to remove by filtration. Cationic polyelectrolytes are reported to improve filtration also via reducing the residual Mn concentrations [36]. As for the permanganate residual, reducing agents such as Na_2SO_3 are proven to be beneficial. However, organic matter deposits on the filter grains material may also exhibit a substantial permanganate demand and act as its reducer [50]. A careful permanganate dosing pattern, e.g. maintaining pink colour in two thirds of the sedimentation tank, may allow for a full utilisation of the oxidizing potential of the permanganate [61], and reduce the necessity of additional chemical permanganate reduction by allowing the filters to satisfy their permanganate demand.

Although application of permanganate may occasionally have a negative impact on the THM formation potential [55], this has little relevance for Dutch drinking water production in which chlorination is not extensively practised. Importantly, presently no direct or indirect hazardous by-products are known to be caused by the permanganate application.

4.1.7 Research rationale

Based on the circular shape and size characteristics of the single cell *M. aeruginosa*, one of the research goals is to test the existence of a stoichiometric relation between their concentration and the coagulant dose necessary for their optimal destabilisation and down-stream DAF removal. The relatively low DAF removal efficiency of algae (*M. aeruginosa*) of 50-70% achieved in the model water research (Chapter 3), as well as the significant increase (up to 99%) of the algae removal via coagulation pH control (pH<7) and cationic polyelectrolyte C-573 coagulant aid, suggest the existence of an operational window for process operation improvement. For the purpose of DAF process enhancement the application of polyelectrolytes of different charge characteristics and in different modes, i.e. as sole coagulant or coagulant aid in combination with iron coagulant, is to be assessed. In the latter case, polyelectrolytes will be applied in order to test the possibility of raising the Fe(III) coagulant efficiency. The process efficiency is to be assessed in terms of the polyelectrolytes' charge and weight characteristics. Another option for DAF process enhancement which is to be studied is conditioning with oxidants, i.e. ozone and $KMnO_4$. The feasibility of the following treatment enhancement options is to be addressed in the research: (i) combined treatment by (iron) coagulant and oxidant conditioning, (ii) (iron) coagulant, polyelectrolyte coagulant aid, and oxidant conditioning, and (iii) polyelectrolyte coagulant and oxidant conditioning. Different coagulant, oxidant and polyelectrolyte dose combinations will be tested in the context of maximising DAF process efficiency and minimising chemicals consumption. The research will also address and investigate some oxidant specific aspects like: oxidants imparted microflocculation phenomena, point of (metal) coagulant dosing relative to the oxidant, temporal algae concentration variation and reduction of $KMnO_4$ by Na_2SO_3.

The research aims at providing deeper and more comprehensive insights into the factors that affect the removal efficiency of algae and the governing process mechanisms. The impact of different treatment combinations is to be assessed by turbidity, particle count, residual coagulant, UV absorbance at 254 nm, and algae and floc surface charge measurements, while computer image analysis will provide insight into (micro)flocculation phenomena. Scanning electron microscopy (SEM) is to be used to assess the impact of ozone on the algal cell structure. The research will also consider and assess the potential risks related to the application of these treatment enhancement technologies. The outcome should be the provision of a more solid knowledge basis for their eventual application in the context of solving short and long term algae treatment problems.

4.2 MATERIALS AND METHODS

4.2.1 Experimental set-up

The laboratory bench-scale experimental set-up consisted of a jar test apparatus with incorporated DAF facilities (Fig. 3.1). The recirculation ratio in the DAF unit was 7%, while the saturator pressure was 500 kPa. The experimental procedure included coagulation at a pH 8, temperature of 20 °C and G of 1,000 s^{-1} for 30 s. Based on previous experimental work the flocculation G was set at G_f= 10, 30 or 70 s^{-1} and the flocculation time at t_f= 10 or 15 min. $FeCl_3 \cdot 6H_2O$ was used as coagulant. The dose range for the optimisation of algae removal and testing for the existence of stoichiometry between algae concentration and optimal removal coagulant dose was 0-15 mg Fe(III)/L. Based on the analysis from Chapter 1, the selected cyanobacterium species was a semi-continuously, laboratory cultured, single cell form of *M. aeruginosa* (see Appendix 3.1 for details on culturing conditions) spiked into water originating from the Biesbosch storage reservoirs (North Brabant, the Netherlands) to different initial concentrations of ≈ 5,000, 10,000, 15,000, 20,000 and 25,000 ±cells/mL. The optimal coagulant dose is defined as the dose at which highest particle (algae), turbidity and coagulant removal takes place.

Four brands of polyelectrolytes were tested as coagulants and as coagulant aids. Two of these were of synthetic (cationic Superfloc C-573 and anionic Superfloc A-100) and two of organic (cationic Wisprofloc-P and non-ionic Wisprofloc-N) nature. Superfloc C-573 is a quaternary amine based, low molecular weight, cationic polyelectrolyte of 50% active matter, used in the 1.0-10.0 mg/L dose range as sole coagulant and 0.05-1.0 mg/L as coagulant aid. Superfloc A-100 is a high molecular weight polyacrylamide based anionic polyelectrolyte of 100% active matter applied in the dose range of 0.2-10.0 mg/L as sole coagulant and 0.05-1.0 mg/L as coagulant aid. The Wisprofloc-P is a soluble potato starch derivative cationic polyelectrolyte applied in a dose range of 0.25-5 mg/L as coagulant or coagulant aid. The Wisprofloc-N is of similar origin and characteristics as the Wisprofloc-P, however it is non-ionic. Polyelectrolytes were prepared in accordance to manufacturers recommendations. This was accomplished by dissolving the liquid or powder polyelectrolyte into demineralised water before each set of experiments (to avoid problems caused by polyelectrolyte aging and inactivation). The preparation of working and stock solutions was done in plastic utensils to avoid loss of polyelectrolyte due to wall adhesion. The use of polyelectrolytes as sole coagulants was tested in a dose range of 0.2-3 mg/L and included rapid mixing at a G value of 500 s^{-1} for 30 s, while the flocculation followed the same pattern as in the case of metal coagulant application. The application of polyelectrolytes as coagulant aids was done in the dose range of 0.1-1.0 mg/L, under the same coagulation G and t conditions, following the application of the previously optimised primary metal coagulant dose (Chapter 3).Tests were limited to polyelectrolyte combinations with the optimal Fe coagulant dose due to time limitations.

For ozone production in the conditioning experiments the Trailigaz LABO LO, France, ozone generator was used. Ozone was produced under standard ozone generator conditions of 220 V, 0.6 A, and 0.6-0.7 bar pressure. The raw water hardness (as $CaCO_3$) to TOC ratio was in the range of 65-70 and thus >25, a value suggested to warrant ozone induced coagulation/flocculation improvement. Most of the TOC was in the dissolved form (DOC≈80% of TOC). Based on the model water TOC concentration for the 10,000 cells/mL algae concentration which was measured prior to each ozone experiment (mean value of 2.4 mg C/L), and the prescribed ozone dose range for improved coagulation/flocculation of 0.4-0.8 mg O_3/mg C [47], three ozone doses were applied. A lower (0.2 mg O_3/mg C), within the prescribed range (0.5 mg O_3/mg C), and a higher (0.9 mg O_3 /mg C) ozone dose were applied, while in some specifically mentioned cases a high dose of 1.3 mg O_3/mg C was also applied. The ozone dosage (transferred ozone) was derived from the ozonation time which was varied from 1-4.5 min. The pH of the raw (model) water subject to ozonation was previously set to 7.5 by HCl addition.

A 5.5 L ozone resistant glass jar was used as a batch ozone reactor. The time gap between ozonation and subsequent coagulation/flocculation was kept at 2 min, in order to comply with previous positive experiences with short time gaps [36]. The pH change after ozonation was 7.5±0.2 and it served as coagulation pH without additional corrections. Ozone was applied in combination with the previously determined optimal coagulant dose for a concentration of ≈10,000 cells/mL of *M. aeruginosa* (3 mg Fe(III)/L), as well as with a 50% lower (1.5 mg Fe(III)/L) and a 50% higher (4.5 mg Fe(III)/L) coagulant dose. The same strategy was applied for a ≈5,000 cells/mL *M. aeruginosa* concentration (ozone combined with 1, 2 or 3 mg Fe(III)/L). The cationic polyelectrolytes Superfloc C-573 and Wisprofloc-P were tested as sole coagulants (0.3-1.0 mg/L), or as coagulant aids (0.1-1.0 mg/L) to Fe(III) coagulant, combined with a low (0.2 mg O_3/mg C) and high (0.9 mg O_3/mg C) ozone dose.

For permanganate conditioning experiments a 0.0057 M stock solution was prepared at weekly intervals. The applied permanganate dose ranged from 0.1-2.0 mg $KMnO_4$/L, while the pH was previously set at 8. The permanganate was applied at a G value of 400 s^{-1} for a period of 1 min, followed by 30 min of slow mixing at a G value of 10 s^{-1}. This rather long contact time before coagulant application was applied in order to simulate conditions under which permanganate is dosed as soon as possible within the treatment process. The positive effect of this strategy is the long time available for permanganate reactions before coagulant application, but in practice some MnO_2 would precipitate on the tank and channel walls. Thus, previous to all experiments, the precipitation rate of permanganate and formation of MnO_2 was assessed within the adopted 30 min contact time. Measurements of both dissolved Mn(VII) and particulate Mn(IV) were conducted for each experiment. The optimisation of the $KMnO_4$ dose was based on the visual determination technique (no pink colour observable in the DAF effluent). It was combined with a previously optimised Fe(III) coagulant dose at coagulation pH 8. The impact of the conditioning pH (6, 7, 8, and 9) on DAF efficiency was also tested. The flocculation G (10, 30, 70 and 130 s^{-1}) and t (10, 15, 20 and 30 min) were varied in search for the existence of an optimal Gt range. An overview of the experimental conditions for each of the particular sets of experiments is given in Appendix 4.1.

4.2.2 Analytical techniques.

Process efficiency was evaluated by turbidity (Sigrist L-65, Switzerland) and particle (alga) count of particles >2.75 μm (HIAC-Royco PC-320, USA). A Nikon Optiphot (Japan) microscope was used for sample inspection. Residual coagulant (Fe_{total}) and manganese were measured by atomic absorption spectrometry at 248.3 nm for iron (NEN 6460) [62] and 279.5 for manganese (NEN 6466) [63]. Electrophoretic mobility was measured with a Tom Lindström AB-Repar apparatus. Transferred ozone was determined by measuring the O_3 concentration in the inlet and outlet air/ozone gas mixture by the iodometric method [64]. Computer image analysis was performed with the Mini-Magiscan (IAS 25/IV25 Joyce-Loebl Ltd., UK) [13].

Scanning electron microscopy (SEM) was performed with a JEOL 6300 F microscope; samples were previously filtered (0.22 μm) to remove surrounding water but leave algae cells intact by avoiding dehydration. Prior to SEM samples were frozen in liquid nitrogen and mounted into a cryo-transfer unit (CT 1500 HF, Oxford Instruments, UK) under high vacuum ($1x10^{-6}$ Pa), temperature of -85°C and sputter coated (Denton) with 3 nm platinum. SEM observation of the coated specimen was carried out at 5KV and -180°C.

Statistical analysis of data was performed with the Statistica software package. One and two way ANOVAs followed by multiple comparison and regression analysis among treatment.

4.3 RESULTS AND DISCUSSION

4.3.1 Stoichiometry between algae concentration and metal coagulant dose

Stoichiometric relationships between alga concentration and coagulant dose were found to apply for spherically shaped, single cell μ-algae [11] . This suggests that these algae can be considered as behaving as inorganic particles to which the common coagulation and flocculation theory applies. This does not apply, however, for algae which have particular morphological features such as bristle-like appendages (*Stephanodiscus hantzscshii*) or a mucilaginous sheath layer around the cells (*Dictyosphaerium*).The size and shape of the single cell form of *M. aeruginosa* suggest a linear stoichiometric relation may exist. DAF jar tests with model water (spiked with different concentrations of laboratory cultured algae), in which the coagulant dose was optimised confirmed the existence of such a relationship (Fig. 4.2). The regression line is a polynomial, suggesting that at a certain point (high algae concentrations above 20,000 cells/mL) there is no need for further coagulant dose increase. Possibly, a critical concentration of coagulant (8-9 mg Fe (III)/L) has been reached which allows for sweep coagulation, i.e. destabilisation to occur without further increase of the coagulant dose. According to classical coagulation theory the coagulant dose is inversely proportional to the concentration of aquatic colloids at high colloid concentration, as is here the case at and above 20,000 cells/mL [16].

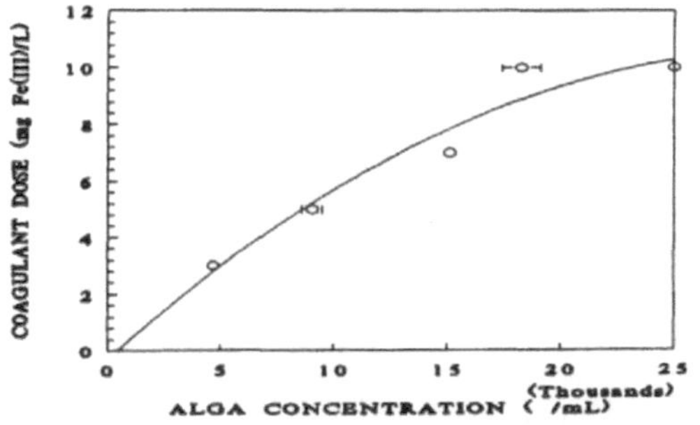

Fig. 4.2 Stoichiometric relation between algae (*M. aeruginosa*) concentration and optimal coagulant ($FeCl_3$) dose, including standard deviations for the initial algae concentrations.

Even at the highest coagulant dose, EM (electrophoretic mobility) did not show particle charge neutralisation (or reversal) in the presence of the algae. However, adding coagulant to water without algae resulted in charge neutralisation (of $Fe(OH)_3$ particles) and reversal at a dosage of 15 mg Fe(III)/L (Fig.4.3). The charge reversal in the water without algae could be attributed to gradual elimination of HCO_3^- by H^+, generated by Fe(III) hydrolysis. High Ca^{2+}concentrations (here 150-200 mg as $CaCO_3$) accompanied by a very low HCO_3^- level tend to shift the IEP of $Fe(OH)_3$ from pH 6-7 to slightly alkaline values, as is the case here [14]. The algae spiked water, on the other hand, would have a substantially different composition with respect to ions and organics, because it contains the growth medium of the algae culture (see Appendix 3.1). Although it would be very difficult to identify the responsible compounds the net effect appears to result in a shift in IEP precluding charge reversal at pH 8.

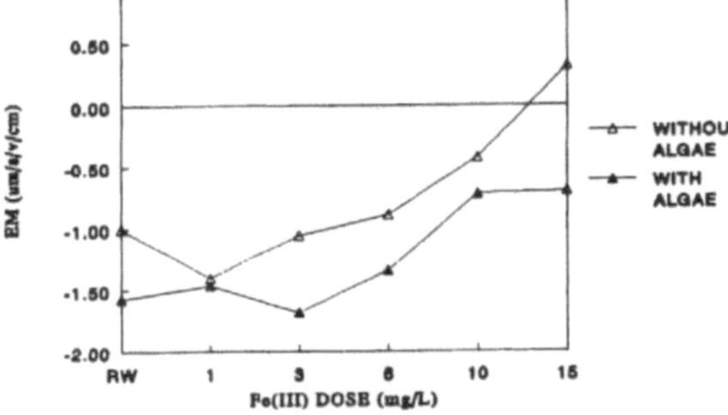

Fig. 4.3 Electrophoretic mobility (EM) of water with algae (≈10,000 cells/mL) and free of algae subject to different coagulant doses (pH 8).

4.3.2 Polyelectrolytes in coagulation

a. Polyelectrolytes as coagulant aid

Earlier work on the importance of the coagulation pH (Chapter 3) resulted in significantly improved coagulation and down-stream DAF efficiency at pH≤6. This is likely to be owed to a combination of partial adsorption coagulation, which occurs at pH below the IEP, and sweep coagulation [3]. In practice, however, it is very often not feasible to correct pH. However, the role of adsorption coagulation suggests cationic polyelectrolytes may be also effective to achieve adsorption coagulation. The cationic polyelectrolyte Superfloc C-573 was proven to be a coagulant aid highly beneficial to DAF (and sedimentation) efficiency, resulting in 94.5% particle (alga) removal, improved turbidity removal and reduced residual coagulant content. In further research different types of organic and inorganic polyelectrolytes of different chemical composition, charge and weight characteristics were considered. Figs. 4.4 and 4.5 represent particle (alga) and turbidity removal efficiency for different types and doses of polyelectrolyte coagulant aid.

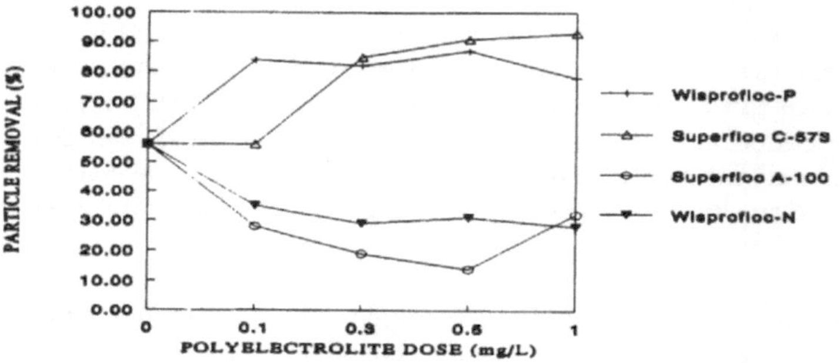

Fig. 4.4 Particle (alga *M. aeruginosa*) removal efficiency by DAF for different types and doses of polyelectrolyte. Conditions : ≈10,000 cells/mL, initial turbidity of 3.85 FTU, 3 mg Fe(III)/L, pH 8, G=30 s^{-1}, t_f=10 min, R=7%, p=500 kPa.

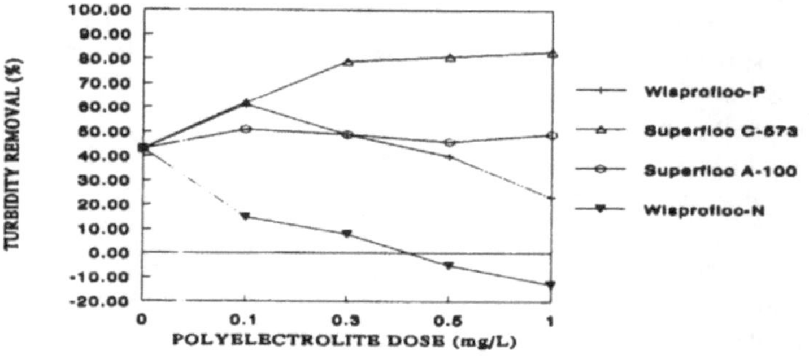

Fig. 4.5 Turbidity removal efficiency by DAF for different types and doses of polyelectrolyte. Conditions : same as Fig. 4.4.

The synthetic cationic polyelectrolyte Superfloc C-573 was the most efficient. These results confirm other positive experiences with polyelectrolyte coagulant aids and *Scenedesmus* spp. algae removal [65]. However they contradict other findings in which synthetic cationic polymers were used as coagulant aid together with Fe(III) coagulant for the removal of *M. aeruginosa* (100,000 cells/mL concentration) and were found inefficient [66]. Other low molecular weight cationic polymers (Catfloc T2 and Catfloc TL, at dosages of 2-3 and 5 mg/L in combination with 30 and 16.5 mg Fe(III) coagulant respectively) were efficient for DAF algae removal [67]. Superfloc C-573 owes its efficiency to the very strong charge characteristics which resulted in particle (algae) charge neutralisation as verified by EM measurements (Fig. 4.6). Similar charge neutralisation effects were noted for the organic cationic polyelectrolyte Wisprofloc-P, however, of lower accompanying process efficiency. The same polyelectrolyte was again inefficient for the DAF treatment of *Stephanodicus* spp. (36,000 cells/mL) [66], while it efficiently reduced DAF effluent turbidity (5 FTU before, and 1.2 FTU after its application) and residual alum (from 400 to 20 μg/L) in other circumstances [68]. This indicates additional factors as relevant for the noted process efficiency. Two such factors are the resultant floc size (distribution) and density. The floc volume distribution has been indicated as dependant on the polyelectrolyte configuration, since it is critical for their adsorption to particulates (and media in filtration) [69].

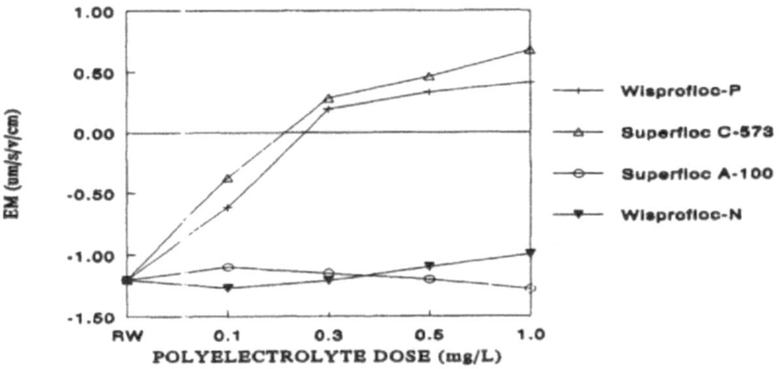

Fig. 4.6 Electrophoretic mobility (EM) of particles for different types and doses of polyelectrolyte. Conditions : 3 mg Fe(III)/L, pH 8, $G=30\ s^{-1}$, $t_f=10$ min, R=7%, p=500 kPa.

One way ANOVAs for the effect of different dose of Superfloc C-573 coagulant aid on the particle (alga) and turbidity removal efficiency showed that both were statistically significantly affected by treatment with coagulant only, as compared with coagulant and coagulant aid (P=0.00000 for particle count and P=0.012 for turbidity). One way ANOVA for the initial particle (alga) count showed that the initial conditions were similar for all experiments (P=0.99), while for turbidity it was not the case (P=0.00054). The difference which was found for initial turbidity between the case of coagulant only and that of coagulant and coagulant aid, is due to the difficulty to prepare identical algae cultures (see Chapter 3, Fig. 3.2). However, including initial turbidity and particle count as covariates in the analysis, showed that both the turbidity and particle removal efficiency were significantly affected by the polyelectrolyte (P=0.00000 for particle count and P=0.000084 for turbidity).

The optimal polyelectrolyte dose in terms of DAF (and sedimentation) efficiency was in the range of one tenth of the coagulant dose. Particle size analysis after flocculation with the Superfloc C-573 as coagulant aid [3, 13] showed that the increase of polyelectrolyte dose from 0.5 to 1.0 mg C-573/L resulted in the production of floc material of denser structure (Fig. 3.8, Chapter 3). The long polyelectrolyte chain structure of the Wisprofloc-P is suggested responsible for the observed formation of floc material larger than in case of the Superfloc C-573. Although no particle size analysis was done in case of the Wisprofloc-P, the floc size visibly increased with the polyelectrolyte dose, opposite to the case of Superfloc C-573. Furthermore, the Wisprofloc-P based floc material tended to stick to the walls of the jars, as well as to other surfaces like the stirrer. The algae were destabilised and flocculated, however, the floc quality (size and density) was not appropriate for DAF. The polyelectrolyte dose increase had a pronounced effect on the turbidity, which rose with the Wisprofloc-P dose, contrary to the case of Superfloc C-573 (see Fig. 4.5). The increased turbidity levels suggest high residual polyelectrolyte concentrations. The absence of a residual polyelectrolyte characterisation technique, precludes discerning the extent to which they affect the residual turbidity, as well as to assess the potential health hazards.

The anionic polyelectrolyte Superfloc A-100 and the non-ionic Wisprofloc-N proved inefficient under the tested circumstances. The DAF effluent deteriorated in comparison with the case when Fe(III) was used as sole coagulant, the removal efficiency for both polyelectrolytes being below 50%. In the case of Superfloc A-100, this was accompanied with a slight increase of turbidity removal, whereas the turbidity increased in the case of Wisprofloc-N. Similar observations pertain to residual iron, which was relatively low in the case of Superfloc A-100 (removal rate of 75-85%). The EM measurements show no significant changes; particles become slightly more negative in the case of Superfloc A-100, and less negative in the case of Wisprofloc-N. This suggests that under the investigated circumstances the bridging mechanism typical for anionic and non-ionic polyelectrolytes, had a minor effect as compared to that of the charge neutralization occurring in the case of the cationic polyelectrolytes. It is further suggested that the algae concentration in our model water was insufficient to allow for successful bridging to occur between algae. Anionic polyelectrolytes are adsorbed by particles through intervention of metal ions such as calcium. Under the model water circumstances it could be that the availability of calcium in the range of 150 mg $CaCO_3$/L enhanced adsorption and hence removal efficiency compared to the non-ionic Wisprofloc-N.

Cationic polyelectrolytes are suggested to be effective coagulant aids for algae laden water coagulation, and treatment by DAF. The charge and weight of the polyelectrolytes determine the process efficiency.

b. Polyelectrolytes as single coagulants

The application of polyelectrolytes as sole coagulants is not common in European water treatment, although it often leads to efficiency comparable to metal coagulants. This restriction is mainly due to health concerns. The most efficient polyelectrolyte as sole coagulant in this study was the cationic polyelectrolyte Superfloc C-573. The particle (alga) removal efficiency increased with the polyelectrolyte dose, the highest tested dose of 3 mg Superfloc C-573/L resulting in 81% removal efficiency (Fig. 4.7), which was higher than the maximum removal efficiency obtained by iron coagulant without coagulant aid under optimal coagulation and flocculation conditions

(71 %). The same holds for turbidity : the polyelectrolyte caused a removal efficiency of 74 % as compared to 56 % obtained by iron coagulant [3] (Fig. 4.8). Two mechanisms are suggested to be responsible for the polyelectrolytes efficiency : (i) improved coagulation of the algae, due to polyelectrolyte adsorption followed by charge neutralization, and (ii) improved attachment of the positively charged floc material to the negatively charged air bubbles [70]. The steep rise in the DAF removal efficiency obtained for the Superfloc C-573 polyelectrolyte suggests that a dose higher than 3 mg/L may have further increased efficiency.

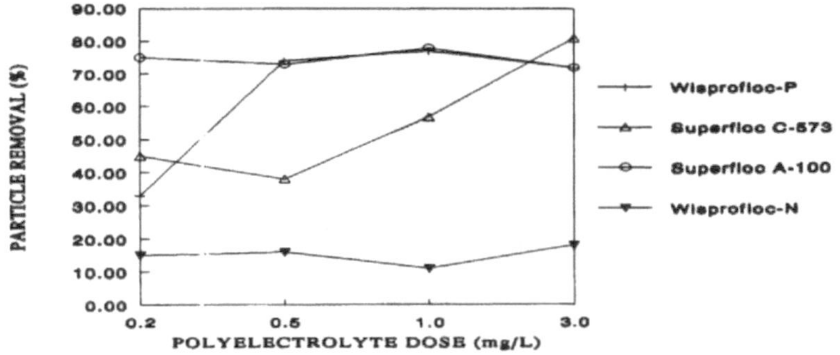

Fig. 4.7 Particle removal efficiency (DAF) for different types and doses of polyelectrolyte as sole coagulant. Conditions: ≈10,000 cells/mL, initial turbidity 3.8 FTU, pH 8, G=30 s^{-1}, t_f=10 min, R=7%, P=500 kPa.

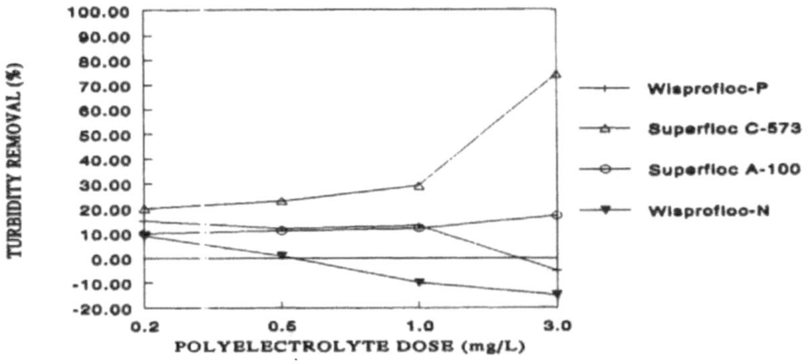

Fig. 4.8 Turbidity removal efficiency (DAF) for different types and doses of polyelectrolyte as sole coagulant. Conditions: ≈10,000 cells/mL, initial turbidity 3.8 FTU, pH 8, G=30 s^{-1}, t_f=10 min, R=7%, P=500 kPa.

Superfloc C-573 emerges as an attractive polyelectrolyte when used as a sole coagulant; this is further supported by the absence of carry-over of metal hydroxide flocs, lower amount of salts, and lower sludge volume. On the other hand, this sludge may be more difficult to dewater and dispose (environmental concerns). Furthermore, accurate determination of the residual is difficult, raising health concerns.

The second cationic polyelectrolyte Wisprofloc-P as a sole coagulant resulted in relatively high particle removal in the range of 75 % (Fig. 4.7). However, turbidity removal was inferior, actually resulting in an increase of the original turbidity of the raw water in the case of the highest tested dose (Fig. 4.8). Characteristic for this polyelectrolyte is its high molecular weight (long polyelectrolyte chain structure), and weaker electric charge compared to Superfloc C-573. This resulted in the formation of comparatively voluminous and weaker flocs which tended to settle or stick to the glass surface of the jars and the mixing equipment. Destabilisation is suggested to have occurred as a combination of adsorption, as well as bridging mechanisms. The relatively weak charge density of the polyelectrolyte could be insufficient to promote a flat configuration on the alga surface, thus leaving its long polyelectrolyte chains protruding into the solution and bridging with other algae or negatively charged sites on existing flocs. EM measurements showed that Wisprofloc-P treatment was accompanied by only slight charge neutralization of the algae (Fig. 4.9).

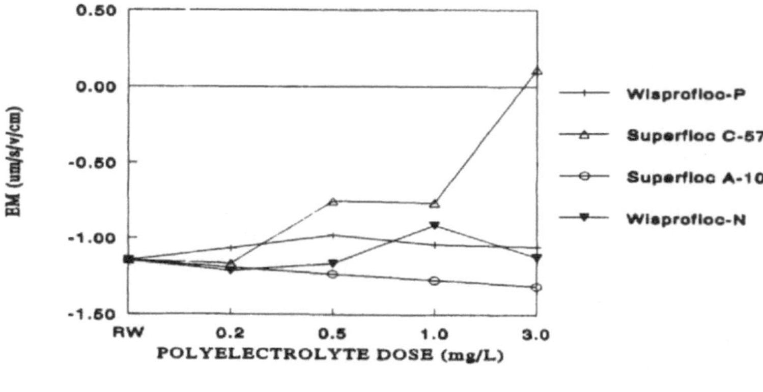

Fig. 4.9 Electrophoretic mobility of particles for different types and doses of polyelectrolyte as sole coagulant. Conditions : 10,000 cells/mL, pH 8, $G=30\ s^{-1}$, $t_f=10$ min, R=7%, P=500 kPa.

Based on the particle removal efficiency, it is suggested that most of the existing algae were flocculated, however the character and quality of these flocs were inferior for efficient DAF. Furthermore, all the applied doses resulted in high residual turbidities presumably of polyelectrolyte origin. These observations suggest that the Wisprofloc-P polyelectrolyte is preferably applicable in cases where sufficient destabilisation has already been imposed e.g. by metal coagulants, in other words as a low-concentration coagulant aid. Similarly, it may be expected to be efficient when the concentration of the particulate and colloidal matter is relatively high, thus providing ample adsorption coagulation opportunity.

The bridging mechanism proved to be efficient in the case of the anionic polyelectrolyte Superfloc A-100 also, resulting in particle removal efficiency in the range of 75%, however, with low turbidity removal of 10-15% presumably ascribed to the residual polyelectrolyte. The presence of metal ions such as Ca^{2+} (raw water hardness of 150 mg/L expressed as $CaCO_3$) in the water provided good conditions for the occurrence of efficient particle coagulation and flocculation. The negative charge of the flocculated material obviously further increased with increase of the anionic polyelectrolyte dose. This suggests that the particle removal in DAF occurred primarily as the result of particle-bubble interception, owed to the large bubble concentration relative to the particle (floc) concentration. The negative charge of the polyelectrolyte did not allow for its more complete incorporation into the floc matter. Furthermore, the negatively charged non-flocculated part of the polyelectrolyte was more difficult to remove by the negatively charged air bubbles. This resulted in inappropriate polyelectrolyte presence and coupled turbidity in the effluent.

Finally, the non-ionic polyelectrolyte Wisprofloc-N resulted in the poorest removal efficiency, both in terms of particle (algae) and turbidity removal. The bridging mechanism involved in the particle destabilisation in case of non-ionic polyelectrolytes occurs via hydrogen bonding and proved inefficient under the test circumstances. The particle charge (EM) showed a slight decrease of the original negative charge (Fig. 4.9). Flocculation occurs to some extent as visually observed, however inefficient and incomplete.

The relatively good model water quality, in which moderate concentrations of algae were the only significant contributor to the suspended matter, created conditions that were presumably not appropriate for the particle bridging coagulation by the anionic or non-ionic polyelectrolytes. Although generally recognized as more efficient, in particular for the removal of microorganisms such as algae, cationic polyelectrolytes are not a guarantee for efficient coagulation and down-stream removal efficiency. Practical experience can allow to roughly predict the effect of a certain polyelectrolyte on the process efficiency. However, the optimal process efficiency and polyelectrolyte choice should be tested case by case. Under the tested circumstances, the application of the synthetic cationic polyelectrolyte Superfloc C-573 proved to be an effective and competitive coagulant at pH 8.

4.3.3 Ozone conditioning

Ozone conditioning of the algae resulted in significant DAF efficiency improvement. The results from one series of ozone conditioning experiments are presented in Fig. 4.10. The ozone dosing was based on similar research [47]. A range of low (0.2 mg O_3/mg C, or 0.48 O_3/L), medium (0.5 mg O_3/mg C, or 1.2 mg O_3/L) and high (0.9 mg O_3/mg C, or 2.16 mg O_3/L) ozone doses were tested in combination with the optimal iron coagulant dose of 3 mg Fe(III)/L), and with 1.5 mg Fe(III)/L and 4.5 mg Fe(III)/L. The results show that the low and medium ozone doses in combination with the 1.5 mg Fe(III)/L dose resulted in slightly deteriorated DAF efficiency, however, the combinations with the optimal (higher) iron dose resulted in all cases in significantly improved removal efficiency (80-97%), approaching 2 log particle removal for the high ozone dose. Full scale ozoflotation (combined ozonation and DAF in one unit) of Thames Water at Walton Works (UK) with PAC (polyaluminum chloride) as coagulant, resulted in 20-88% chlorophyll reduction in case of 0.5 and 4 mg O_3/L respectively, compared to 20% achieved without ozone [71]. In the case of 4 mg O_3/L the down-stream algae filtration removal efficiency was 99%, compared to 55% without ozone application. Other ozoflotation facilities in France achieved in 55-85% algae removal for an ozone dose of 1.6 mg O_3/L, the variation depending on the predominant algae species, the most abundant and difficult to remove being the blue-green *Gomphosphaeria* spp. [72]. Ozone induced flotation at very high doses of 15-50 mg O_3/L has been noted to reduce total suspended solids, including algae from a waste-water stabilization pond by 98% [73].

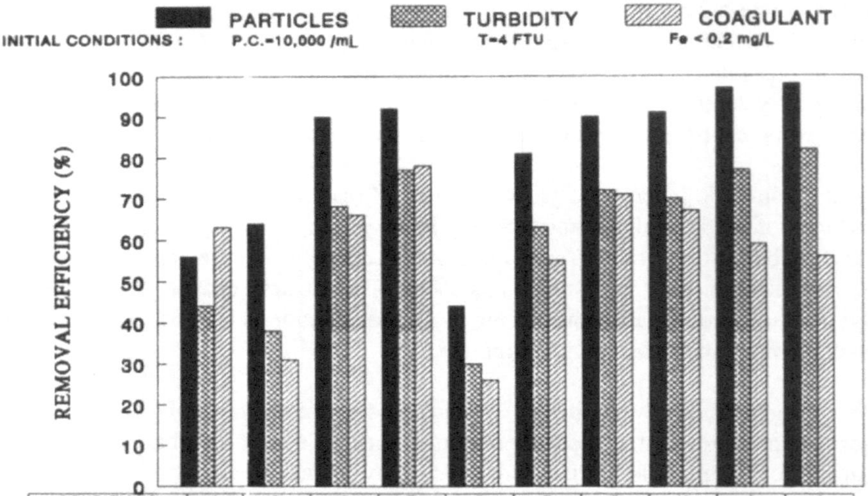

CONDITIONS	A	B	C	D	E	F	G	H	I	J
mg Fe(III)/L	3	1.5	3.0	4.5	1.5	3.0	4.5	1.5	3.0	4.5
mg O3/mg TOC	-	0.2	0.2	0.2	0.5	0.5	0.5	0.9	0.9	0.9

Fig. 4.10 Removal efficiency of particles, turbidity, and coagulant for different combinations of ozone and coagulant doses.

While DAF reduced algae by 1 log unit, intermediate ozone application after DAF reduced downstream filtrate algae concentration by 2-3 log units [74]. Pilot plant direct filtration efficiency of 1 log algae removal was raised to approximately 3 log removal after ozonation, providing also longer filter run-time [36].

In our case particle and turbidity removal efficiency increased with increase of the coagulant dose for each ozone dose, while the combinations of ozone and iron tended to be consistently more efficient in the case of the low and high ozone dose as compared to the medium one. This finding is related to the complicated chemistry of ozone and the induced release of EOM and possibly IOM, which is algae species and age specific [71, 72, 75]. The extent of the reaction of EOM and IOM with ozone depends on the dose. This phenomenon is further discussed in Section 4.3.3.a.

The results from the one way ANOVA for the case of 3 mg Fe(III)/L in combination with different ozone doses are summed-up in Fig. 4.11. The results suggest that ozone significantly affects the particle (alga) removal efficiency (P<0.00001) and not significantly the turbidity removal efficiency (P=0.48). The notation a, b and ab denote similarities between treatments based on multiple comparison analysis, suggesting that all ozone doses perform significantly better than the iron coagulant alone, however with no significant efficiency differences between themselves. One way ANOVA for the initial particle count shows that the initial conditions (with respect to particle - algae count) were significantly different for different ozone treatments (P=0.0047). However, including initial particle count as a covariate in the ANOVA still suggests that the particle removal efficiency is significantly affected by different ozone treatments. One way ANOVA for the initial turbidity shows that the initial conditions (with respect to turbidity) were not significantly different (P=0.12). Including the initial turbidity as a covariate in the ANOVA suggests that ozone treatment had a significant effect on the turbidity removal efficiency.

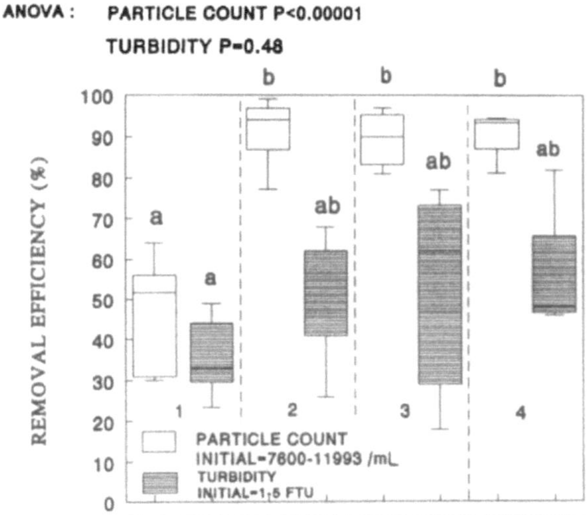

CONDITIONS	FeIII (mg/L)	mg O3/mg TOC
1 (DAF)	3	-
2 (DAF)	3	0.2
3 (DAF)	3	0.5
4 (DAF)	3	0.9

Fig. 4.11 One way ANOVA for the combination of 3 mg Fe(III)/L iron coagulant and ozone treatment.

Multiple regression analysis for the initial particle count and ozone treatment (as independent variables) shows that they both significantly affected the particle removal efficiency (with ozone as a dependent variable r^2=0.56, *P*=0.0015; and with initial particle count *P*=0.028). Similar results were achieved by multiple regression analysis for turbidity and ozone treatment (r^2=0.47, *P*=0.0018 for ozone; and *P*=0.00034 for the initial turbidity).

In case of a higher coagulant dose of 5 mg Fe(III)/L the same conclusions were reached (Fig. 4.12.). In order to assess the influence of the different coagulant dose in combination with ozone, a two way ANOVA was performed on the 3 mg and 5 mg Fe(III)/L cases. Although the results suggest a tendency of slightly increased DAF efficiency for the higher coagulant dose, especially with regard to particle removal (estimated 5%), no significant differences between treatment were found, both for particle count (*P*=0.098 for the coagulant dose, *P*=0.85 for the ozone dose, and *P*=0.999 for the combined influence of ozone and iron dose) and turbidity (*P*=0.68 for the coagulant dose, *P*=0.80 for the ozone dose and *P*=0.70 for the combined influence of ozone and iron dose). Two way ANOVAs for the initial particle count and the turbidity showed that the initial experimental conditions (in terms of both particle count and turbidity) were not significantly different. Multiple regression analysis for the particle removal efficiency (r^2=0.57) shows that it depended significantly on the ozone dose (*P*=0.0011) and the initial particle count (*P*=0.017), while it depended insignificantly on the iron coagulant dose (*P*=0.082). Multiple regression analysis for the turbidity removal efficiency (r^2=0.2) similarly shows that it was significantly affected by the ozone dose (*P*=0.036) and the initial turbidity (*P*=0.013), while it was insignificantly affected by the iron coagulant dose (*P*=0.165).

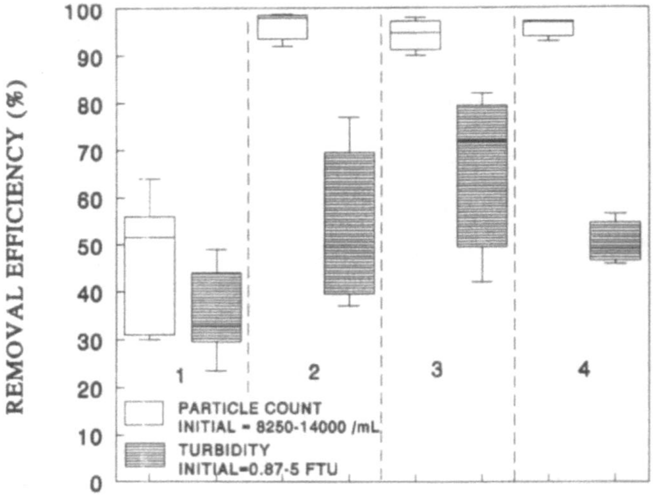

CONDITIONS	FeIII (mg/L)	mg O3/mg TOC
1 (DAF)	5	-
2 (DAF)	5	0.2
3 (DAF)	5	0.5
4 (DAF)	5	0.9

Fig. 4.12 Particle and turbidity removal efficiency for the combination of 5 mg Fe(III)/L iron coagulant and ozone treatment.

Experimental series with model water that contained a 50% lower initial algae concentration yielded similar efficiency results (Fig. 4.13).

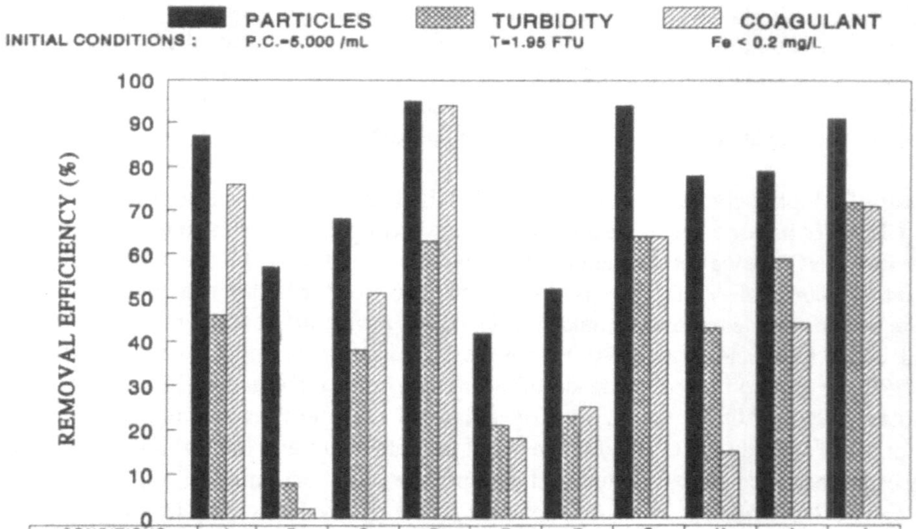

CONDITIONS	A	B	C	D	E	F	G	H	I	J
mg Fe(III)/L	2	1.0	2.0	3.0	1.0	2.0	3.0	1.0	2.0	3.0
mg O3/mg TOC	-	0.2	0.2	0.2	0.5	0.5	0.5	0.9	0.9	0.9

Fig. 4.13 Removal efficiency of particles, turbidity and coagulant for different combinations of ozone and coagulant doses. Initial algae concentration ≈ 5,000 cells/mL.

The same trend of low and high compared to medium ozone dose efficiency was observed again, however, ozone application resulted in a general low efficiency in combination with the low and medium coagulant dose. Although the combinations of ozone with high coagulant dose resulted in improved process efficiency compared to the zero- ozone case, ozone application in this case seemed to raise the coagulant demand. Based on comparison of these results and the results for the 10,000 cells/mL case, it can be argued that a critical dissolved organic matter concentration has to be present, and hence a minimal concentration of algae that are assumed to be the cause for a substantial part of such dissolved matter, in order to benefit from ozone conditioning [26, 45]. Higher initial particle concentration also provides more opportunity for the effectuation of the bridging mechanism which is directly related to : (i) collision frequency between the particles and (ii) the EOM secreted by the algae.

a. Mechanisms of ozone induced flocculation

Increased process efficiency by ozone has been attributed to ozone induced changes of pH or calcium precipitation during ozonation [76]. This implies that the same effects induced by ozonation could be induced by decreasing the pH with strong acids, or by carbon dioxide stripping and thus causing precipitation of supersaturated calcium carbonate. The pH of the water used for preparation of the model water was 7.9±0.2. Prior to the ozonation the pH was corrected to 7.5 by HCl addition. However, after ozonation the pH variation was 7.5 ± 0.2, suggesting that these arguments do not hold in this case.

Ozone conditioning can be regarded responsible for the creation of stress conditions in the algae environment, which results in the intensified excretion of EOM. The particle count of our ozonated model water (particle sized <10 μm) at a relatively low dose of 0.2 mg O_3/mg C (or 0.48 mg O_3/L) prior to further treatment, was decreased by 30 %, while the concentration of particles larger than 10 μm increased (Fig. 4.14). This could be explained by algal cell lysis and IOM leakage that initiated spontaneous micro-flocculation. Among algae characteristics that have been suggested to affect process efficiency are the extra cellular and intra cellular organic matter (EOM and IOM). Their concentration, composition and characteristics (molecular weight) vary for different algae species [77], growth phase and growth conditions [78]. A substantial amount of the total

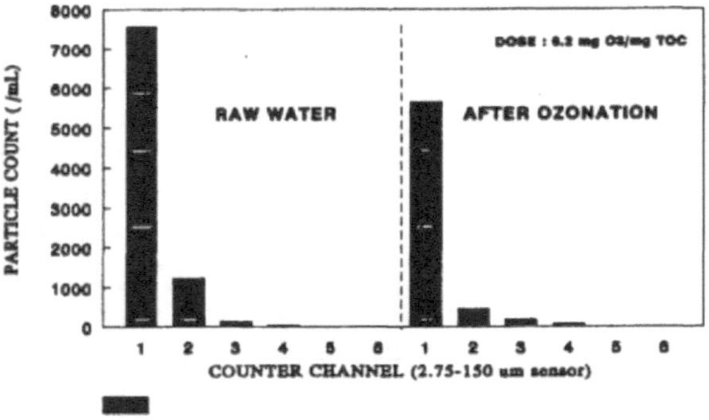

Fig. 4.14 Particle count before and after ozonation; classification for different size fractions (CH1:2.75-5 μm; CH2:5-10 μm; CH3 :10-20 μm; CH4 : 20-30 μm; CH5:30-50 μm, CH6: 50-150 μm).

assimilated carbon can be released in the water in the form of EOM (up to 50%), while stress growth conditions (e.g. related to availability of light and nutrients) may increase this amount up to 95% [79]. Major EOM constituents have been analyzed for different species [80] and can generally be characterized as neutral and acidic polysaccharides and nonsaccharide acidic macromolecular compounds. Under certain water quality and coagulation conditions (DOC<3 mg/L, ratio between algogenic organic matter (DOC) and particle surface concentration <75 mg/m^2, and ferric salts used as coagulant at pH 6.5), the EOM, especially its neutral and acidic polysaccharides component, have been considered as anionic polymer coagulant aids. The EOM coagulant aid effect is asserted through : (i) attachment to particulate matter via ligand complexes, (ii) chemisorption and further coagulation of particles by the bridging mechanism, or (iii) by particle capture in a gel like structure. The model water DOC content was relatively low (≤2 mg C/L). Furthermore, this water had been subjected to ozonation in a full-scale treatment process (3 mg O_3/L) before being abstracted and used in the experiments. So it can be expected that the initially present organic matter had already been substantially modified before spiking it with the algae. Therefore, the ozone that was added during the experiment reacted preferentially with the spiked algae and caused the release of EOM and IOM.

On the other hand, the significance of fine organic matter matrices or meshes of filaments which embed very small organic and inorganic colloids in the water, has been recognized as a very important determinant of their coagulation and sedimentation behaviour in natural surface water [81]. The higher molecular weight EOM constituents (>100,000 D) play a significant role in the formation of these organic matrices. Thus, the organic matrices which embed released EOM biopolymer, may facilitate the inter-particle bridging mechanism to a certain degree. On the other hand, excessive EOM release may hinder the flocculation process by coating of the particles and increasing their negative charge, as well as densening the organic matrice structure to an extent which makes particle-metal coagulant contact and flocculation more difficult.

The issue of possible algal cell lysis and release of IOM into the water still remains controversial [27, 43, 82]. This release depends primarily on the ozone dose, but also on the alga species and the presence of other natural organic matter (NOM) in the water, which may preferentially react with ozone. SEM analysis of ozonated algae spiked in untreated reservoir water (Fig. 4.15) and containing ozone scavengers, strongly suggests algal lysis and leakage of inner cell constituents occurring at an ozone dose of 0.57 mg O_3/mg C (or 2.0 mg O_3/L). Measurement of $UV_{254\,nm}$ absorbance further support this. Although weakly correlated with the DOC concentration [27, 36], it may be a useful indicator of process efficiency. It is indicative of the presence of organic compounds which contain a hydrocarbon ring in their structure (e.g. aromatic and pigmented compounds). The verified algae cell lysis strongly implies the involvement of IOM, together with EOM, in the observed flocculation improvement.

The low, medium and high ozone doses (0.2, 0.5, 0.9 mg O_3/mg TOC) resulted in different DAF removal efficiency, both for the 10,000 cells/mL (Fig. 4.10) and the 5,000 cells/mL algae concentration. Although these differences were statistically not significant (Fig. 4.11), the low and the high ozone dose tended to perform better than the medium dose. In view of the previous discussion on the effect of ozone on the algae cell structure and EOM and IOM leakage (Fig. 4.14 and 4.15), and their effect on removal efficiency, $UV_{254\,nm}$ absorbance of the water was measured for different ozone doses and times after ozonation (Fig. 4.16).

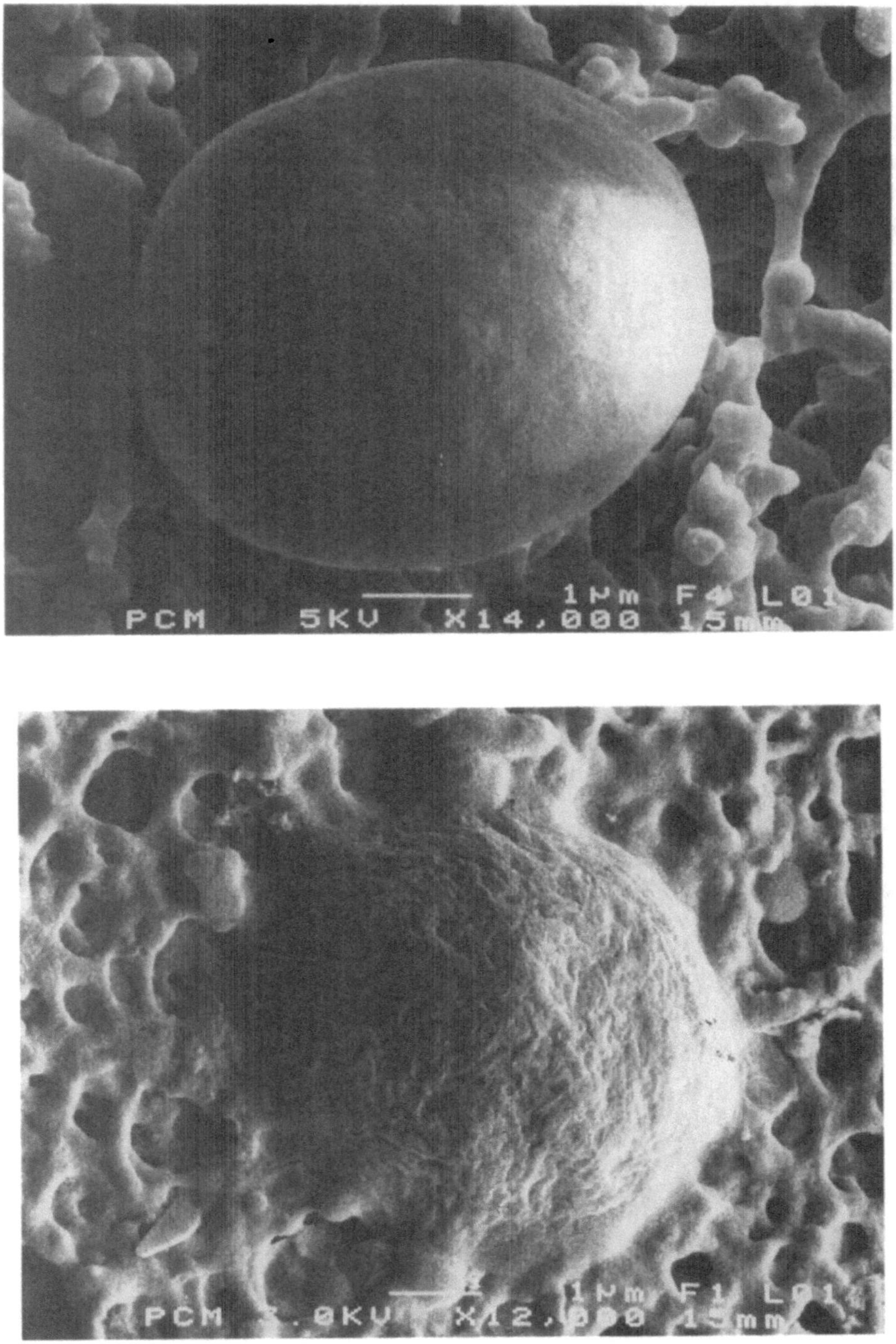

Fig. 4.15 Scanning electron microscopy (SEM) of *M.aeruginosa* spiked in reservoir water before and after ozonation at a dose of 0.57 mg O_3/mg C (2 mg O_3/L).

The low ozone dose (0.2 mg O_3/mg C) steadily increased the absorbance over in time, suggesting gradual release of algae EOM and IOM into the water. The medium ozone dose (0.5 mg O_3/mg C) caused a higher increase of the absorbance, suggesting a higher EOM and IOM content in the water. This dose resulted in modification of the released organic matter and a steady decrease of the absorbance with time. The 0.9 mg O_3/mg C and 1.3 mg O_3/mg C doses, both resulted in a decrease of the raw water absorbance, suggesting that the ozone concentration was high enough to not only cause the release of algae EOM and IOM, but also to substantially modify them through oxidation and reduction of their aromatic character.

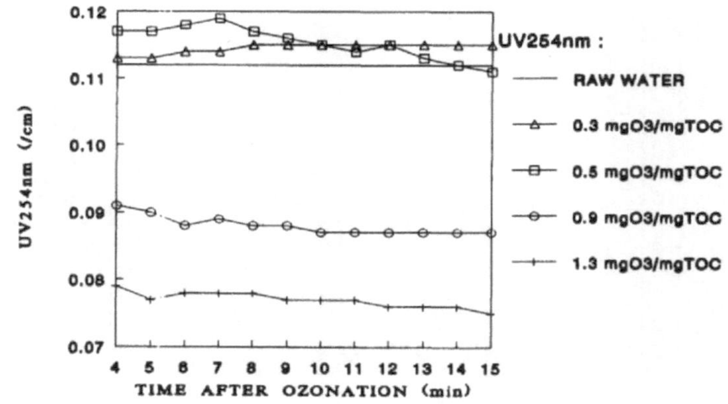

Fig. 4.16 UV_{254nm} absorbance in the water after ozone dosing, as a function of the ozone dose and time.

It has been reported that ozonation decreases the molecular weight of the present organic matter, which becomes more acidic and hydrophilic than the parent organic material [46], while the particle charge is reduced (Fig. 4.17).

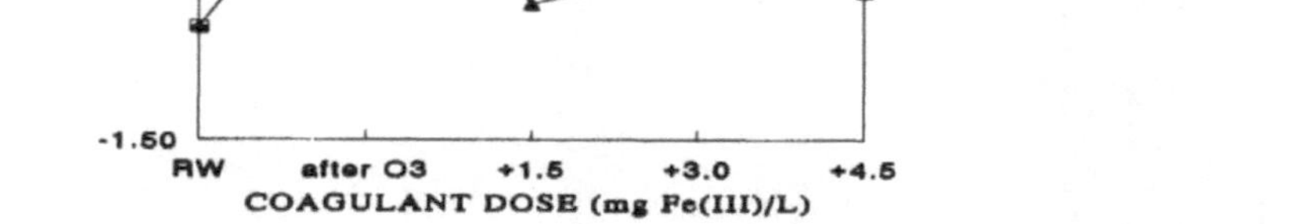

Fig. 4.17 Electrophoretic mobility (EM) of raw, ozonated and flocculated water as a function of ozone and coagulant dose.

The increased acidic functional group content is reported to provide for higher sorbability to metal-hydroxy species formed during the metal coagulant hydrolysis. This may result in the formation of more stable (i.e. of increased negative charge) particulate and colloidal material than before ozonation, which can impart an additional coagulant demand [6, 82]. This is observed in Fig. 4.10 for the high ozone dose of 0.9 mg O_3/mg C where the increase of the coagulant dose resulted in increased residual coagulant, suggesting the formation of difficult to remove soluble complexes of the modified organic matter and the added iron coagulant.

The higher the ozone dose, the larger the amount of released EOM, but also the larger the possibility of cell lysis and IOM release. Furthermore, the higher the ozone dose, the more significantly the NOM and EOM/IOM were modified by oxidation. While the low ozone dose seemed to cause moderate amounts of released EOM and IOM which benefitted flocculation, the medium dose resulted in EOM, but also in more substantial IOM release, which seemed to impair flocculation (Figs. 4.10 and 4.16). The high removal efficiency for the larger doses of ozone is suggested to be the result of larger EOM and IOM release, which however, is compensated by extensive reaction of the excess ozone with these compounds. Therefore, the application of low ozone doses in the range of 0.2 mg O_3/mg TOC was found most suitable for algae conditioning purposes under the bench-scale jar test conditions.

b. The effect of ozone on particle size distribution

Results (Fig. 4.18) suggest that ozone decreased the volume of small particles and increased the larger particle size fraction, which partly coincides with the particle counter results (Fig. 4.14).

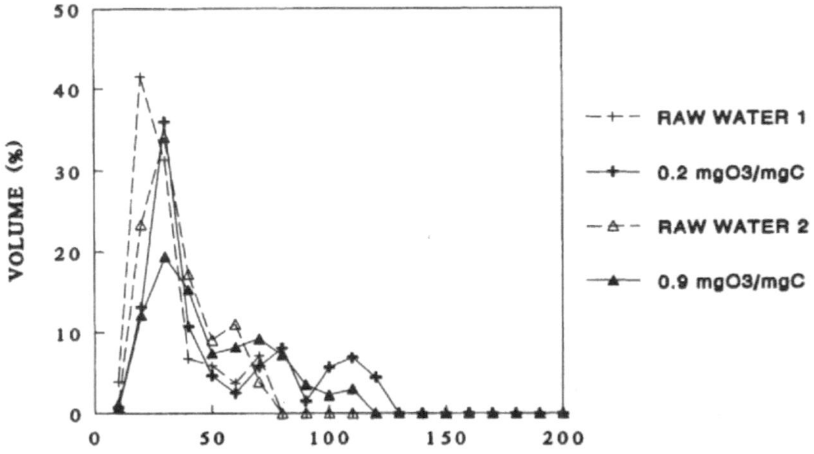

Fig. 4.18 Particle volume distribution as a function of ozone dose (0.2 and 0.9 mg O_3/mg TOC).

The increase of the larger particle size volume fraction is not very obvious from the graph, since the impact of a larger particle on the volume distribution is much higher than that of a smaller size one. This suggests that : (i) cell lysis took place to a certain degree, and (ii) consequent spontaneous micro-flocculation has been induced due to EOM and IOM. Cell lysis may have contributed to the rise of concentration of particles in the size range below the detection limit of the computer image analyser (1.9 μm), however, it is also likely that a portion of these is entrapped in the floc material which spontaneously formed due to EOM and IOM release. An increase of the overall particle volume has also been registered, suggesting the production of hydrophilic floc material with water entrapped within a hydrous floc structure. This is more pronounced in case of the lower ozone dose, presumably due to continued reactions with the organic matter of excess ozone in the case of the higher ozone dose. The influence of further coagulant addition and flocculation, as well as the down-stream DAF, on the particle size and volume distribution depend on the process conditions.

The effect of ozone on the particle size distribution still remains controversial. Similar to our results, it has been stated that ozone application shifts the particle size distribution towards larger sizes, thus indicating the initiation of spontaneous microflocculation. In addition, ozone is known to induce the formation of colloidal matter from dissolved organic matter and it enhances its subsequent removal [26, 29, Section 4.3.3]. However, others stated that ozone did not have a significant effect on the particle size distribution, whilst resulting in an increase of the smaller size fractions [25, 83]. Ozonation has also been found to decrease the overall particle volume and mass and result in lower filter loading [84]. These effects of ozone, however, appear to depend on the ozone dose and may result in contradictory findings; higher doses tend to be associated with reduction of the amount of particles in the larger size range, and with lower particle volume and mass [85].

One of the probable reasons for the inconsistencies is the use of non-standardised observation equipment (different types of light-blocking or laser particle counters, image analysis systems, etc.) which often have a limited applicability. One such limitation is related e.g. to the use of the (light-blockage) particle counter in this research. The high shear forces which are applied during the passage of the sample through the orifice of the counting cell may cause break-up of aggregates, which makes it a less reliable qualitative assessment method of flocculated and ozonated water. In view of the limitations of the available particle counting techniques we applied the Mini-Magiscan (IAS 25/IV25 Joyce-Loebl Ltd., UK) computer image analysis system, which proved to be especially useful for the characterisation of ozonated and flocculated water. It provides minimum sample disturbance and may be regarded as more reliable for such purposes.

c. Combined ozone and polyelectrolyte treatment (alone or combined with metal coagulant)

The combination of ozone and polyelectrolyte treatment has been assessed before, especially for cationic polyelectrolytes [34, 43, 84]. Process efficiency was expected to improve due to the effects of the ozone induced biopolymers and the cationic polyelectrolyte charge. The positive effect of the latter is assumed to be related to the reduction of the negative surface charge of the algae and the improvement of the particle collision kinetics. The cationic polyelectrolyte Superfloc C-573 and Wisprofloc-P were tested as sole coagulants and as coagulant aids in combination with ozone conditioning.

The efficiency of the Superfloc C-573 polyelectrolyte as a sole coagulant proved here to depend on the ozone dose (0.2 and 0.9 mg O_3 /mg C, or 0.5 and 2.25 mg O_3 /L) (Fig. 4.19, F and G).

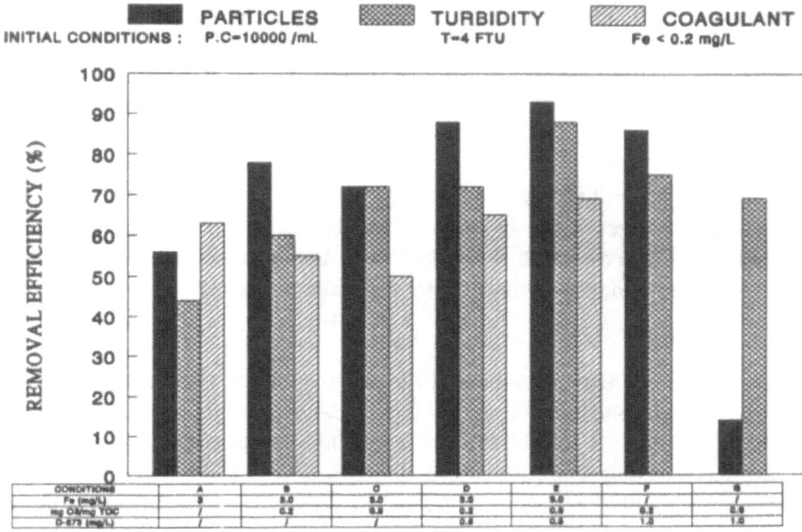

Fig. 4.19 Removal efficiency for the combination of ozone dose conditioning with polyelectrolyte Superfloc C-573 as coagulant aid or as sole coagulant.

The process efficiency for the same polyelectrolyte dose (1.0 mg C-573/L) was lower for the high ozone dose (0.9 mg O_3/mg C) than for the low one (0.2 mg O_3/mg C), especially regarding the particle removal. Considering the relatively clean model water circumstances in which the coagulant demand was exerted mainly by the organic matter, the increase of the ozone dose results in changes in the NOM structure which eventually result in lower MW and more acidic and hydrophilic compounds. The influence of these changes on coagulant demand depends on their extent. In case of metal coagulant application the increased acidity tends to increase the affinity of the organics to the metal hydroxide, thus increasing metal coagulant demand, while the MW decrease tends to produce the opposite effect. The same applies for polyelectrolyte coagulants, since most of them selectively precipitate high MW and NOM with low functional group content. Ozonation causes the opposite effect [6, 83] : ozone decreases the particle surface charge, thus decreasing the amount of polymer needed to neutralize it and theoretically decreasing the polymer demand. However, the amount of precipitated TOC under such conditions decreases due to the preference of the polymers for organics with high MW and low functional group content.

Thus, the high ozone dose is suggested to disfavour particle flocculation through the applied polymer coagulant and to have resulted in lower DAF efficiency. The high ozone dose reduced the initial particle (alga) count by 25%, reflecting algal lysis and EOM and IOM leakage. Under these conditions, the particle (alga) count after DAF remained at the same level reached after ozonation, suggesting that the EOM and IOM interfered with the coagulation (the algae were not properly destabilised and flocculated) and that the polyelectrolyte was ineffectively utilised. It is also suggested that ozone and its hydroxyl radical may oxidise some of the polyelectrolyte molecules resulting in inefficient particle coagulation; this is more pronounced for the higher ozone dose. In the case of the low ozone dose the initial algae count was reduced by only 7%, suggesting only minor IOM leakage accompanying the EOM release and probably less changes in their original structure. These conditions (Fig. 4.19, F) resulted in more efficient coagulation and flocculation of the algae and in higher particle and turbidity removal than the ozone + metal coagulant combination (Fig. 4.19, B and C). In both cases this was accompanied by a decrease of the $UV_{254\,nm}$ absorbance; the DOC content was more substantially reduced in the case of the lower ozone dose.

In the case where polyelectrolyte was used as coagulant aid (D and E), the overall NOM removal efficiency has been suggested to depend on the concentration ratio of metal coagulant to polyelectrolyte [6, 83]. The overall effect of the ozonation on the coagulation in the case of dual coagulants (metal coagulant + polyelectrolyte) depends on the individual characteristics of each coagulant.

The results show that the characteristics of the polyelectrolyte play a specific role (Fig. 4.20). Ozonation in the case of a medium Superfloc C-573 polyelectrolyte coagulant aid dose range (D and E; 0.3 and 0.5 mg C-573/L) tended to perform slightly better than the metal coagulant only (A), even when the coagulant dose was doubled (B). Unlike the case where the same polyelectrolyte was applied as a sole coagulant, the DAF efficiency tended to improve with an increase of the ozone dose (data not presented here), suggesting that in this case the cationic polyelectrolyte induced adsorption phenomena compound the conditioning effect of the ozone and particle destabilisation by the metal coagulant. This discussion is supported by EM measurements, with the particle surface charge decreasing with the polyelectrolyte dose. Although the Wisprofloc-P polyelectrolyte proved beneficent for the flocculation and DAF removal of the algae in the ozone + Fe(III) coagulant scheme, its weaker charge density and resultant high residuals are suggested responsible for the accompanying high residual turbidities (F, G and H).

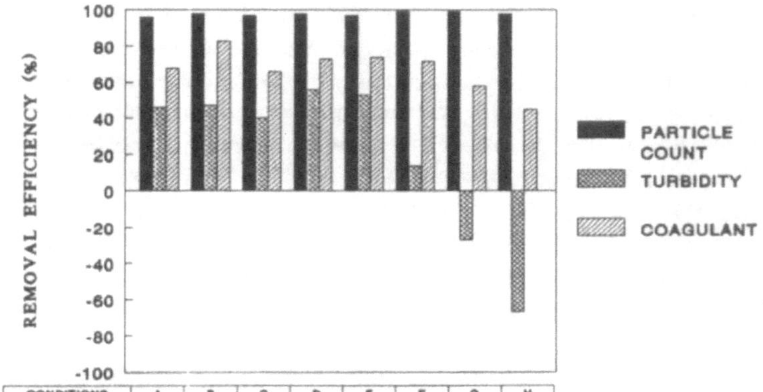

CONDITIONS	A	B	C	D	E	F	G	H
Fe (mg/L)	2.5	5.0	2.5	2.5	2.5	2.5	2.5	2.5
mg O3/mg TOC	0.2	0.2	0.2	0.2	0.2	0.2	0.2	0.2
C-573 (mg/L)	/	/	0.1	0.3	0.5	/	/	/
W-P	/	/	/	/	/	0.1	0.3	0.5

Fig. 4.20 Removal efficiency as a function of low ozone dose conditioning in combination with cationic polyelectrolyte Superfloc C-573 and Wisprofloc-P coagulant aid.

Fig. 4.21 shows the results from the two way ANOVA for iron coagulant alone, the combination of iron and Superfloc C-573 as coagulant aid, the combination of ozone conditioning and iron coagulant, and ozone conditioning, iron coagulant and Superfloc C-573 coagulant aid.

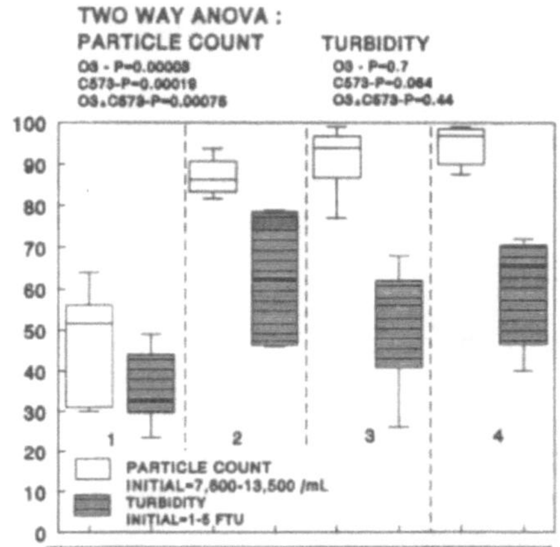

CONDITIONS	FeIII (mg/L)	mg O3/mg TOC	C-573 (mg/L)
1 (DAF)	3	-	-
2 (DAF)	3	-	0.3
3 (DAF)	3	0.2	-
4 (DAF)	3	0.2	0.3

Fig. 4.21 Two way ANOVA for the different DAF treatment options.

The particle removal efficiency is statistically significantly affected by ozone, Superfloc C-573 coagulant aid, and by the combination of both. The turbidity removal efficiency, on the other hand is not significantly affected by these parameters, however, introducing the initial turbidity in the ANOVA as a covariate suggests that turbidity removal was significantly affected by the polyelectrolyte and the combination of polyelectrolyte and ozone, but not by ozone alone (P=0.32 for O_3, P=0.002 for polyelectrolyte and P=0.049 for the combination of polyelectrolyte and ozone). Two way ANOVA for the initial turbidity and particle count showed that the initial turbidity was not significantly different, but that initial particle count was significantly different for the different experiments. Multiple comparison analysis shows that for particle removal efficiency there is a significant difference between the iron coagulant treatment only (Fig. 4.21; 1) and the other treatment options (2, 3 and 4), however, the results for the latter did not significantly differ from each other. Similarly, if initial particle count is included in the analysis as a covariate, the results for the iron coagulant only and the other treatment options were again significantly different (P<0.001). The multiple comparison analysis for turbidity removal efficiency showed that there was no significant difference between the different treatment options 1 to 4, however, introducing the initial turbidity as a covariate in the analysis suggests again that polyelectrolyte and the polyelectrolyte + ozone conditioning significantly affect the DAF removal efficiency. Multiple regression analysis for the particle removal efficiency (as the dependent variable, r^2= 0.69) showed that the application of ozone (P=0.04) and polyelectrolyte (P=0.00036) significantly affected it, while the initial particle count (P=0.58) did not. Multiple regression analysis for the turbidity removal efficiency (as the dependent variable, r^2= 0.35) showed that the polyelectrolyte (P=0.0025) and the initial turbidity (P=0.011) significantly affected it, while the ozone (P=0.14) did not.

The influence of the ozone dose on the process efficiency proved to be larger in the case of application of cationic polyelectrolytes as sole coagulants compared to the case when it was used as a coagulant aid to Fe(III) coagulant. This implies that process flexibility is lower and that if this option would be applied in practice the process would need to be optimised more frequently. Dual coagulant systems provided more flexibility in this sense, as the polyelectrolyte had a synergistic effect on particle destabilisation, to that of the metal coagulant. Optimisation of the polyelectrolyte coagulant aid dose is likely to lower required coagulant doses. The positive effects of polyelectrolyte application in this case, such as reduced coagulant demand and lower filter headloss, must be balanced with the possible negative effects, e.g. higher polyelectrolyte residuals and accompanying turbidity, and higher organic matter, presumably AOC, in the effluent.

d. Point of coagulant addition

The point of coagulant addition relative to the ozone can influence process efficiency. Our experiments with a low dose of ozone (0.2 mg O_3/mg C or 0.48 mg O_3/L, with 3 mg Fe(III)/L) showed that increase of the time lag between the ozonation and the coagulation from 1.5 to 9 min, slightly increased DAF removal efficiency (Fig. 4.22). This is contrary to previous results from research on the application of ozone as an algae conditioner in the context of direct filtration [36]. There, an increase of the time lapse between ozonation and coagulant addition from 1-10 min was found to negatively influence process efficiency.

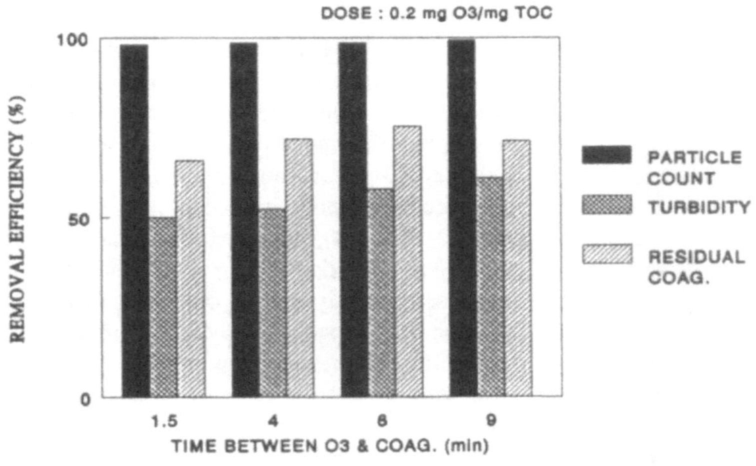

Fig. 4.22 DAF removal efficiency as a function of the time gap between ozonation and coagulation.

In both cases similar ozone doses were used, however, in the direct filtration case raw Biesbosch Reservoirs water had been used for the experiments [36], while in our case the model water was prepared by using Biesbosch Reservoirs water after it had already been treated with ozone in the full-scale treatment. This implies that the organic matter present before treatment had already been oxidised and modified. Our model water was thus, characterised by very small particulate matter concentration except for the spiked algae, while the largest portion of the TOC was DOC (80%≥DOC≥85%). In such circumstances, the spiked algae were presumably the only significant source of particulate and dissolved organic matter which was prone to react with the newly applied ozone, and affect the process efficiency.

The observed differences between our results and that of others [36], can partly be explained with the raw water quality differences. In our case, the removal efficiency for the shortest time lag of 1.5 min was already high (>90%), suggesting very fast reactions of the ozone with the algae within this time after ozone application. In both cases the mg $CaCO_3$/mg TOC ratio was 60-70, which favours ozone effects on flocculation [47]. At this ratio and the used ozone doses, the efficiency rose with the elapsed time before coagulant addition, suggesting that ozone continued to react with the algae. The applied ozone dose was probably not high enough to further extend organic matter changes and increase coagulant demand, which is in accordance with other findings [84]. Although not tested in this experiment, the application of a higher ozone dose (e.g. in the range of 1.0 mg O_3/mg C) may produce a different effect, in which the additional ozone is likely to further oxidise the organic molecules into compounds with a MW and acidic functional group content that may impart an additional coagulant demand.

4.3.4 $KMnO_4$ conditioning

Till now the use of potassium permanganate for algae conditioning in the context of DAF hasn't been considered. The benefits which arise from its application include taste and odour removal, a bactericidal effect, the reduction of the THM formation potential and the absence of potentially harmful by-products.

$KMnO_4$ dose optimisation for a constant iron coagulant dose in model water (Fig. 4.23) shows that an optimal dose of 0.7 mg $KMnO_4$/L more than doubled the DAF particle removal efficiency achieved with the iron coagulant only. However, the accompanying turbidity removal was not much affected, presumably due to the high residual Mn values (Fig. 4.24). The optimal permanganate dose coincided with the visually determined permanganate demand (no pink colour observable after DAF). Excessive permanganate dosing caused colouration of the water and lowering of the process efficiency. The influence of an increase of the permanganate dose on the residual Mn fractions is seen in Fig. 4.24. Both, the residual dissolved MnO_4^- and colloidal MnO_2 concentrations increased with the permanganate dose, which is in accordance with reports by Middlemans and Ficek [49].

For the same $KMnO_4$ dose the permanganate (dissolved Mn) decreased non-linearly with time (Fig. 4.25). 60% of the overall $KMnO_4$ reduction which takes place within the 30 min available contact time before coagulation, occurred during the first 5 min. The precipitation rate of colloidal MnO_2 for the 0.7 mg $KMnO_4$ dose and the model water conditions was 0.135 mg Mn/L, or 0.214 mg MnO_2/L in the 30 minutes of contact time. Fig 4.25 suggests that although the $KMnO_4$ contact time in the experiments was rather long, the results from the $KMnO_4$ conditioning experiments would be valid for contact times as short as 20 or even 15 minutes, because very little permanganate reduction took place after the 15 min.

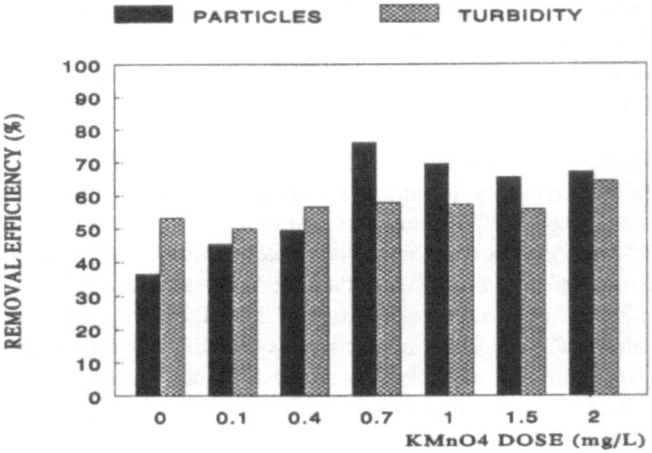

Fig. 4.23 Optimisation of the $KMnO_4$ dose for a constant dose of coagulant (5 mg Fe(III)/L), based on removal of particles and turbidity. Conditions : $KMnO_4$ contact time 30 min, contact and coagulation pH 8, G=70 s^{-1}, t_f=15 min, R=7%, P=500 kPa.

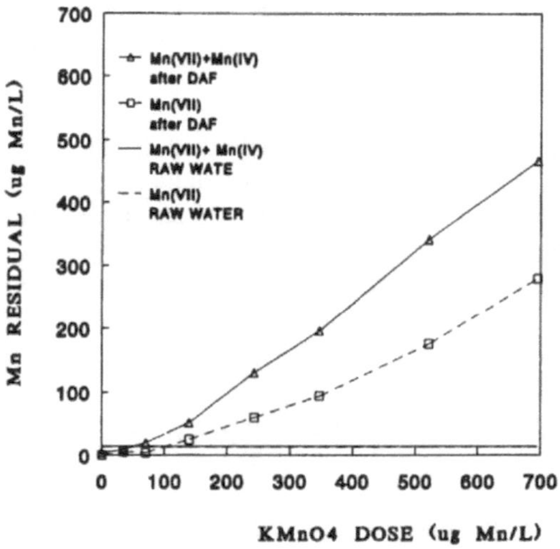

Fig. 4.24 Residual Mn concentration as a function of the $KMnO_4$ dose. Conditions : $KMnO_4$ contact time 30 min, 5 mg Fe(III)/L, pH 8, G=70 s^{-1}, t_f=15 min, R=7%, P=500 kPa.

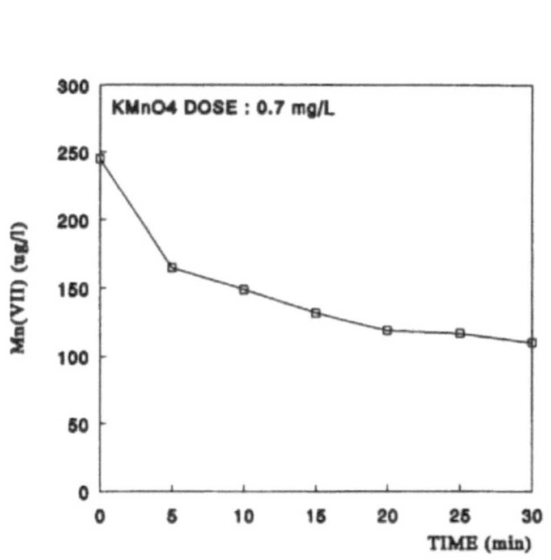

Fig. 4.25 $KMnO_4$ reduction for model water (≈10,000 cells *M. aeruginosa*/mL, pH 8), as a function of contact time.

The decrease of the permanganate concentration (dissolved Mn) after DAF was linear with increase of the conditioning (and coagulation) pH (Fig. 4.26), due to increase of permanganate reactivity with increase of pH.

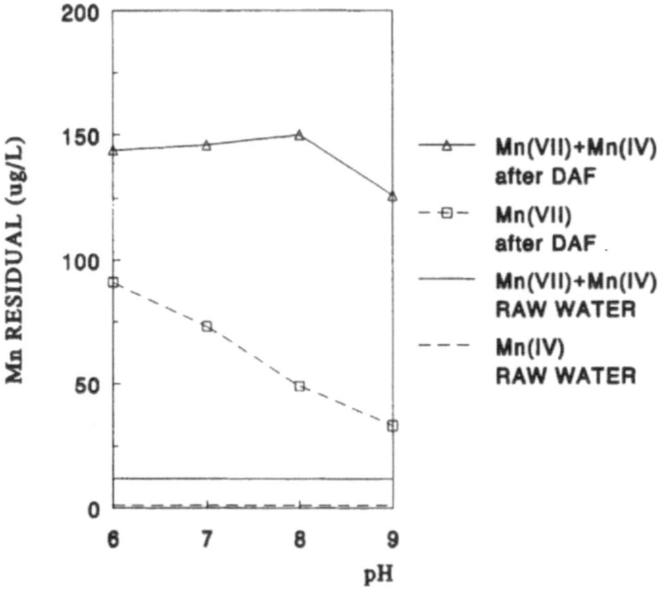

Fig 4.26 Total and dissolved residual Mn as a function of $KMnO_4$ conditioning and coagulation pH. Conditions : 0.7 mg $KMnO_4$/L, G=70 s^{-1} , t=15 min, R=7%, P=500 kPa.

Fig. 4.26 suggests that the accompanying high particle removal efficiency achieved at coagulation pH 6 conditions is to be ascribed predominantly to adsorption coagulation and charge neutralisation. These mechanisms are typical for such pH conditions below the IEP [3, 11, 12], compared to sweep coagulation occurring at the pH 7, 8 and 9 [86]. The high(er) permanganate concentration at pH 6 suggest that the permanganate reduction at this pH was substantially lower than at pH 7, 8 and 9. This again implies that at pH 6 less oxidative reactions with the algae and EOM took place, consequently less MnO_2 was produced and hence, lower flocculation rates should occur. At pH 6 one would expect efficient removal of the negatively charged MnO_2 by adsorption coagulation with the predominantly positively charged iron-hydroxo species [9, 10]. However, the principal mechanisms in which metal species and organic ions are adsorbed, are cation sorption and direct ion attraction. The cation exchange properties of MnO_2 are related to the functional groups at its surface, which strongly depend on pH [50, 60]. Thus, as a decrease of pH would result in a decrease of its negative charge and of its cation exchange capacity, this would result in lower adsorption of positively charged iron-hydroxo species onto MnO_2.

DAF efficiency is closely related to the applied energy input (G value) and flocculation time (t). In the case of permanganate conditioning, the removal efficiency of the residual colloidal MnO_2 forms an additional criterion of optimal flocculation. For our model water a Gt value of approximately 60,000 (G=70 s^{-1} and t=15 min) was found appropriate for efficient and economical flocculation preceding DAF (Fig. 4.27).

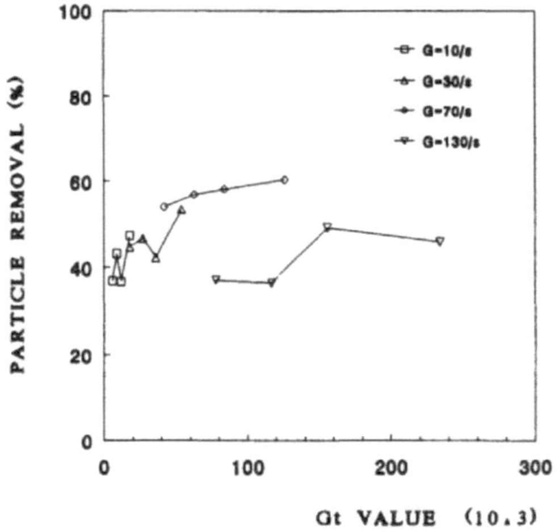

Fig. 4.27 Particle removal efficiency as a function of Gt (and G); model water with $KMnO_4$ conditioning.

This is in accordance to findings for algae laden water treatment by DAF [66] and practically applied Gt values [87] without $KMnO_4$ conditioning, although much lower Gt values (15,000-16,000) have also been implicated as favourable [65]. Prolonged flocculation times did not substantially increase process efficiency. Higher G values are suggested to have resulted in floc break-up, while lower ones did not provide sufficient contact opportunities per unit of time for efficient flocculation. In all tested cases discussed previously, the total residual Fe coagulant and

Mn concentrations were higher than their respective maximum allowable concentration (MAC) of 200 and 50 μg/L. Further process optimisation or modification may be required, as well as down-stream filtration.

a. Mechanisms of $KMnO_4$ induced flocculation

The results from the one way ANOVA (Fig. 4.28) for different treatment options including iron coagulant alone, combined $KMnO_4$ conditioning and iron coagulant, and combined $KMnO_4$ conditioning, iron coagulant and polyelectrolyte Superfloc C-573 coagulant aid, suggest that the particle removal efficiency was significantly affected by the different treatments. The same applies for the turbidity. Multiple comparison analysis for the particle removal efficiency suggests that the option including the polyelectrolyte (Fig. 4.28; 3) performed statistically significantly better than the other two treatment options (Fig. 4.28; 1 and 2). The same holds for turbidity. Although the $KMnO_4$ conditioning tended to raise the DAF efficiency where the iron coagulant was used, the obtained improvement was statistically insignificant. The standard deviation of the removal efficiency obtained with experiments was comparatively high (Fig. 4.28).

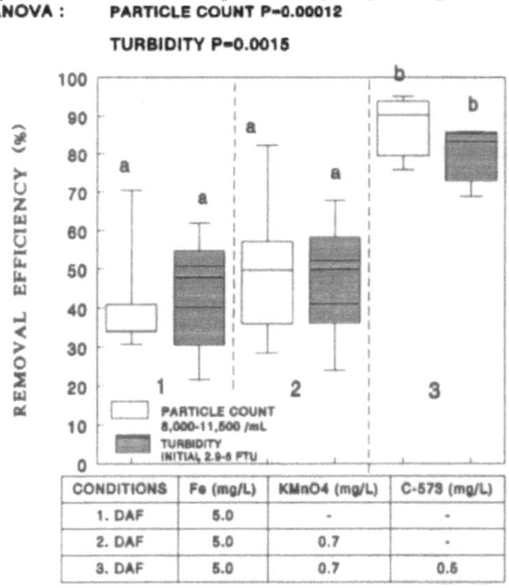

CONDITIONS	Fe (mg/L)	KMnO4 (mg/L)	C-573 (mg/L)
1. DAF	5.0	-	-
2. DAF	5.0	0.7	-
3. DAF	5.0	0.7	0.5

Fig. 4.28 Results from the ANOVA for different treatment combinations including $KMnO_4$ conditioning and combined $KMnO_4$ and polyelectrolyte treatment.

Although the algal concentration was kept constant, periodically a different permanganate demand was exerted, indicating that the organic matter content of the water nevertheless was varying. The TOC/DOC content of an algae batch was shown to vary at different growth stages, as well as different time after the algae were sampled from the original culture. Our attempts to correlate the DOC concentration with the achieved process efficiency proved not fully successful. This suggested that the DOC composition was a factor that affected the process efficiency as well. In the absence of significant amounts of other organic matter in the water, the spiked algae with their EOM and possibly IOM were the only significant contributor to the particulate and dissolved organic matter susceptible to $KMnO_4$ oxidation. The effect of the algal growth stage on the EOM content composition and structure, and on flocculation process efficiency, is well documented [77, 78, 82, 85]. On the other hand, the longer the time gap between the algae sampling from the original culture and their actual addition to the model water, the higher the possibility of change of the quantity and composition of EOM. Although no characterisation (e.g. measurements of the MW) of the present EOM fractions for different experiments was done, this is suggested as one of the major sources of noted results inconsistency.

The overall effect of permanganate conditioning consists of its oxidative activity, and the flocculation and adsorption characteristics of the colloidal MnO_2. Similar to ozone, permanganate application to algae laden water causes stress for the algae resulting in algal EOM release, which may serve as biopolymer [36]. The 0.7 mg $KMnO_4$ dose caused a 10% reduction of particle count suggesting IOM leakage as well. DOC and UV_{254nm} increased by 6% and 65 % respectively. Furthermore, the permanganate has been known to oxidise organic matter which may otherwise bind the iron and form soluble complexes, imparting an additional coagulant demand. It has also been suggested that the released biopolymer may initiate microflocculation even before coagulant addition [36]. In surface water rich in NOM fine organic matter matrices or meshes of filaments exist [81], both EOM and MnO_2 precipitates can be captured within the meshes, thus enhancing a bridging flocculation mechanism. The precipitated MnO_2 increases the overall particle concentration and adsorbs multivalent cations such as Ca^{2+}, organic molecules, and can be adsorbed on alga cells [36]. Although of high density (5,026 kg/m^3), the hydrous colloidal nature of MnO_2 provides conditions which do not necessarily favour settling of the algae. The MnO_2 adsorption on alga cells was accompanied by a slightly higher negative charge of the algae, as verified by EM measurements. The physical coverage of the algae by MnO_2 and the resulting increased negative charge contradict the requirements for their efficient flocculation. Yet, this effect could be balanced by the oxidising effect of the permanganate and the release of biopolymer which can behave as a coagulant aid. This leads to an overall better coagulation and flocculation of particles.

b. Combined $KMnO_4$ and polyelectrolyte treatment

The residual Mn after $KMnO_4$ conditioning and DAF was systematically higher than the 50 µg/l MAC. The negative surface charge of the colloidal MnO_2 at the tested pH 8, suggests the use of cationic polyelectrolytes can assist in its destabilisation and removal. The Superfloc C-573 and Wisprofloc-P polyelectrolytes were tested for the purpose. They proved efficient as coagulant aids to Fe(III) coagulant in a DAF scheme (Fig. 4.29), supporting the predominant sweep coagulation (see Chapter 3) by the neutralisation of the surface charge of MnO_2. The results from the one way ANOVA (Fig. 4.28) suggested that although the $KMnO_4$ conditioning tended to raise the DAF efficiency where the iron coagulant was used, the only statistically significant improvement was obtained if cationic polyelectrolyte was introduced in the scheme.

As discussed in Section 4.3.5, the efficiency improvement achieved by the combined Fe(III) coagulant and polyelectrolyte treatment, especially in the case of the higher charged Superfloc C-573, was significant (Fig. 4.20). The incorporation of $KMnO_4$ conditioning in the scheme tended to raise the process efficiency further. This is valid in particular for the turbidity removal efficiency which improved by approximately 10-20%, although the introduction of the permanganate in effect increases particle concentration due to resulting MnO_2 production.

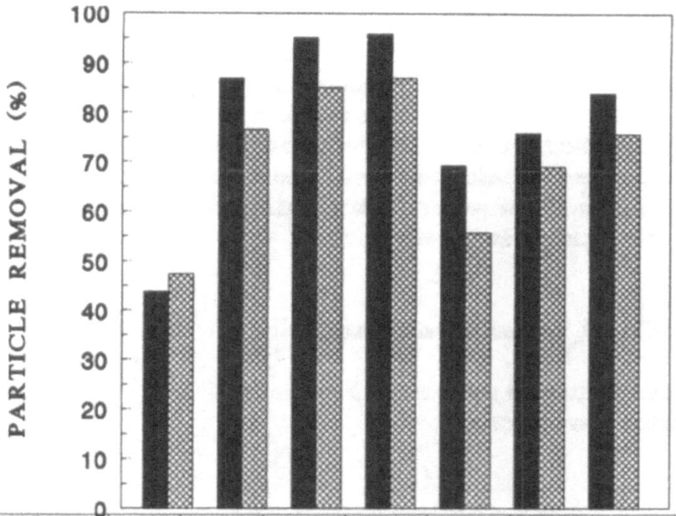

CONDITIONS	A	B	C	D	E	F	G
Fe (mg/L)	5	5	5	5	5	5	5
KMnO4 (mg/L)	-	-	0.7	0.7	-	0.7	0.7
C573 (mg/L)	-	0.5	0.3	1.0	-	-	-
W-P (mg/L)	-	-	-	-	0.5	0.3	1.0

Fig. 4.29 $KMnO_4$ conditioning in combination with iron and cationic polyelectrolytes as coagulant aid.

Residual coagulant and manganese levels confirm that the lower final turbidity in the case of the $KMnO_4$-coagulant-polyelectrolyte scheme is caused by improved coagulant removal (Fig. 4.30).

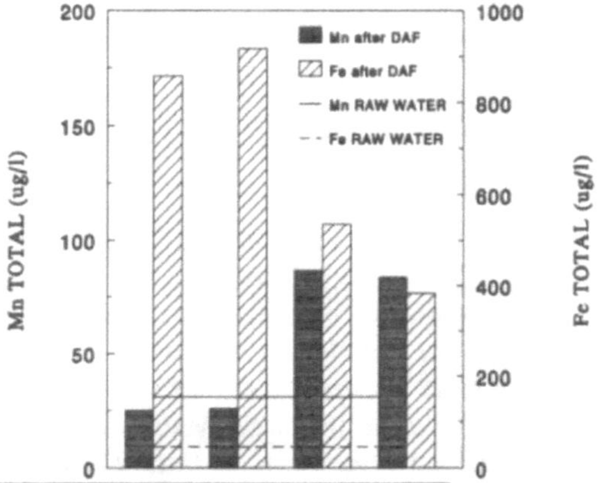

CONDITIONS	A	B	C	D
Fe (mg/L)	5.0	5.0	5.0	5.0
KMnO4 (mg/L)	-	-	0.7	0.7
C-573 (mg/L)	-	0.5	0.3	1

Fig. 4.30 Residual Fe and Mn after DAF with different treatment combinations of $KMnO_4$ conditioning, Fe(III)coagulant and Superfloc C-573 polyelectrolyte treatment. Conditions : pH 8, G=70 s^{-1}, t_f=15 min, R=7%, P=500 kPa.

Permanganate is assumed to react with the organic matter preventing it from complexing the iron coagulant into stable organo-iron compounds. This oxidative action ensures more complete coagulant utilisation for particle destabilisation. In this case it eventually resulted in better coagulant removal due to adsorption effects related to the cationic polyelectrolyte. The same adsorption coagulation effects are responsible for the reduction of the residual Mn below the 100 μg/l, suggesting that cationic polyelectrolytes may also be utilised successfully for the purpose of reducing the residual Mn. There were no indications that the applied $KMnO_4$ dose of 0.7 mg/L affected the polymer structure and its efficiency.

c. The effect of $KMnO_4$ on particle size distribution

Fig. 4.31 represents the particle volume distribution after $KMnO_4$ conditioning with a 0.7 mg/L dose and after different contact time.

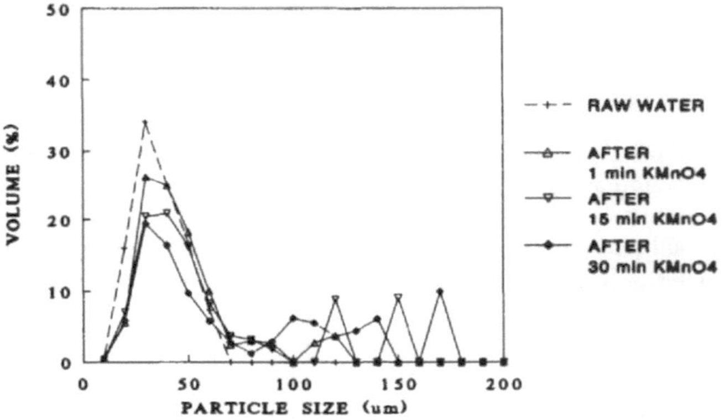

Fig. 4.31 Effect of a 0.7 mg $KMnO_4$/L dose and contact time on the particle volume distribution.

The stress which results from the permanganate conditioning results in the excretion of algal EOM, as well as cell lysis and IOM leakage. Even at very short contact time of 1 minute there is a noticeable reduction of the volume of the smaller-size particles (predominantly algae) and an increase of volume of the larger-size particles. This indicates that spontaneous micro-flocculation was initiated by the EOM and IOM. Longer contact times did not have a more pronounced effect on the algal volume fractions, however, the larger-size particles were further modified resulting in more voluminous floc material with longer contact time. The particle distribution measurement did not allow to detect particles smaller than approximately 2 μm, which would encompass the colloidal (0.3-0.4 μm) MnO_2. As more permanganate will be reduced with time (Fig. 4.25), the MnO_2 and the submicron particle concentration will increase. The increase with its concomitant high adsorptive capacity principally results in an increase of the overall process efficiency. The MnO_2 fraction will be removed in the final filter, resulting in acceptable effluent Mn concentrations [22].

d. Reduction of $KMnO_4$ residual by Na_2SO_3

Chemical reduction of the permanganate Mn residual by Na_2SO_3 has been attempted with limited success in direct filtration [36]. The background motive for this approach was to reduce the dissolved permanganate Mn (VII) and simultaneously convert it into colloidal Mn(IV)O_2, which can be removed by the final filtration. The experiments here (at 0.35 and 0.7 mg Na_2SO_3 /L) lowered total Mn, after DAF and prior to filtration, to levels in the range of 70 μg/l. These concentrations are low for the purpose. The stoichiometric amount of Na_2SO_3 necessary for the reduction of 1 mg $KMnO_4$/L is approximately 1.2 mg Na_2SO_3/L; a considerable time is needed to achieve complete reduction. Since flotation time is 15-30 minutes, full reduction by Na_2SO_3 may not be completed within the available flotation time. Increasing the dose of Na_2SO_3 in excess to the stoichiometric amount necessary for the reduction of the residual permanganate can speed up the permanganate reduction.

Finally, the chemical reducing activity of the down-stream filter should also be considered as a contributor to lower permanganate concentrations in the effluent.

4 .4 CONCLUSIONS

Coagulant dose stoichiometry was established for single cell cyanobacteria *Microcystis aeruginosa* in the concentration range of 5,000-20,000 cells/mL. A further 25% rise of algae concentration did not raise the coagulant dose requirement, suggesting that a critical concentration of algae and of coagulant dose had been reached allowing for full utilisation of the sweep coagulation of the particles. This is supported by the fact that even the highest tested coagulant dose of 15 mg Fe(III)/L did not result in algae charge neutralisation, although particle charge neutralisation and charge reversal took place for the same dose in the absence of algae. It can be assumed that the IEP of the colloids in the absence of algae was shifted towards the neutral coagulation pH, resulting in particle charge reversal. The presence of the algae (and the constituents of their culture medium), prevented the IEP shift towards the neutral pH region and their charge neutralisation.

Polyelectrolyte as coagulant aid is a treatment process option which proved to be efficient in particle destabilisation. The efficiency for the two cationic polyelectrolytes (Wisprofloc-P and especially Superfloc C-573) was significantly higher than that of the metal coagulant alone under sweep coagulation conditions. A combination of 3 mg Fe(III)/L and 0.3 mg Superfloc C-573/L achieved 87% particles (algae) removal, compared to 84 % achieved with 3 mg Fe/L + 0.1 mg Wisprofloc-P, and 47% achieved by the 3 mg Fe/L only. The turbidity removal for these treatment options was 79%, 61% and 35% respectively. Both polyelectrolytes induced adsorption coagulation which supported the sweep coagulation; this was verified by EM measurements. The efficiency of the Superfloc C-573 and Wisprofloc-P depends on their electric charge and molecular weight characteristics. The more highly charged Superfloc C-573 proved to be more efficient and reliable than the 'heavier' but lower charged Wisprofloc-P, both in terms of algae and turbidity removal. The latter tended to form sticky floc material that kept being attached to any available surfaces. The bridging mechanism was prevalent in the case of the non-ionic and anionic polyelectrolytes; it proved to be of inferior efficiency as compared to the cationic polyelectrolytes' charge neutralisation mechanism. The abundant presence of Ca^{2+} in the model water which enhanced the adsorption of the anionic polyelectrolyte Superfloc A-100 onto the particles, is thought to be the reason for its slightly higher efficiency as compared to the non-ionic polyelectrolyte Wisprofloc-N.

Polyelectrolyte as sole coagulant is a treatment option which may result in similar (or slightly higher) DAF particle removal efficiency as compared to that of sole metal coagulants. This applies to cationic polyelectrolytes where particle adsorption, and in the case of Superfloc C-573 charge neutralisation, were again more efficient than the bridging mechanism associated with the non-ionic and anionic polyelectrolytes. The most important prerequisite for high particle and turbidity removal is the charge characteristic of the polyelectrolyte, which has an influence on the resulting floc size and structure. Polyelectrolyte doses >0.5 mg/L of the higher charged Superfloc C-573 resulted in smaller and more compact floc material compared to the same doses of the Wisprofloc-P polyelectrolyte, which again tended to produce large and sticky floc material. Hence, the difference in achieved DAF maximum particle and turbidity removal efficiency, which amounted to 81% and 77% particle removal for 3 mg Superfloc C-573 and 0.5 mg Wisprofloc-P, coupled to turbidity removal efficiency of 74% for the first, and only 12% for the second polyelectrolyte. The additional benefit of cationic polyelectrolytes application in the case of DAF is the increase of particle-bubble adsorption due to attraction forces between the negatively charged air bubbles and positively charged floc material.

The final choice of polyelectrolyte, whether as a coagulant or coagulant aid, also depends on the raw water characteristics, including the suspended and colloidal particles concentration, the TOC/DOC content and composition, the presence of ions such as calcium, etc. The cationic polyelectrolytes were overall more efficient and reliable in the context of DAF. However, the decision to apply them and the final type selection will also depend on the residual concentration after the final filter. The long-term health hazard related to consumption of water with polyelectrolyte residual, and the absence of real-time analytical technique calls for a careful approach.

Ozone conditioning is a pre-treatment option which proved efficient for improving the coagulation and flocculation before down-stream DAF. With the prerequisite that a range of raw water characteristics and process parameters can be addressed (e.g. 0.4-0.8 mg O_3/mg C, hardness to TOC ratio >25 mg $CaCO_3$/mg TOC, low bromide concentration, ozonation pH preferably <7.5 or even 7), the application of ozone may be considered a viable option for significantly increasing DAF efficiency. Ozone conditioning, initial particle count and initial turbidity were found to statistically significantly affect DAF efficiency. The resultant DAF particle and turbidity removal efficiency for different ozone doses (0.2, 05 and 0.9 mg O_3/mg C) did not significantly vary. The particle removal for these ozone doses combined with 3 mg Fe/L amounted to 91%, 89% and 90.5%, coupled to 50%, 52% and 56% turbidity removal respectively. However, the higher the ozone dose, the more significantly the organic matter (mostly algae, EOM and IOM) was modified, as measured by UV_{254nm} absorbance. An increase of the coagulant dose in combination with the same ozone dose did not improve DAF efficiency significantly, although a positive trend was noted. A coagulant dose of 5 mg Fe/L combined with the 0.2, 0.5 and 0.9 mg O_3/mg C doses, resulted in 96%, 94% and 96% particle removal, and 54%, 65% and 51% turbidity removal efficiency.

The principal mechanisms involved in the ozone induced flocculation effects are related to its strong oxidising activity. Low ozone dosages of 0.2 mg O_3/mg C (or 0.48 mg O_3/L) reduced the initial particle count, indicating reaction with the ozone and modification of the algal cell wall layer, as well as partial rupture and lysis of the algae. Ozone application may be regarded as stress conditions for the algae, causing EOM release and possibly IOM leakage in the case of ruptured alga cells. This was verified by SEM observation, as well as by UV_{254nm} absorbance measurement, which is sensitive to the presence of dissolved organic matter with hydrocarbon rings in its structure, such as the pigmented compounds originating from lysed algae cells. The released EOM and IOM act as natural coagulant aid (biopolymer) and enhance flocculation; they may cause spontaneous micro-flocculation as verified by particle count and computer image analysis data. The micro-flocculation was stimulated by the ozone induced decrease of the initial algal surface charge and corresponding mutual electrostatic repulsion.

Larger doses of ozone (0.5 mg O_3/mg C) provided an opportunity for more pronounced oxidation and modification of the algae cells, resulting in further decrease of their initial concentration. This was accompanied by prolonged ozone activity, and the chemical modification of the EOM and IOM into compounds of lower MW and higher acidic group content, more hydrophilic organic matter compounds, and a resulting overall negative influence on the flocculation and the DAF efficiency. The highest ozone dose (0.9 mg O_3/mg C) had a severe impact on the algae population in terms of their EOM and IOM. The ozone further oxidized and split the released EOM and IOM compounds into ones with lower MW, which resulted in the formation of difficult to remove organo-iron complexes and increased residual coagulant in the effluent.

A similar tendency of slightly lower efficiency (compared to the lower and the higher ozone dose) was found to occur at a medium dose for a two times lower initial algae concentration. In this case the ozone appeared to induce an increase of the coagulant demand compared to the optimal dose for the coagulant without conditioning. This suggests that a minimum organic matter content has to be present in order to benefit from ozone conditioning. In the model water the main portion of organic matter prone to oxidative modification, was in the form of algae and their EOM and IOM. In the case of the higher algae concentration this resulted in improved flocculation through the enhancement of bridging flocculation. In reservoir water with often considerable amounts of colloidal and particulate inorganic and organic matter, the ozone conditioning may have different outcomes per location.

The use of polyelectrolytes as sole coagulant combined with ozone conditioning proved to be a relatively attractive option in view of the possible achievement of relatively high algae removal efficiency together with the absence of problems related to the complexing of metal coagulants by dissolved organic matter. This option mainly pertains to cationic polyelectrolytes and its efficiency is related to charge reduction effects. A combined 0.2 mg O_3/mg C and 1.0 mg Superfloc C-573/L treatment resulted in 86% particle removal and 75% turbidity removal. Although there was a slightly lower particle removal efficiency compared to the ozone + Fe(III) coagulant combinations, there was an improvement of the turbidity removal by more than 20%. This is related to the absence of the residual Fe coagulant in the effluent.

The cationic polyelectrolytes destabilise the ozone conditioned algae, which have a lower negative charge, by adsorption coagulation and partial charge neutralisation. The non-ionic and anionic character of the ozone induced EOM and IOM results in their adsorption onto the polyelectrolyte-floc material, resulting in overall good flocculation conditions. However, this option proved sensitive to ozone overdosing, which resulted in deteriorated effluent quality. It is suggested that high ozone doses may not only affect the algal structure and result in higher amounts of, and more extensively oxidised EOM and IOM, but they may also oxidise some of the polyelectrolyte molecules, thus affecting the polyelectrolyte structure and lowering its efficiency.

The use of polyelectrolytes as coagulant aid combined with ozone conditioning tended to further increase the ozone-metal coagulant scheme efficiency, in the case of cationic polyelectrolytes. The combination of 0.2 mg O_3/mg C, 2.5 mg Fe/L coagulant, and 0.2 mg Superfloc C-573/L resulted in 95% particle removal and 59% turbidity removal, compared to a similar case in which a 0.1 mg Wisprofloc-P coagulant aid resulted in 99% particle removal, but very low turbidity removal of 14%.

The efficiency of (cationic) polyelectrolytes in this cases is based on aiding the particle flocculation by adsorption coagulation and charge neutralisation of particles which had already been destabilised by the ozone conditioning and the metal coagulant induced sweep coagulation. Consequently, the algae coagulation does not solely depend on the polyelectrolyte as in the previously discussed option, but instead the flocculation of already destabilised particulate matter is being aided by the cationic polyelectrolyte. Thus, more flexibility and reliability is provided for achieving higher process efficiency. Again, the polyelectrolyte charge and MW play an important role in the process, in view of their possible modification by the ozone.

Statistical analysis confirmed that DAF preceded by combined ozone conditioning and metal coagulant application, combined metal coagulant and cationic polyelectrolyte coagulant aid, and combined ozone conditioning, metal coagulant and cationic polyelectrolyte coagulant aid, in all cases outperformed DAF with metal coagulant application only. Ozone and polyelectrolyte had a statistically significant effect on treatment efficiency. Ozone and polyelectrolyte treatment had a significant effect on the particle removal efficiency irrespective of the initial particle count, while initial turbidity and polyelectrolyte treatment had a significant effect on the turbidity removal efficiency irrespective of the ozone treatment. No significant efficiency differences were detected between the above combined treatments.

$KMnO_4$ conditioning is attractive because of its oxidative capacity without the hazardous by-products formation, such as bromate formation in the case of ozonation. The $KMnO_4$ induced flocculation efficiency improvement is based on a range of process mechanisms which are often similar to the ones encountered with ozonation. $KMnO_4$ also causes algae stress and the release of EOM and IOM biopolymers, as verified by particle count, $UV_{254\ nm}$ absorbance and DOC measurements. The released EOM and IOM act as biopolymers and promote spontaneous micro-flocculation even before any coagulant addition, as verified by particle size analysis. Under normal pH conditions the colloidal (0.3-0.4 μm size) hydrous MnO_2 is formed. The negatively charged MnO_2 exhibits a high adsorption capacity for organic and inorganic particles and adds to the concentration of suspended particles, which reinforces the positive flocculation effect of any oxidant. Although the attachment of MnO_2 to algae tended to increase their negative surface charge presumably creating unfavourable coagulation conditions, the overall algae removal efficiency was similar to or higher than the one achieved with the iron coagulant only. The optimal permanganate dose (0.7 mg $KMnO_4$/L) for the model water conditions coincided with the visually determined permanganate demand (no pink colour observable after DAF). It resulted in 49 % particle removal and 47% turbidity removal, compared to 40% and 44% achieved by the Fe(III) coagulant alone. The $KMnO_4$ conditioning option proved significantly less efficient than the ozone conditioning, especially in terms of particle removal. The relatively high standard deviation of the efficiency results, as well as the occasional change of the permanganate demand, suggested that the organic matter concentration and composition, play an important role in defining the permanganate demand, its chemical effect and the resulting DAF efficiency. The optimal flocculation Gt range was related to the flocculation and removal of the colloidal MnO_2 and was 60,000 (typically $G=70\ s^{-1}$, $t=15$ min).

The Mn and coagulant concentrations in the DAF effluent at the optimal permanganate dose were higher than their MAC values. This suggested that cationic polyelectrolytes could reduce the electrostatic repulsion between the oppositely charged Mn and Fe colloids. The DAF particle and turbidity removal efficiency in the case of cationic polyelectrolyte Superfloc C-573 coagulant aid application resulted in significantly better efficiency than was achieved in the case of iron coagulant only, or in the case of the combination of $KMnO_4$ conditioning and iron coagulant. A particle and turbidity removal efficiency of 95% and 86% respectively, were achieved with 0.5 mg Superfloc C-573 as coagulant aid, applied in combination with permanganate and Fe(III) coagulant. A 5-25% improvement of turbidity removal efficiency was also noted in comparison with the iron coagulant and cationic polyelectrolyte combination. The observed improvement was achieved via improved removal of both the Fe and Mn residuals. The option of lowering the residual Mn by the reducing agent Na_2SO_3 proved potentially viable. In any case, other studies proved that residual MnO_2 can be efficiently removed in a final filtration step.

Appendix 4.2 synthesizes the DAF particle and turbidity removal efficiency achieved under different treatment combinations.

REFERENCES

1. Bernhardt H. and Clasen J., 1992. Studies on removal of planktonic algae by flocculation and filtration. Technical papers, 2. *Water Malaysia '92, 8th ASPAC-IWSA Regional Water Supply Conference*, Kuala Lumpur, Malaysia.
2. Zabel T., 1992. The advantages of dissolved air flotation for water treatment. *JAWWA*, **77**, pp. 42-46.
3. Vlaški A., van Breemen A.N. and Alaerts G.J., 1995. Optimisation of Coagulation Conditions for the Removal of Cyanobacteria by Dissolved Air Flotation or Sedimentation. *J. Water SRT-Aqua*, **45**, 5, pp. 253-261.
4. Berger M. and Carnahan R.P, 1991. Fouling prediction in reverse osmosis processes. *Desalination*, **83**, pp.3-33.
5. Edzwald J.K., 1993. Coagulation in drinking water treatment : particles, organics and coagulants. *Wat. Sci. Tech.*, **27** (11).
6. Edwards M. and Benjamin M.M., 1992.Effect of preozonation on coagulant-NOM reactions. *JAWWA*, **84**, pp. 63-72.
7. Randtke S.J., 1988. Organic contaminant removal by coagulation and related process combinations. *JAWWA*, **80**, pp.40.
8. Davis J.A., 1982. Adsorption of natural dissolved organic matter at the oxide/water interface. *Geochim. Et cosmochim. Acta*, **46**, pp. 2381.

9. Stumm W and Morgan J., 1962. Chemical aspects of coagulation. *JAWWA*, **54**, pp. 971-992.
10. Bratby J., 1980. *Coagulation and flocculation*. Upland Press Ltd., England.
11. Bernhardt H. and Clasen J., 1991. Flocculation of micro-organisms - *J.Water SRT-Aqua*, **40**, 2, pp. 76-87.
12. Ives K.J., 1959. The significance of surface electric charge on algae in water purification. *J. Biochem. And Microb. Techn. Eng.*, **1**, 1, pp. 37-47.
13. Vlaški A., van Breemen A.N. and Alaerts G.J., 1996. The role of particle size and density determining dissolved air flotation and sedimentation efficiency. *4th IAWQ-IWSA International Conference on The Role of Particle Characteristics in Separation Processes*, Jerusalem, Israel.
14. Bernhardt H., Schell H. And Lusse B., 1986. Criteria for the control of flocculation and filtration processes in water treatment of reservoir water. *Water Supply*, **4**, pp. 99-116.
15. Stumm W. and Sigg L., 1972. Kolloidchemisch grundlagen der phosphorelimination in fallung, flockung und filtration. *Z Wasser-Abwasser-Forsch*, **12**, pp. 73-83.
16. Packham R.F., 1965. Some studies of the coagulation of dispersed clays with hydrolysed salts. *J. Coll. Sci.*, **20**, pp. 81-92.
17. Gregory J., 1983. Chemistry and technology of water soluble polymers. Ed. Finch C.A., Plenum Press, pp. 307-320.
18. Kawamura S., 1991. Effectiveness of natural polyelectrolytes in water treatment. *JAWWA*, **83**, pp. 73-81.
19. Habibian M.T. and O'Melia C.R., 1975. Particles, polymers and performance in filtration. *J. Envir. Eng. Div.- ASCE*, **101**, pp. 567-583.
20. Tenney M.W., Eichelberger W.F. Schuessler R.G. and Pavoni J.L., 1969. Algal flocculation with synthetic polymeric flocculants. *Applied Microbiology*, **18**, pp. 965-971.
21. Tilton R.C., Murphy J. And Dixon J.K., 1972. The flocculation of algae with synthetic polymeric flocculants. *Wat. Res.*, **6**, pp. 155-164.
22. Petruševski B., van Breemen A.N. and Alaerts G.J., 1995. Effect of permanganate pre-treatment and coagulation with dual coagulants on particle and algae removal in direct filtration. *Proc. IAWQ-IWSA Workshop on Removal of Microorganisms from Water and Wastewater*, Amsterdam, the Netherlands.
23. Malley J.P., 1994. The use of selective and direct DAF for removal of particulate contaminants in drinking water treatment. *Joint IAWQ-IWSA Specialised Conference on Flotation Processes in Water and Sludge Treatment*, Orlando, Fl., USA.
24. Janssens J.G., van Hoof F. and Dirickx, 1986. Study on preozonation and prechlorination applied in the treatment of surface waters from impounding reservoirs. *Proc. AWWA Annual Conference*, Part 2, Kansas City, Mo., USA.
25. Jodlowski A., 1990. *Preoxidation effect on algae coagulation related to direct filtration water treatment*. Internal Report, Delft University of Technology, Delft.
26. Langlais B., Reckhow D.A. and Brink D.R., editors, 1991. *Ozone in Water Treatment Applications and Engineering*. Lewis Publishers, Chelsea, Mi, USA.
27. Sukenik A., Teltch B., Wachs A.W., Shelef G., Nir I. And Levanon D., 1987. Effect of oxidation on microalgal flocculation. *Wat. Res.*, **21**, 5, pp. 533-539.
28. Singer P.C. and Chang S.D., 1988. Impact of ozone on the removal of particles, TOC and THM precursors. *AWWA Research Foundation Report*, AWWA, Denver, Co., USA.
29. Jekel M.R., 1983. The benefits of ozone treatment prior to flocculation processes. *Ozone Sci. And Eng.*, **5**, pp. 21-35.

30. Vlaški A., van Breemen A.N. and Alaerts G.J., 1997. Algae laden water treatment by dissolved air flotation (DAF). *Proc. CIWEM International Conference Dissolved Air Flotation - an Art or a Science ?*, London, United Kingdom.
31. Krasner W.S., Glaze W.H., Wienberg H.S., Daniel P.A. and Najm I.N., 1993. Formation and control of bromate during ozonation of waters containing bromide. *JAWWA*, **85**, pp.73-81.
32. Kruithof J.C., Meijers R.T., and Schippers J.C., 1993. Formation, restriction of formation and removal of bromate. *Wat. Supply*, **11**, 3/4, pp. 331-342.
33. Kruithof J.C. and Meijers R.T., 1993. Presence and formation of bromate in Dutch drinking water treatment. *Proc. AIDE/IWSA International Workshop : Bromate and Water Treatment*, Paris, France.
34. Schalekamp M., 1986. Pre- and intermediate oxidation with ozone, chlorine and chlorine dioxide. *Wat. Supply*, **4**, pp. 499-522.
35. Reckhow D.A., Singer P.C. and Trussell R.R., 1986. Ozone as a coagulant aid. *AWWA National Conference, Sunday Seminar on Ozonation : Recent Advances and Research Needs*, USA.
36. Petruševski B., 1996. Algae and particle removal in direct filtration of Biesbosch Water, Influence of algal characteristics, oxidation, and other pre-treatment conditions. Ph.D. thesis, published by A.A. Balkema, Rotterdam, the Netherlands.
37. Von Gunten U. and Hoigné J., 1993. Bromate formation during ozonation of bromide containing waters. *Proc. IOA World Congress*, San Francisco, USA.
38. Siddiqui M.S. and Amy G.L., 1993. Factors affecting DBP formation during ozone-bromide reactions. *JAWWA*, **85**, pp. 63-72.
39. Orlandini E, Kruithof J.C., van der Hoek J.P., Siebel M.A. and Schippers J.C., 1997. Ozonation in Biological activated carbon filtration : it's effect on bromate, Assimilable Organic Carbon and the Ct value. *J. Water SRT-Aqua*, **46**, 1, pp. 20-30.
40 Kooij van der D., 1992. Assimilable organic carbon as an indicator of bacterial regrowth. *J. AWWA*, **84**, 2.
41. Orlandini E., Siebel M.A., Graveland A. and Schippers J.C., 1994. Pesticide removal by combined ozonation and GAC filtration, *Water Supply*, **14**, pp. 99-108.
42. Ma J. and Li G., 1993. Laboratory and full scale plant studies of permanganate oxidation as an aid in coagulation. *Wat. Sci. Tech.*, **27**, 11, pp. 47-54.
43. Edzwald J.K. and Paralkar A., 1992. Algae, coagulation, and ozonation - *Chemical Water and wastewater treatment II*, Klute R. and Hahn H.(Eds.), Springer-Verlag, Germany, pp. 263-279.
44. Ginocchio J.C., 1981. Effect of ozone on the elimination of various algae by filtration. *Ozonation Manual for Water and Wastewater Treatment*, edited by Masschelein W.J., J.Wiley.
45. Singer P.C., 1990. Assessing ozonation research needs in water treatment. *J. AWWA*, **82**, pp.78-88.
46. Dowbiggin B. And Singer P.C., 1989. Effects of natural organic matter and calcium on ozone-induced particle destabilisation. *JAWWA*, **81**, pp. 77-85.
47. Chang S.D. and Singer P., 1991. The impact of ozonation on particle stability and removal of TOC and THM precursors. *JAWWA*, **84**, pp. 63-72.
48. Ficek J.K., 1978. Potassium permanganate for iron and manganese removal and taste and odour control. *Water treatment plant design*, edited by Sanks R.L., Ann Arbor Publishers Inc., Mi., USA.

49. Middlemans E.R. and Ficek J.K., 1986. Controlling the potassium permanganate feed for taste and odour treatment. AWWA Annual Conference, Denver. Co., USA.
50. Emanuel A.G., 1965. The chemistry and application of potassium permanganate in water treatment. *South West America Water Works Association Meeting*, Carus Chemicals Co., La Salle, Ill., USA.
51. Willey B.F. and Jennings H., 1963. Iron and manganese removal with potassium permanganate. *JAWWA*, **65**, pp. 729-735.
52. Fitzgerald G.P., 1964. Evaluation of potassium permanganate as an algicide for water cooling towers. *I&EC Product Research and Development*, **3**, 2, pp. 82-85.
53. Kemp H.T., Fuller P.G. and Davidson R.S., 1966. Potassium permanganate as an algicide. *JAWWA*, **59**, pp. 255-263.
54. Bernhardt H., and Lusse B., 1989. Elimination of zooplankton by flocculation and filtration. *J. Water SRT-Aqua*, **38**, 1, pp. 23-31.
55. Ficek J.K. and Boll J.E., 1980. Potassium permanganate : an alternative to prechlorination. *Aqua*, **7**, pp. 153-156.
56. Colthurst J.M. and Singer P.C., 1982. Removing trihalomethane precursors by permanganate adsorption. *JAWWA*, **74**, pp. 78-83.
57. Ma J., and Graham N., 1996. Controlling the formation of chloroform by permanganate preoxidation-destruction of precursors. *J. Water SRT-Aqua*, **45**, 6, pp. 308-315.
58. Willey B.F., Jennings H. And Muroski F., 1964. Removal of hydrogen sulphide with potassium permanganate. *JAWWA*, **50**, pp. 475-479.
59. Ficek J.K. Potassium permanganate "The unique water treatment oxidant'. Carus Chemical Company, La Salle, Ill., USA.
60. Posselt H.S., Anderson F.J. and Weber W.J., 1968. Cationic sorption on colloidal hydrous manganese dioxide. *Env. Sci. And Techn.*, **2**, 12, pp. 1087-1093.
61. Weltch W.A., 1962. Potassium permanganate in water treatment. Conf. Paper, *Chesapeake Section Meeting*, Wilmington, Ill., USA.
62. Nederlands Normalisatie Instituut, 1982. NEN: 6460: Water - Determination of iron content by atomic absorption spectrometry (flame technique).
63. Nederlands Normalisatie Instituut, 1982. NEN: 6466: Water - Determination of manganese content by atomic absorption spectrometry (graphite furnace technique).
64. APHA, AWWA and WPCF, 1985. *Standard methods for examination of water and wastewater*, Am. Publ. Health Assoc., Washington D.C., USA.
65. Janssens J.G., 1990. The application of dissolved air flotation in drinking water production, in particular for removal of algae. *DVGW Wasserfachlichen Aussprachetagung*, Essen, Germany.
66. Rees A.J., Rodman D.J. and Zabel T.F., 1979. Water clarification by flotation - 5, Technical report TR114, Water Research Centre, UK.
67. Wobma P, Bellamy B., Pernitsky D., Kjartanson K., Adkins M. And Sears K., 1997. Effects of dissolved air flotation on water quality and filter loading rates. Conference proceedings, *CIWEM (Chartered Institution of Water Institution and Environmental Management) International Conference on Dissolved Air Flotation in Water Treatment - an art or a science?*, London, UK, April 1997, printed by Formara Ltd., Southend on Sea, Essex, UK.
68. Ponton G., 1997. Experince of DAF plants in West of Scotland Water. Conference proceedings, *CIWEM (Chartered Institution of Water Institution and Environmental Management) International Conference on Dissolved Air Flotation in Water Treatment -*

an art or a science?, London, UK, April 1997, printed by Formara Ltd., Southend on Sea, Essex, UK.

69. Tanaka T.S. and Pirbizari M., 1986. Effects of cationic polyelectrolytes on the removal of suspended particulates during direct filtration. *JAWWA*, **72**, pp. 57-65.
70. Okada K. and Akagi Y., 1987. Methods and apparatus to measure the ζ potential of bubbles. *Jour. Chem. Eng. Japan*, **20**, pp.11-15.
71. Bourbigot M-M., Martin N., Faivre M., Le Corre K. and Quennell S., 1991. Efficiency of ozoflotation-filtration process for the treatment of the River Thames at Walton Works. *J. Water SRT-Aqua*, **40**, 2, pp. 88-96.
72. Baron J., Martin Ionesco N. And Bacquet G., 1997. Combining flotation and ozonation - the FLOTTAZONE process. Conference proceedings, *CIWEM (Chartered Institution of Water Institution and Environmental Management) International Conference on Dissolved Air Flotation in Water Treatment - an art or a science?*, London, UK, April 1997, printed by Formara Ltd., Southend on Sea, Essex, UK.
73. Betzer N., Argaman Y. and Kott Y., 1980. Effluent treatment and algae removal by ozone-induced flotation. *Water Research*, **14**, pp. 1003-1009.
74. O'Connell J.K., Phillips N.R. and Lutz C.A., 1997. Pilot testing and implementation of full-scale dissolved air flotation, intermediate ozonation, and high rate filtration for public water supply in the United States of America - case study. Conference proceedings, *CIWEM (Chartered Institution of Water Institution and Environmental Management) International Conference on Dissolved Air Flotation in Water Treatment - an art or a science?*, London, UK, April 1997, printed by Formara Ltd., Southend on Sea, Essex, UK.
75. Markham L., Porter M. and Schofield T., 1997. Algal and zooplankton removal by dissolved air flotation in Severn Trent Ltd. surface water treatment works. Conference proceedings, *CIWEM (Chartered Institution of Water Institution and Environmental Management) International Conference on Dissolved Air Flotation in Water Treatment - an art or a science?*, London, UK, April 1997, printed by Formara Ltd., Southend on Sea, Essex, UK.
76. Edwards M. and Benjamin M.M., 1991.A mechanistic study of ozone-induced particle destabilisation. *JAWWA*, **83**, pp. 96-105.
77. Hoyer O., Lusse B. and Bernhardt H., 1985. Isolation and characterisation of Extracellular Organic Matter (EOM) from algae. *Z. Wasser-Abwasser Forsch.*, **18**, pp. 76-90.
78. Lusse B., Hoyer O. and Soeder C.J., 1985. Mass cultivation of planktonic freshwater algae for the production of Extracellular Organic Matter (EOM)., *Z. Wasser-Abwasser Forsch.*, **18**, pp. 67-75.
79. Fogg G.E., Nalewajko, Czeslawa, Watt W.D., 1965. Extracellular products of phytoplankton photosynthesis. *Proceedings Royal Society (B)*, **162**, pp. 517-534.
80. Bernhardt H. et al., 1985. Reaction mechanisms involved in the influence of algogenic organic matter on flocculation., *Z. Wasser-Abwasser Forsch.*, **18**, pp. 18-30.
81. Fillela M, Buffle J. And Leppard G.G., 1992. Characterisation of submicron colloids in freshwaters : evidence for their bridging by organic structures. *Proceedings from the IAWQ/IWSA Joint Specialised Conference on Control of Organic Material by Coagulation and flocculation processes*, pp. 71-80.
82. Paralkar A. And Edzwald J.K., 1996. Effect of ozone on EOM and coagulation., *JAWWA*, **88**, pp.143-154.

83. Mathonnet S., Casselas C., Bablon G. and Bontoux J., 1985. Impact of preozonation on the granulometric distribution in suspension. *Ozone Science & Engineering*, **7**, pp. 107-120.
84. Edwards M., Benjamin M.M. and Tobiason J.E., 1994. Effect of ozonation on NOM using polymer alone and polymer/metal salt mixtures. *JAWWA*, **86**, pp. 105-116.
85. Edzwald J.K. and Paralkar A., 1992. Algae, coagulation and ozonation. *Chemical water and wastewater treatment II*, Klute R. And Hahn H. (Eds.), Springer-Verlag, pp. 24-35.
86. Alaerts G.J. and Van Haute A., 1982. Stability of colloid types and optimal dosing in water flocculation. In : Pawlowski L., ed. *Physicochemical Methods for Water and Wastewater Treatment. Amsterdam*, the Netherlands : Elsevier, pp.13-29.
87. AWWA Committee Report, 1982. Survey of polyelectrolyte coagulant use in the United States. *JAWWA*, **74**, pp. 600-608.

APPENDIX 4.1 Experimental conditions for different treatment combinations.

CONDITIONS	O_3 ($mgO_3/mgTOC$)	$KMnO_4$ (mg/L)	Contact time(min)	Contact pH	Fe(III) (mg/L)	Poly. (mg/L)	Coag. G (s^{-1})	Coag. t (sec)	Coag. pH	Flocc. G (s^{-1})	Flocc. t (min)	R (%)	P (kPa)
STOICHIOMETRY EXP.	/	/	/	/	0 - 15	/	1,000	30	8.0	70	10	7	500
POLYELECTROLYTE (C.A.)	/	/	/	/	3.0	0 - 1.0	+ 500	+ 30	8.0	30	10	7	500
POLYELECTROLYTE (S.C.)	/	/	/	/	/	0 - 3.0	500	30	8.0	30	10	7	500
O_3 CONDITIONING	0.2; 0.5; 0.9	/	/	7.5	1.5; 3.0; 4.5	/	1,000	30	7.5±0.2	10	10	7	500
O_3 COND. + POLY (C.A.)	0.2; 0.9	/	/	7.5	2.5; 3.0	0-0.5	+ 500	+ 30	7.5±0.2	10	10	7	500
O_3 COND. + POLY (S.C.)	0.2; 0.9	/	/	7.5	/	0 - 1.0	500	30	7.5±0.2	10	10	7	500
$KMnO_4$ CONDITIONING	/	0 - 2.0 (0.7)	30	8.0	5.0	/	1,000	30	8.0	70	15	7	500
$KMnO_4$ COND. + POLY. (C.A.)	/	0.7	30	8.0	5.0	0 - 1.0	+ 500	+ 30	8.0	70	15	7	500

(C.A.) : Coagulant aid
(S.C.) : Sole coagulant
Poly. : Polyelectrolyte
Coag. : Coagulation
Flocc. : Flocculation

Comment :
+ 500 : Coagulation G value in case of polyelectrolyte as coagulant aid, applied after 30 s at G of 1,000 s^{-1} for the $FeCl_3$
+ 30 : Coagulation time in case of polyelectrolyte as coagulant aid, applied after 30 s at G= 1,000 s^{-1} for the $FeCl_3$

APPENDIX 4.2 **Particle and turbidity removal efficiencies for different treatment combinations.**

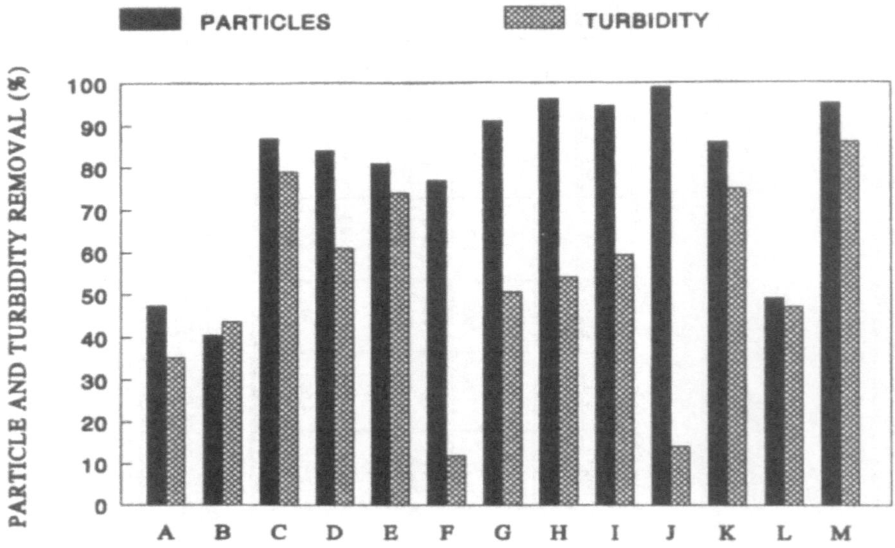

Fig. 4.32 DAF particle and turbidity removal efficiency for different treatment combinations.

CONDITIONS	Fe(III) (mg/L)	Superfloc C-573 (mg/L)	Wisprofloc-P (mg/L)	O_3 (mg O_3/mg TOC)	$KMnO_4$ (mg/L)
A	3.0	/	/	/	/
B	5.0	/	/	/	/
C	3.0	0.3	/	/	/
D	3.0	/	0.1	/	/
E	/	3.0	/	/	/
F	/	/	0.5	/	/
G	3.0	/	/	0.2	/
H	5.0	/	/	0.2	/
I	3.0	0.2	/	0.2	/
J	2.5	/	0.1	0.2	/
K	/	1.0	/	0.2	/
L	5.0	/	/	/	0.7
M	5.0	0.5	/	/	0.7

Chapter 5

ALGAE LADEN WATER TREATMENT BY DISSOLVED AIR FLOTATION (DAF) - PILOT PLANT RESULTS

A.Vlaški, A.N.van Breemen, and G.J.Alaerts (1997). CIWEM (The Chartered Institution of Water and Environment Management) International Conference : Dissolved Air Flotation Technology in Water Treatment - An Art or a Science ?, London, April 1997.

ABSTRACT : Based on previous bench-scale dissolved air flotation (DAF) studies with model water (tap or reservoir water spiked with laboratory cultured *Microcystis aeruginosa*), DAF + (post) filtration pilot plant research was conducted using eutrophied reservoir water with a naturally occurring bloom of the same alga. The DAF removal of the algae was optimised, and the DAF + filtration process was compared with a full-scale sedimentation + filtration process scheme. Process variables included coagulant dose, coagulation pH, flocculation time and energy input, flocculation mode (one-, two- and three-stage flocculation), the application of organic and synthetic polyelectrolytes as coagulant aids, DAF recirculation ratio, etc. The application of ozone and $KMnO_4$ was tested in the context of algae conditioning and improved coagulation/flocculation and down-stream DAF + filtration effectiveness. The assessment of the process efficiency and kinetics was based on particle count and computer image analysis, in addition to commonly applied analytical techniques such as turbidity, residual coagulant, dissolved organic carbon (DOC), and electrophoretic mobility (EM). Similarly, the MFI (Modified Fouling Index) was measured. Results indicate the possibility of producing water of high quality after filtration, though the achievement of an appropriate MFI value (<5 s/L^2) remains problematic. Suggestions are given for the appropriate approach to achieve it, including further optimisation of the agglomeration (coagulation/flocculation) process, application of conditioning oxidants like ozone or $KMnO_4$, and (post)filtration optimisation.

5.1 INTRODUCTION

Within the context of the research on the relevance and removal of cyanobacteria in water treatment (Chapters 1, 3 and 4), pilot plant investigations were conducted on location of the 'Princess Juliana' water treatment plant of the Rhine-Kennemerland Water Transport Company (WRK III) in Andijk, the Netherlands. The main goal of this investigation was to study the removal of naturally occurring cyanobacteria by dissolved air flotation (DAF). The eutrophic status of the IJssel Lake which replenishes the raw water reservoir, causes in late summer or early autumn cyanobacteria blooms of mostly *Microcystis aeruginosa*, a highly versatile and competitive species. Fig. 5.1 shows representative algae related parameters for the reservoir water quality for 1995. Two seasonal peaks of total chlorophyll are observed, one in early spring (ascribed to diatoms, mostly *Melosira* spp.) and one in late summer or early autumn (ascribed to the cyanobacterium *Microcystis* spp.). Much higher algae counts are registered during cyanobacteria blooms, these representing more than 99% of the overall algae count during these periods, and contributing to high chlorophyll values and algal volume. This situation results in numerous treatment problems, as well as a deteriorated product water quality. The problems include increase of required coagulant dose, shorter filter runs due to filter blockage, increased backwash water needs, algae as the source of trihalomethane formation (during short periods of NaOCl application to prevent mussels growth, a practice which has recently been abandoned), increased MFI (Modified Fouling Index) [1] and AOC (Assimilable Organic Carbon) [2] values of the product water [3], and algae passing through treatment in excessive quantities [4].

The integral water treatment process of the WRK III treatment plant relies on an efficient combination of reservoir water quality management, and water treatment. The 20 m deep reservoir with a retention period of 20 days (for an installed production of 12,000 m^3/h) is periodically mixed with air to avoid stratification. $FeSO_4$ (average 5 g Fe(II)/m^3) is dosed in the

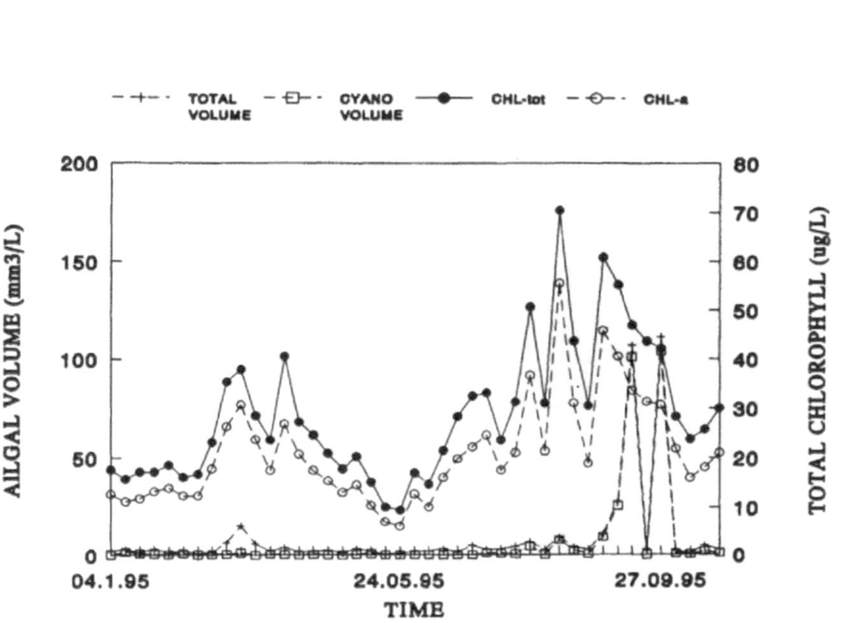

Fig. 5.1 Algae related water quality parameters (total algal volume, volume of cyanobacteria, total chlorophyll (CHL-tot) and chlorophyll-*a* (CHL-a) at the inlet of the WRK III reservoir (Andijk).

reservoir mostly to promote phosphorus precipitation and algae growth reduction, as well as settling of algae within the reservoir itself. This technique is able to more than halve the original IJssel Lake algae concentration, which periodically reaches several hundred μg of chlorophyll/mL. The treatment process comprises microstrainers (one street with 35 μm, the two others with 200 μm mesh) (Fig. 5.2), coagulation/flocculation (coagulant $Fe_2(SO_4)_3$ with a dose 15-20 g Fe(III)/m^3, maximum 43 g Fe(III)/m^3; cationic polyelectrolyte Wisprofloc-P as coagulant aid with dose 0.1-0.2 g/m^3, maximum 0.5 g/m^3; and with pH correction by dosing $Ca(OH)_2$ (Fig. 5.3), up-flow lamella sedimentation (Fig. 5.4), and up-flow filtration (one street with a two-layer filter with barite and granite, and the two others with a single sand bed) (Fig. 5.5). The plant delivers water to the steel industry 'Hoogovens' and the Provincial Waterworks of North-Holland which further uses the water for ground infiltration. The infiltration line additionally comprises granular activated carbon filtration (GAC). The primary purpose of this concept is replenishment of the dunes fresh ground water aquifers.

The large coagulant doses are applied mainly to cope with the high organic load of the raw water (DOC level of the raw water during the experimental period ranged 5.0 - 14.4 g C/m^3). They result in good product water, however, during the algae bloom periods they have to be substantially increased. This results in high production costs, which periodically even double.

Fig. 5.2 The microstrainer units of the WRK III treatment plant at Andijk.

Fig. 5.3 The flocculation units of the WRK III treatment plant at Andijk.

Fig. 5.4 The lamella sedimentation units of the WRK III treatment plant at Andijk.

Fig. 5.5 The up-flow rapid sand filtration units of the WRK III treatment plant at Andijk.

The most critical product water quality parameters related to the ground water recharge are the MFI and the AOC. The first defines the clogging capacity of the water, while the second defines its bacteriological after-growth potential. Recently, the required product water MFI value was set at 5 s/L^2. Together with the ambitious guideline of 10 mg AOC/m , which roughly corresponds to the low value of 0.3 mg chlorophyll-*a*/m^3[5], this poses increasingly difficult demands on the existing production, especially during cyanobacteria blooms (Fig. 5.6). The granular activated carbon (GAC) filtration is applied before infiltration in order to reduce pesticides with an additional beneficial side effect of AOC reduction.

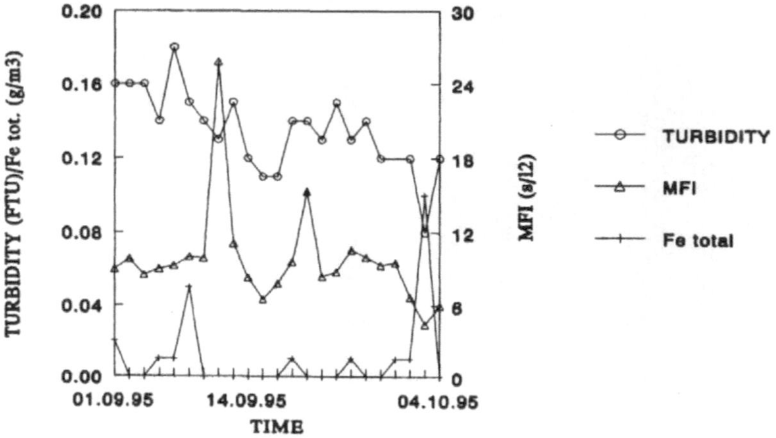

Fig. 5.6 Turbidity, residual Fe (total) and MFI (modified fouling index) for the WRK III filtrate.

This study investigates the applicability of the DAF process for algae removal on pilot plant scale on this reservoir water. It addresses DAF pilot plant optimisation (alone, and in combination with subsequent up-flow filtration), and comparison with the existing sedimentation + filtration full-scale treatment. For this purpose, the same coagulant used in the full-scale plant is considered ($Fe_2(SO_4)_3 \cdot 9H_2O$). The same applies for the coagulant aid Wisprofloc-P, while based on positive results from earlier bench-scale experiments with model water (Chapters 3 and 4) Superfloc C-573 is also considered.

Commonly, conditioning of the water by microstraining (usually of mesh size ≥35 μm) is used as an initial barrier for coarse suspended matter in water treatment plants, thus lowering the load on subsequent treatment. This is also the case at the WRK III treatment plant. Microstrainers are particularly efficient in removing zoo-plankton and larger filamentous and chain-forming algal species; the retention of smaller algae varies as a function of their size and shape [6].

Although the potential benefits of oxidants such as chlorine, chlorine dioxide, ozone, hydrogen peroxide and potassium permanganate on the agglomeration phase have been recognised (Chapter 4) [7, 8, 9, 10, 11], their application has been restricted due to the formation of hazardous by-products, notably trihalomethanes, haloacetic acids and haloacetonitrils by chlorine application, chlorite by chlorine dioxide application, and bromate by ozone application to waters rich in bromide [12]. This study aims at investigating ozone and $KMnO_4$ particle (algae) conditioning in the context of more efficient DAF on a pilot-plant scale, considering the potential disadvantages related to their application.

5.2 METHODS AND MATERIALS

The research was conducted on the premises of the 'Princess Juliana' (WRK III) treatment plant, situated on the shores of IJssel Lake in Andijk, the Netherlands. Raw water is subjected to settling within the raw water reservoir after dosing of $FeSO_4$. For experiments, water was abstracted from a location behind the microsieves. This is the same water which is further pumped to the full-scale sedimentation + filtration treatment. The research was conducted over a two year period. In the first year (September-November, 1995) it coincided with a cyanobacteria bloom of *M. aeruginosa* (Fig. 5.1), a common species for this period of the year. The second year (August-October, 1996) was characterised by a slightly lesser algae activity, but algae were still present in bloom [13] concentrations and chlorophyll-*a* levels reached 60 μg/L (average 29.3 μg/L).

5.2.1 Pilot plant experimental set-up

The experimental work was conducted with a commercial DAF pilot plant (PURAC, Sweden) (Fig. 5.7).

Fig. 5.7 View of the micro-strainer unit and the DAF pilot plant container with the filter column in the front, at the WRK III site.

The maximum capacity of the plant was 13 m^3/h, however, it was operated at Q=4-6 m^3/h in order to avoid air bubbles being carried over into the effluent. The DAF unit itself comprised a 1 m^3 stainless steel tank (Fig. 5.8) with four needle valves located at the bottom of the inlet for the recycled flow re-introduction.

5.8 The DAF pilot plant tank with the raw water inlet, chemicals dosing points and flash mixing (Kenics type, Purac, Sweden) on the left and the sludge pump at the bottom.

A variable level effluent outlet and automatic periodical sludge scraping were incorporated in the pilot plant (Fig. 5.9). Two options for coagulation were provided, a Kenics type in-line mixer and a flash mixing unit, while one-, two- or three-stage flocculation could be applied, each provided with variable speed mixers. A fully automated system provided the pumping, saturation and re-introduction of the recycled flow (Fig. 5.10). The product water was partly discharged, while the remaining part (0.4 m^3/h) was carried over to a transparent PVC up-flow filter column (Fig. 5.11).

The pilot filter was run in up-flow mode and had a multiple layer filling (simulating the full scale plant filters) with 30 cm gravel (d=15-25 mm), 30 cm barite (d=5-10 mm), 50 cm larger size granite (d=1.2-2.0 mm) and 75 cm fine granite (d=0.5-0.95 mm) layers. The filtration rate was kept constant at 10 m/h, simulating full-scale plant conditions. The filter was backwashed after each run according to the full-scale filters backwash procedure, using air (from 0-70 m/h in the first 15 min), air (70 m/h) and water (the backwash rate rising from 0-20 m/h in the next 15 min) and water alone (in the next 30 min, the backwash rate rising from 20-108 m/h in the first 10 min of this period, the latter value being applied in the following 20 min period). After a series of experiments it was concluded that the DAF reached steady state very fast, approximately 45 minutes after start-up. In cases when DAF alone was evaluated, sampling for each experimental run was done after one hour from start-up. In cases when the DAF + filtration line was investigated, samples were taken after two hours from start-up. The effect of a longer filtration time was also investigated, and in this case samples were taken on an hourly basis for a maximum period of eight hours. During the pilot plant investigations period, the full-scale plant was not operating at its design capacity of 12,000 m^3/h, but dropped from an initial 7,120 to 5,560 m^3/h, with an accompanying filtration rate of approximately 10 m/h (September-November, 1995). Although no adequate base-line values existed, it may be assumed that the low production rate had a beneficial effect on the effluent quality during that period.

Fig. 5.9 The automatic sludge scraping unit of the DAF tank.

$Fe_2(SO_4)_3 \cdot 9H_2O$ was used as coagulant in a dosage range of 0-20 g Fe(III)/m^3. pH in coagulation was set at pH 8 using $Ca(OH)_2$, as in the full-scale process. The organic cationic polyelectrolyte Wisprofloc-P and the synthetic cationic polyelectrolyte Superfloc C-573 were tested in the concentration range of 0.1-0.7 g/m^3. The Kenics static mixer was used for the rapid mix with a calculated G value of ≈3,500 s^{-1} for a flow of 4.6 m^3/h. Different flocculation time, energy input (G value) and flocculation sequence (one-, two- and three-stage flocculation) were tested; the values were partly dictated by the technical provisions of the pilot plant. In the discussion section the applied flocculation time is denoted as t_{fi}, while the flocculation energy input is denoted as G_i,, the subscript 'i' signifying the flocculation stage.

Fig. 5.10 The automated DAF saturation unit of the pilot plant.

The recirculation ratio (R) in the DAF unit was set at 5, 7 and 10%, while the saturator pressure (P) was 500 kPa. The water temperature decreased from 16.7°C to 10.1°C in 1995, and from 16.4°C to 12.9°C in 1996, over the experimental period.

For ozone production, the Trailigaz LABO LO ozone generator was used (Fig. 5.12). A 4.5 m vertical, transparent PVC column of 135.5 mm diameter and allowing a contact time of 3.5 min (at Q_{max}=1.14 m^3/h) was used as a continuous-flow ozone reactor. Potassium permanganate (0.0021M working solution) was dosed immediately before pH correction and coagulant addition, allowing for very short contact times (< 1 s) before coagulation and flocculation. This approach was preferred to avoid the extensive deposition of MnO_2 precipitates on the tank and walls.

Fig. 5.11 The multiple layer up-flow rapid filtration unit of the DAF + filtration pilot plant.

Fig. 5.12 The ozone production and metering units, part of the pilot plant set-up.

5.2.2 Analytical techniques

The process efficiency evaluation was based on turbidity (Sigrist L-65, Switzerland) and particle count of particles of size >2.75 μm (HIAC-Royco PC-320, USA). The notation T and PC used in the graph captions represents the initial turbidity and particle count, respectively. For particle count of particles in the size range of 0.3-5 μm the Particle Measuring Systems Inc./Liquid Batch Sampler LBS-100, and the Microlaser Particle Spectrometer (Boulder, Co., USA) were used. An inverted microscope (Zeiss Axiovert 135, Germany) was used for sample inspection. Residual coagulant (Fe_{total}) and manganese were measured by atomic absorption spectrometry at 248.3 nm for iron (NEN 6460) [14] and 279.5 for manganese (NEN 6466) [15). Electrophoretic mobility was measured with a Tom Lindström AB-Repar apparatus. The transferred ozone concentration was determined by measuring the O_3 concentration in the inlet and outlet air/ozone gas mixture, by the iodometric method [16]. Bromate was measured by ion-chromatography [17]. The MFI measurements were performed using the KIWA N.V. Research and Consultancy standard MFI measurement apparatus (Fig. 5.13) [1]. Finally, computer image analyses were performed with the Mini-Magiscan, IAS 25/IV25 Joyce-Loebl Ltd., UK (Fig. 3.12, Chapter 3).

Fig. 5.13 The KIWA N.V. Research and Consultancy standard MFI measurement apparatus, with the filter column in the background.

The image analysis results are presented in the form of particle volume distributions. The volume distributions represent the volume of recorded particles in size ranges of 10 μm width for a specific sample. For this purpose, the recorded particles or floc material were approximated to spheres. The analyses were performed on fresh as well as on photographed samples [18, 19].

Here, preference was given to the latter, because this enabled multiple registration of relevant and representative situations along the treatment line without the risk of disturbing the process or the quality of the sample. For this purpose, a specially devised flat photo cell was connected at different locations of the pilot plant unit and samples were photographed before and after flocculation, and after treatment. In the case of DAF, the released air bubbles were also photographed and analysed. Professional high resolution black-and-white film AGFA 25 or Kodak TMAX 100 was used for the purpose, while forced development was applied for film processing. 30-100 fields, or more than 500 particles of each frame were processed and analysed, using a ccd-camera (604*288 pixels) mounted to a Nicon Optiphot microscope (at 40x magnification). The size limit of the set-up was 1.9 μm. The software package Genias25 developed by Joyce-Loebl Ltd., UK (Version 2.0, 1990) was used for data processing.

The presented experimental results emerge from duplicate or multiple experiments. Statistical analysis of data was performed with the Statistica software package. One and two way ANOVA followed by multiple comparison and regression analysis among treatment results were performed where appropriate.

5.3. RESULTS AND DISCUSSION

5.3.1. Coagulation conditions

a. Coagulant dose

Bench-scale experiments with model water (tap water spiked with laboratory cultured *M. aeruginosa*) showed that the DAF optimal dose was two to three times lower than the one for sedimentation, whilst the effluent quality in terms of the removal efficiency was comparable (Chapter 3) [20]. Pilot plant DAF without the use of coagulant resulted in poor turbidity (6.3%) and particle (19.5%) removal (Fig. 5.14), emphasizing the need to first destabilise - coagulate suspended material [19, 20, 21, 22, 23]. Increase of coagulant dose increased process efficiency. The highest coagulant dose tested was 20 g Fe(III)/m^3 and efficiency continued to increase (up to 99.3%) without a clear 'optimum' being reached. A dose of 7 g Fe(III)/m^3 consistently produced turbidity removal in the range of 84.9-93.2% (mean value of 89.3%) and particle removal efficiency of 87.7-97.5% (mean value of 94.5%). Multiple regression analysis (results from seven experiments) for particle removal efficiency (r^2=0.48, as a dependent variable) and initial particle count and coagulant dose (as independent variables) yielded results suggesting that the particle removal efficiency was significantly affected by both the initial particle count (P=0.0015) and the coagulant dose (P=0.00072). The same analysis for turbidity removal efficiency (r^2=0.53) showed that it was significantly affected by the coagulant dose (P=0.000001) and insignificantly affected by the initial turbidity (P=0.37).

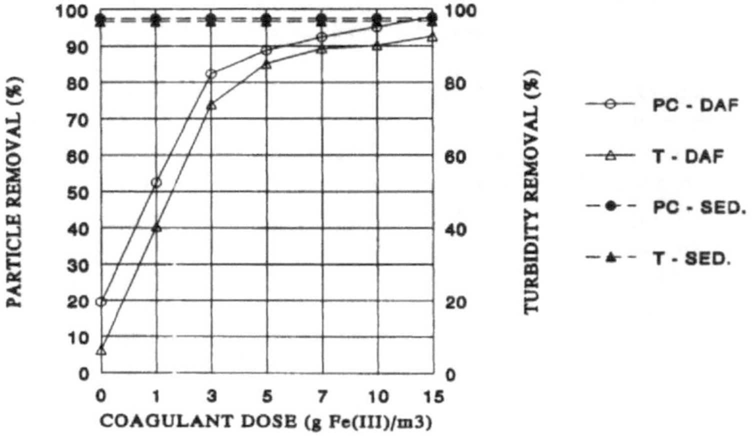

Fig. 5.14 DAF removal efficiency as a function of coagulant dose (duplicate experiment). Conditions: pH 8, two-stage flocculation with t_{f1}=18.2 min, G_1 =50 s^{-1} and t_{f2}=3.8 min, G_2=30 s^{-1}, R=7%, P=500 kPa, T=7.8 FTU, PC= 12,580/mL.

The average DAF efficiency for the coagulant dose of 7 g Fe(III)/m^3 tended to be slightly lower compared to the mean full-scale sedimentation efficiency registered in the same period : turbidity removal of 96.0% and particle removal efficiency of 97.3%. The sedimentation unit, however, was operated with a combined coagulant dose of 20-24 g Fe(III)/m^3 and 0.2-0.5 g W-P/m^3 cationic coagulant aid. The comparatively lower DAF efficiency can partly be attributed to the higher residual (colloidal) iron. The residual (total) iron after DAF decreased with increase of the coagulant dose, reaching 1.05 g Fe/m^3 for the 20 g Fe(III)/m^3 dose, as compared to the average 0.39 g Fe/m^3 after sedimentation. The DOC removal at the 7 g Fe (III)/m^3 dose was also found lower than observed in the full scale plant at 20-25 g Fe(III)/m^3 (7-20% and 25-35%, respectively). It is suggested that the precipitating iron species formed during the sweep coagulation process partially form complexes with the abundantly present organic matter. The high coagulant doses applied before the sedimentation are responsible for the more comprehensive sweep coagulation and consecutive higher degree of turbidity and iron removal compared to the three times lower DAF coagulant dose. The colloidal character and size of the organo-iron complexes coincides with the size range (<1 μm) for which Brownian diffusion plays a major role in the particle-bubble attachment, which renders them more difficult to remove by DAF than particles in the size range larger than 1 μm [22]. At a coagulant dose which lies between 10 and 15 g Fe(III)/m^3 the DAF particle removal efficiency (of particles $\geq$2.75 μm) surpassed the one obtained by the full-scale sedimentation (for identical raw water quality conditions), indicating more efficient particle coagulation. However, this improvement of performance did not apply for the turbidity and residual iron.

The turbidity removal efficiency of the final filter for a coagulant range of up to 20 g Fe(III)/m^3 was 95.5-98.4%, while the particle removal efficiency was 96.8-99.7%, increasing with increasing coagulant dose. It is suggested that the residual Fe coagulant after DAF was mainly in the organo-iron complex form of a colloidal size, which is equally difficult to remove by filtration; thus, the increase of the filtration efficiency could be ascribed mainly to the better water quality obtained after DAF at the higher coagulant doses. The filter effluent turbidity corresponding to these removal efficiencies was 0.1-0.35 FTU; according to the guidelines of the Association of Dutch Water Supply Companies (VEWIN Aanbevelingen, 1993) this is acceptable from the point of view of product water quality (<0.8 FTU if no disinfection is applied and <0.2 FTU in case of disinfection), and is comparable with the full-scale plant filtrate turbidity.

b. Coagulation pH

Due to the high buffering capacity of the raw water and limitations on the available dosing equipment, the lowest tested pH was 6.35. Fig. 5.15 represents the turbidity and particle removal efficiency under different pH conditions. The highest turbidity and particle removal efficiency was obtained at pH 6.35 (93.2% and 97.5%, respectively). The same applied for DOC removal (33%). This is in accordance with the experiments on model water (Chapter 3), as well as with theory, which suggests that adsorption coagulation preferentially occurs in this pH range at a medium to low coagulant dose and supports the otherwise dominant sweep coagulation. Electrophoretic mobility (EM) measurements, however, did not confirm complete charge neutralisation, contrary to results with model water (Chapters 3 and 4) [19, 23]. The lowest DAF turbidity, residual iron and DOC removal were obtained at pH 8. As for particle removal, the lowest efficiency was obtained at pH 9, possibly due to the precipitation of calcite particles under alkaline conditions. These particles have a size similar to that of *Microcystis* (≈ 10 μm, [12]). Final filtrate quality was again influenced most strongly by the DAF effluent quality.

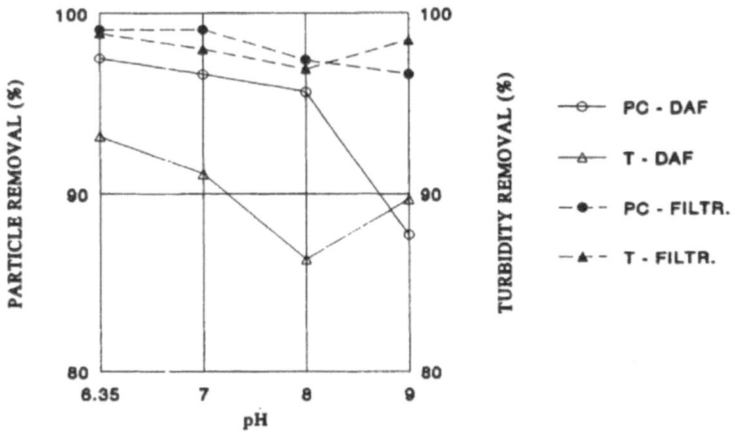

Fig. 5.15 Variation of coagulation pH in DAF + filtration. Conditions : 7 g Fe(III)/m^3, two-stage flocculation with t_{f1}=18.2 min, G_1=50 s^{-1} in the first, and t_{f2}=3.8 min, G_2=30 s^{-1} in the second stage, R=7%, P=500 kPa, T=5.8-6.9 FTU, PC=12,100-18,100/mL.

5.3.2 Flocculation conditions

a. Flocculation time

Research and practice have yielded contradictory conclusions with respect to optimal flocculation time and energy input in case of DAF. Janssens [21] and Malley and Edzwald [24] stated that flocculation times as short as 5 minutes can suffice for efficient down-stream DAF. Our bench-scale experiments on model water on the other hand, have shown that flocculation times longer than 10 minutes did increase DAF process efficiency (Chapter 3), but only by ≈5% [20]. On the other hand, full-scale DAF plants are rarely operated with flocculation times shorter than 20-30 minutes [4, 25].

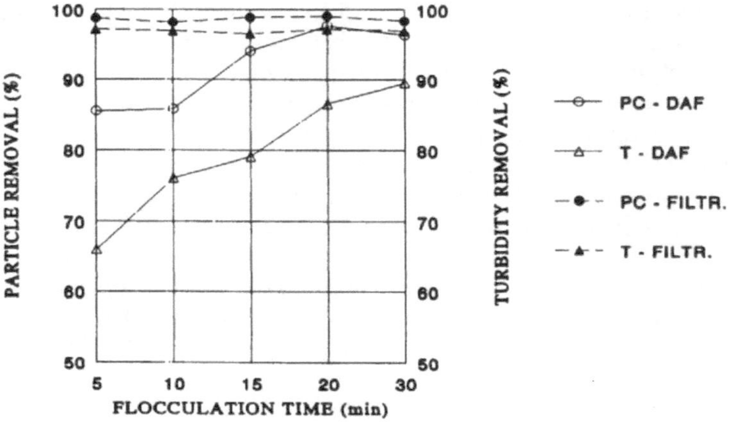

Fig. 5.16 Variation of flocculation time for DAF + filtration. Conditions : 7 g Fe(III)/m^3, G=50 s^{-1}, R=7%, P=500 kPa, T=5.8-6.9 FTU, PC=12,100-18,100/mL.

Results of pilot plant optimisation of flocculation time are presented in Fig. 5.16. No clear optimum was established, the DAF efficiency rising with longer flocculation times. Flocculation times shorter than 15 minutes produced a DAF effluent of inferior quality regarding turbidity, particle count and residual coagulant. The filtrate quality after run time of two hours was similar for different flocculation times, although longer flocculation times will most probably enable longer filter runs (before filter breakthrough or blockage occurs). For further experiments a flocculation time in the range of approximately 20 minutes was chosen, which was partly determined by limitations of the pilot plant.

b. Flocculation energy input

Literature commonly states that relatively high G values (e.g. G=50-100 s^{-1}) produce small, strong and dense flocs, able to resist the shear applied when the DAF recycle stream is re-introduced in the flotation chamber via the nozzles or needle valves, thus favouring DAF efficiency [22]. Our experiments on model water showed that this may need further confirmation (Chapter 3), as the overall DAF efficiency is influenced by a combination of factors, of which the relation between floc size distribution, and floc density and strength are of primary importance.

Two sets of pilot plant experiments were conducted, one with a short flocculation time of 5 min and a different flocculation G value (G=10, 30 and 50 s^{-1}), and a second with a flocculation time of 20 min and different flocculation G value (G=10, 30, 50 and 70 s^{-1}). Turbidity and particle removal efficiency for these experiments are presented in Figs. 5.17 and 5.18.

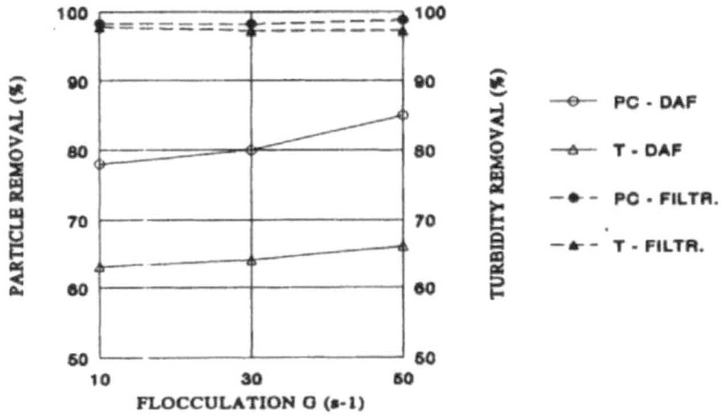

Fig. 5.17 Variation of flocculation G value for flocculation time of 5 min. Conditions : 7 g Fe(III)/m^3, pH 8.4, R=7%, P=500 kPa, T=6-7 FTU, PC=14,800-18,100/mL.

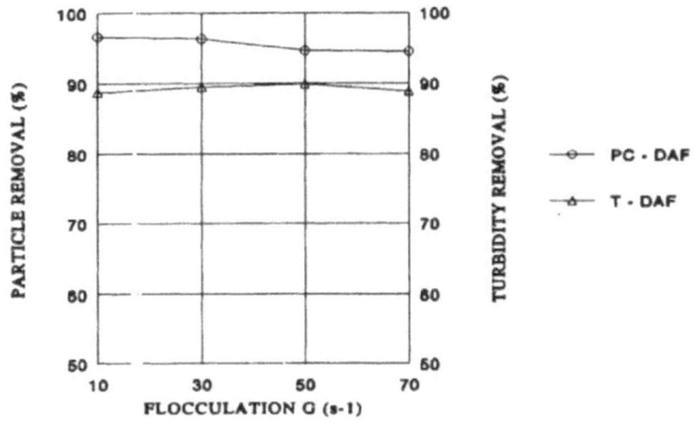

Fig. 5.18 Variation of flocculation G value for flocculation time of 20 min. Conditions : 7 g Fe(III)/m^3, pH 8.4, R=7%, P=500 kPa, T=7.7 FTU, PC=13,460/mL.

The process efficiency for the short flocculation time of 5 min rose with increasing G value (from

78.1-85.5% for particle removal and from 63.6-66.0% for turbidity removal). It is suggested that the contact opportunities during the short flocculation time were insufficient to flocculate the colloidal and particulate matter by the precipitating coagulant. This was supported by the high residual iron of more than 2.5 g Fe(III)/m^3. Increase of the G value resulted in higher DAF efficiency, including a lower residual iron value. The DOC removal was low, typically below 10% (6-7%).

In the case of the longer flocculation time of 20 minutes the turbidity removal efficiency rose slightly with increase of the G value, reaching its 'optimum' (89.9%) at G=50 s^{-1}, while the particle removal efficiency slightly decreased with increase of the G value (from 96.6 to 94.5%). This was accompanied by increased residual Fe, and decreased DOC removal (from 20 to 12%).

Image analysis data of raw and flocculated water with flocculation G=10, 30 and 50 s^{-1} (Fig. 5.19) show that the inverse relation between floc size and flocculation energy input is less pronounced than in the case of flocculation of the model water (Figs 3.6 and 3.7). This difference is suggested to be related to the different model water and reservoir water quality, especially the TOC and DOC concentration and composition, and the organic matter - metal coagulant interactions. It is to be considered that the particles after coagulation of the reservoir water are of colloidal and suspended size, including organo-iron complexes <1 μm, and algae >1 μm, however, of different quality (charge and steric phenomena) and size (fairly monodispersed in case of the model water and heteropdisperse in case of the reservoir water) characteristics, and of significantly higher concentrations than in the model water circumstances.

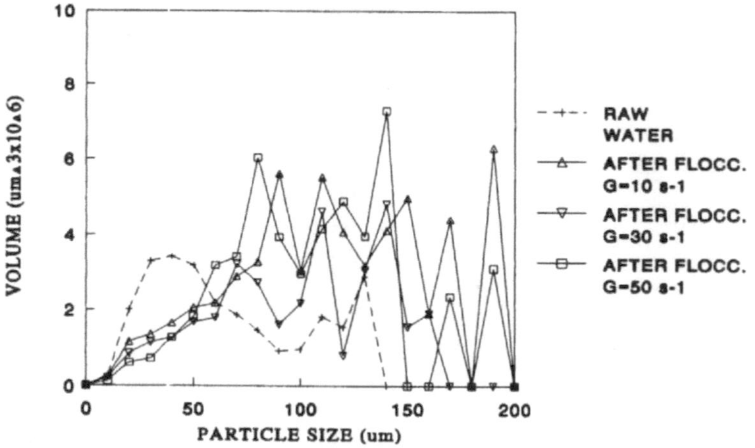

Fig. 5.19 Particle volume distribution as a function of flocculation G value. Conditions: 10 g Fe(III)/m^3, flocculation time t=22 min. pH 7.8-8, R=7%, P=500 kPa, T=6.8-7.9 FTU, PC=18,040-20,680/mL.

According to the heterodisperse curvilinear approach to (turbulent) flocculation modelling [26], the trajectory of one particle approaching another is curvilinear and the velocity gradient G is not of primary significance compared to the rectilinear model of Smoluchowski. Accordingly, Brownian motion and differential settling control the collision opportunities of most combinations of particle sizes. Low water temperature and hence high viscosity lower the collision rate. The energy input G is of relevance only if both particles are larger than 1 μm and have a size ratio of approximately 10. Here (Fig. 5.19), the collision rate between particles in the size range of up to 50 μm seemed to be affected indeed by the energy input, the increase of G resulting in more successful agglomeration.

The reduction of the number frequency of particles of size <50 μm with increase of the G value (not presented here), was more pronounced than the reduction in volume, due to the significantly smaller contribution of the small particles (<50 μm) to the total particle volume as compared to the larger particles and flocs. On the other hand, the residual iron rose and DOC removal decreased with increase of the G, suggesting that the successful collision rate of the smaller colloidal particles, to which the organo-iron complexes belong, was negatively affected by the higher G. The limitation of the image analysis system to particles larger than 1.9 μm impedes further discussion. The flocculation between particles of very different sizes (as in our case) is predicted to be much slower by the curvilinear flocculation model, which could explain the increase of process efficiency with increase of flocculation time (Figs. 5.16, and 5.17 and 5.18). The documented presence of NOM in the form of fine matrices or meshes of filaments in waters of similar quality [27, 28] may have also played a role in the flocculation process of the reservoir water and additionally influenced the floc volume distribution. The role of these structures, however, is difficult to interpret in the context of the different G values.

The particle (floc) distributions obtained for different flocculation G values did not significantly differ from each other (for particles of size >1.9 μm), and the same holds for the DAF efficiency. It is suggested that the slight increase of DAF particle removal efficiency at lower G was partly due to the lighter floc structure obtained at the low flocculation G values.

c. Flocculation mode

The optimal flocculation mode similarly has raised contradictory experiences between practice and theory [4, 20]. Tapered flocculation is often applied in combination with DAF although there is little theoretical support for this choice. Tapered flocculation generally results in a floc size distribution and characteristics favourable for sedimentation. For the research here, three flocculation modes were tested: single-stage flocculation ($t_{f1}=3.8$ min, $G_1=30$ s^{-1}), two-stage flocculation ($t_{f1}=18.2$ min, $G_1=50$ s^{-1}, $t_{f2}=3.8$ min, $G_2=30$ s^{-1}) and three-stage flocculation ($t_{f1}=10.8$ min, $G_1=70$ s^{-1}, $t_{f2}=18.2$ min, $G_2=50$ s^{-1}, $t_{f3}=3.8$ min, $G_3=30$ s^{-1}), determined by the pilot plant flow.

The first mode resulted in noticeably lower process efficiency (84.1% particle removal), especially in terms of residual turbidity (63.4% removal) (Fig. 5.20), accompanied by a high residual iron level of 3.35 g Fe (III)/m^3. Again, it is suggested that the short flocculation time provided for insufficient contact opportunities between particles and for inefficient flocculation. The benefits of the additional third stage of flocculation were minor in terms of turbidity and particle count reduction, though it led to a residual iron level that was significantly lower (0.76 compared to 1.13 g Fe (III)/m^3 for the two-stage mode). The DOC removal efficiency increased with the number of flocculation stages (and thus longer flocculation time) from 11.9-29.3%, the

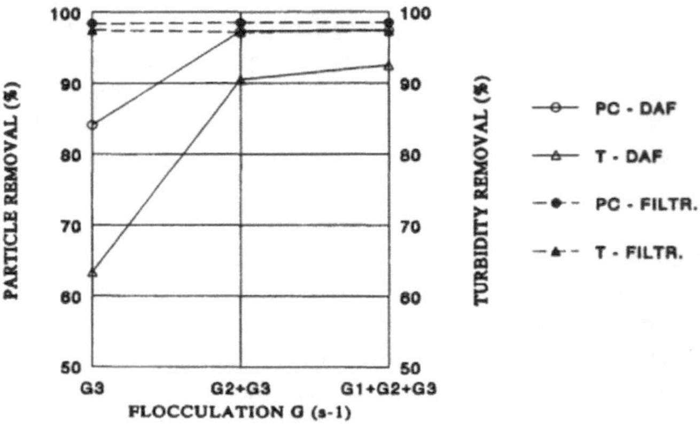

Fig. 5.20 Particle and turbidity removal efficiency as a function of tapered flocculation mode. Conditions : 7 g Fe(III)/m^3, pH 7.8-8.0, R=7%, P=500 kPa, T=6.8-7.9 FTU, PC=18,040-20,680/mL.

higher value being in the range of the full-scale plant efficiency. A comparison of the efficiency results presented in Fig. 5.18 and 5.20 suggests that it is rather the total flocculation time and not the tapered flocculation, that is responsible for the overall DAF efficiency. For further experiments here, the two-stage flocculation mode was chosen with flocculation time and energy input in the first stage t_{f1}=18.2 min, G_1=50 s^{-1}, and t_{f2}=3.8 min, G_2=30 s^{-1} in the second stage.

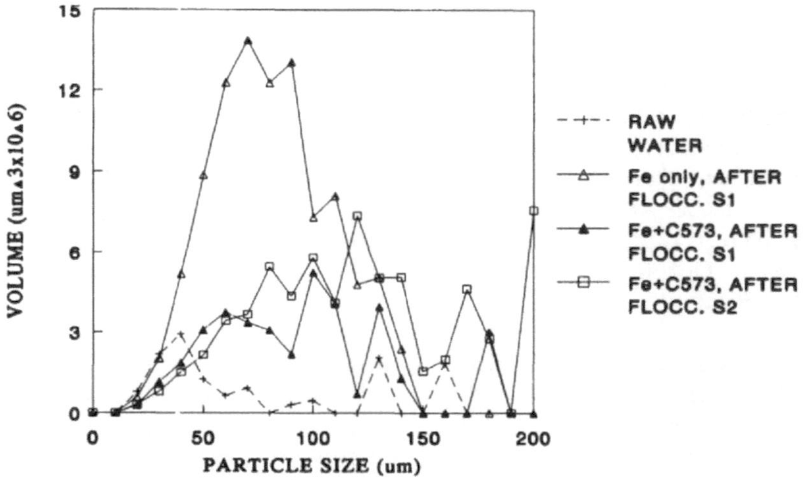

Fig. 5.21 Particle volume distribution as a function of coagulant choice and flocculation stage. Conditions : 15 g Fe(III) vs. 15 g Fe(III) + 0.5 g C-573/m^3, pH 8, tapered flocculation with t_{f1}=18.2 min and G_1=50 s^{-1} in the first stage (S1) and t_{f2}=3.8 min and G_2=30 s^{-1} in the second stage (S2).

The particle volume distribution after each of the first two stages featured a noticeable increase of particle volume in the larger particle size range on account of a decreasing volume of the smaller particles, and the decreasing G value in the second flocculation stage (Fig. 5.21). However, there was little impact on the final DAF effluent quality after tapered flocculation as compared to that after single-stage flocculation with similar G value and flocculation time.

5.3.3 Application of polyelectrolytes as coagulant aids

The application of cationic polyelectrolytes as coagulant aid resulted in increased DAF (and sedimentation) efficiency in earlier bench scale experiments (Chapters 3 and 4) [19, 20]. Other positive experiences [29], as well as the actual operation practice at WRK III which applies the cationic polyelectrolyte Wisprofloc-P in its treatment (0.2-0.5 g/m^3), suggested an opportunity for further research. Based on the bench-scale experiments with model water (Chapters 3 and 4) two cationic polyelectrolytes were chosen for evaluation : the organic Wisprofloc-P and the synthetic Superfloc C-573. The dosage range for both was 0.1-0.7 g/m^3, in combination with 7 g Fe(III)/m^3 (Figs. 5.22 and 5.23).

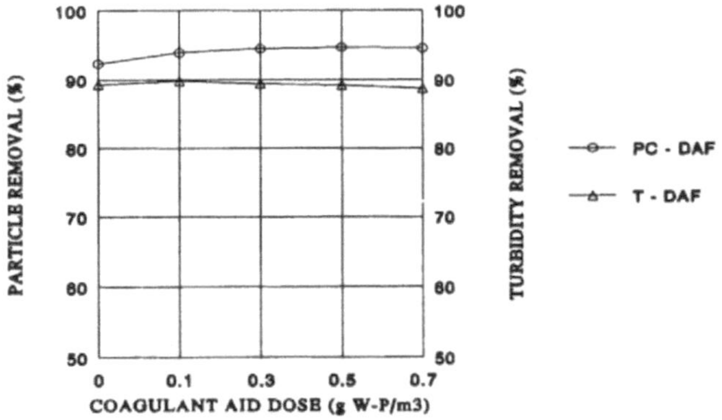

Fig. 5.22 Variation of coagulant aid (Wisprofloc-P) dose. Conditions : 7 g Fe(III)/m^3, two stage flocculation with t_{f1}=18.2 min and G_1 =50 s^{-1} in the first, and t_{f2}=3.8 min and G_2 =30 s^{-1} in the second stage, pH 8, R=7%, P=500 kPa, T=4.6-7.5 FTU, PC=13,730-19,370/mL.

The turbidity removal efficiency slightly decreased with increasing Wisprofloc-P dose (from 89.8 to 88.7%), while the opposite was noted for the particle removal efficiency (from 94.0 to 94.7%) and residual coagulant, suggesting that the polyelectrolyte residual is the probable cause for the turbidity increase.

In the case of Superfloc C-573, slightly better results were obtained (94.6-96.4% particle removal and 90.1-91.4% turbidity removal, see Fig. 5.23); the same was true when the results were compared to the case without coagulant aid (92.4% particle removal and 89.3% turbidity removal). The efficiency did not change noticeably with the coagulant aid dose. This contradicts the bench scale model water experiments (Chapter 3) in which quality improvement after coagulant aid application was very significant (increase of approximately 40% for particle and 30% for turbidity removal efficiency). EM measurements showed that although the particles electric charge decreased (from -0.84 to -0.78 μm/s/V/cm for 7 g Fe(III)/m^3 in combination with 0.5 g Superfloc C-573/m^3, at pH 8), no complete charge neutralisation occurred as in the model water experiments. It is suggested that the abundant DOC present in the WRK III reservoir water exerted an additional demand for polyelectrolyte [30], reducing its availability for coagulation. This, and the relatively low coagulant dose of 7 g Fe(III)/m^3 are probable causes of the comparatively low removal efficiency improvement.

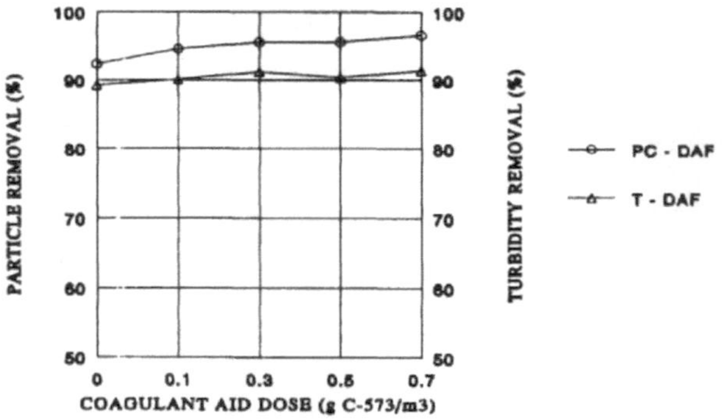

Fig. 5.23 Variation of coagulant aid (Superfloc C-573) dose. Conditions : 7 g Fe(III)/m^3, two stage flocculation with t_{f1}=18.2 min and G_1 =50 s^{-1} in the first, and t_{f2}=3.8 min and G_2 =30 s^{-1} in the second stage, pH 8, R=7%, P=500 kPa, T=6.5-6.7 FTU, PC=14,960-16,080/mL.

One way ANOVA for different treatment options (Fe and Superfloc C-573, with conditions as mentioned in Fig. 5.23) showed that DAF efficiency in the case of Superfloc C-573 showed no significant differences between the treatment options (P=0.98 for particle removal efficiency and P=0.97 for turbidity removal efficiency). Multiple regression analysis for particle removal efficiency (r^2=0.1) showed that neither the initial particle count (P=0.89) nor the polyelectrolyte (P=0.11) was a significant factor in determining DAF efficiency. The same applies for the turbidity.

Fig. 5.21 represents particle volume distribution of flocculated water with and without the addition of coagulant aid. Improved agglomeration was achieved in the former case (0.5 g of Superfloc C-573/m^3 and 15 g Fe(III)/m^3) resulting in a more pronounced larger particle size fraction and a reduction of the smaller particles fraction. The increased concentration of particles in the 30-150 μm size range in the case of only 15 g Fe(III)/m^3, changed when coagulant aid was added, resulting in denser floc structures with a slight increase of particles (flocs) in the >150 μm range. In the DAF pilot plant study, the beneficial effect of the cationic polyelectrolyte application tended to be most evident for doses ≥0.5 g Superfloc C-573/m^3 in combination with coagulant dose ≥15 g Fe(III)/m^3.

However, two way ANOVA for the cases of a low coagulant dose of 7 g Fe(III)/m^3, and combined coagulant dose of 7 g Fe(III)/m^3 and 0.5 g Superfloc C-573/m^3 coagulant aid, versus a high coagulant dose of 15 g Fe(III)/m³, and combined coagulant dose of 15 g Fe(III)/m³ and 0.5 g Superfloc C-573/m³ coagulant aid, showed that only the coagulant dose had a statistically significant effect (P=0.034) on the DAF efficiency (Fig. 5.24).

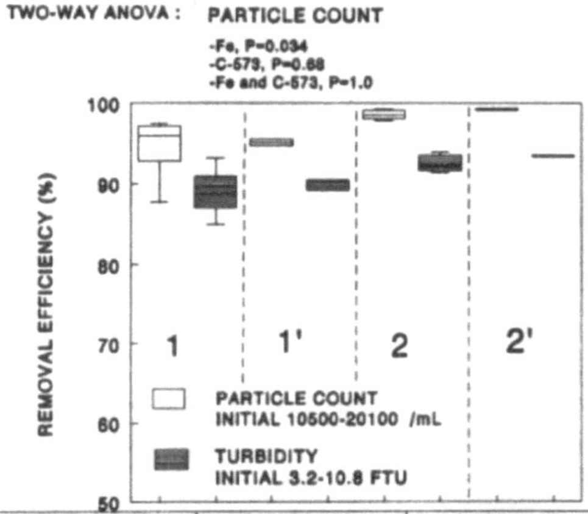

CONDITIONS	Fe(III) (g/m3)	C-573 (g/m3)
1 (DAF)	7	-
1' (DAF)	7	0.5
2 (DAF)	15	-
2' (DAF)	15	0.5

Fig. 5.24 Particle removal efficiency for different dosage combinations of coagulant and coagulant aid (including P-levels from the two-way ANOVA, number of experiments n=14).

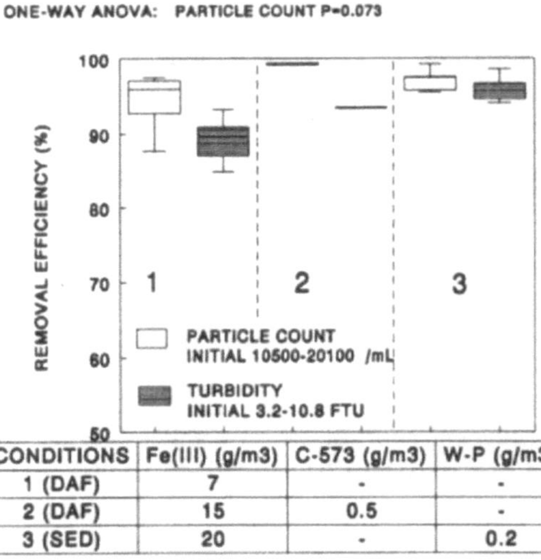

CONDITIONS	Fe(III) (g/m3)	C-573 (g/m3)	W-P (g/m3)
1 (DAF)	7	-	-
2 (DAF)	15	0.5	-
3 (SED)	20	-	0.2

Fig. 5.25 Particle removal efficiency for different dosage combinations of coagulant and coagulant aid, for the pilot plant DAF (DAF) and full-scale sedimentation (SED) (including the P-level from the one-way ANOVA, n=15).

Finally, one way ANOVA for the particle removal efficiency achieved at the cases of the low coagulant dose of 7 g Fe(III)/m^3, combined coagulant dose of 15 g Fe(III)/m^3 and 0.5 g Superfloc C-573/m^3 coagulant aid, and the full-scale sedimentation (20 g Fe(III)/m^3 and 0.2 g W-P/m^3) showed that there was no statistically significant difference (P=0.073) between the three treatment options (Fig. 5.25). The results of both ANOVAs may have been influenced by the limited availability of replicates for the combined 15 g Fe(III)/m^3 and 0.5 g Superfloc C-573/m^3 option.

5.3.4 Particles (algae) conditioning

a. Ozone conditioning

The significance of the results from the pilot plant study on ozone conditioning was constrained by the technical limitations of the ozone pilot column. The maximum flow in the ozone column was 1.1 m^3/h which led to long DAF retention times of approximately 1 hour and a deteriorated effluent quality due to particles settling within the DAF unit. On the other hand, the experiment allowed to quantify the bromate formation under the circumstances of the WRK III reservoir water (Fig. 5.26).

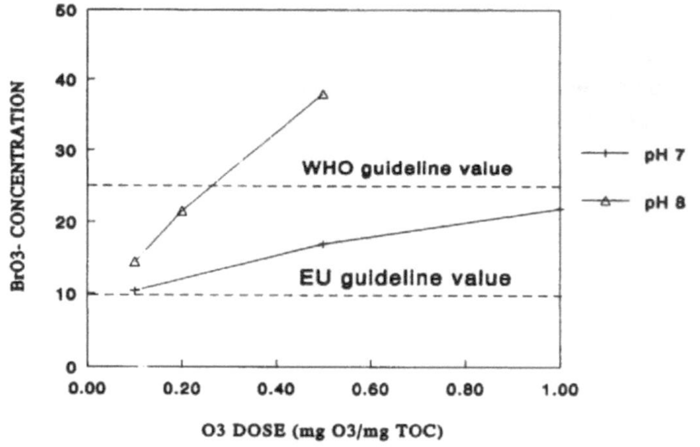

Fig. 5.26 Bromate (BrO_3^-) concentration in DAF product water for different O_3 doses and pH (raw water TOC=6.2 g/m^3).

The relatively high TOC concentration of the raw water (mostly DOC) imposed a need for adequately high ozone doses of typically 1.2-5.5 gr O_3/m^3 [31]. The resulting bromate concentrations of 10-38 μg/L were all (except one) below the WHO provisional guideline value of 25 μg BrO_3^-/L, however, higher than the European Union drinking water guideline of 10 μg/L, and far higher than the envisaged bromate standard for the Netherlands of 0.5 μg/L, or 5 μg/L where ozone is applied for disinfection.

The bromate formation via O_3 and $OH^{\cdot}$ mediated reactions is significantly affected by pH (Fig.

5.26); its concentration increases sharply with pH and ozone dose. Earlier bench-scale experiments confirmed that ozone as conditioner for algae removal by DAF yielded higher removal efficiency compared to treatment with coagulant alone (Chapter 4). Ozone dosages as low as 0.2 mg O_3/mg TOC (0.6 mg O_3/L) in combination with pH 7.5 resulted in significant DAF particle (algae) removal efficiency improvements of 40-50% (Chapter 4, Figs. 4.11 and 4.12). Use of cationic polyelectrolytes as coagulant aid further increased DAF efficiency typically by 5%. Scanning electron microscopy (SEM) of ozonated model water (Fig. 4.15, Chapter 4) showed that ozone affected the outer algal cell wall and damaged part of the algal cells. Leakage of the IOM that then acts together with EOM as natural coagulant aid has been proposed to be one of the main coagulation mechanisms [32, 33, 34, 35].

In the case of WRK III water, the 0.2 g O_3/g TOC corresponded roughly to a dose of 1.2 g O_3/m^3, which at pH 7 resulted in pilot plant bromate levels approaching the EU guideline value of 10 μg/L. This promotes ozone conditioning at lower pH (≤pH 7) as an attractive option in WRK III circumstances, especially in view of the possible accompanying reduction of the clogging capacity of the water [9], and improvement of organic matter (including pesticides) removal by the GAC filtration [36].

b. $KMnO_4$ conditioning

The pilot plant research confirmed the beneficial influence of this oxidant. $KMnO_4$ application proved to be beneficial under the conditions of the model water, however, its influence on the process efficiency was lower and less predictable compared to ozonation (Chapter 4) [35].

The DAF removal efficiency of particles larger than 2.75 µm (the lower size limit of *M. aeruginosa*) at a dosage of 10 g Fe(III)/m^3 was comparable to or higher than the full-scale sedimentation performance with 24 g Fe(III)/m^3 and 0.5 g Wisprofloc-P/m^3 (i.e. 96.3% and 95.9% mean particle removal efficiencies). Additional $KMnO_4$ conditioning at 1.0 g $KMnO_4/m^3$ tended to increase the DAF particle removal efficiency (97.7% mean particle removal), the results after DAF being consistently better than those of the full-scale sedimentation and at some occasions also better than the results after the full-scale filtration (maximum DAF particle removal efficiency of 99.7 %). This enhanced performance should result in lower filter loading by particles larger than 2.75 µm. However, negatively charged MnO_2 colloids are formed that cause manganese concentration in the DAF effluent to over the 50 µg/l MAC, and thus, together with organo-iron complexes contribute to a higher residual turbidity after DAF. In order to further increase the DAF removal efficiency and improve the effluent quality, $KMnO_4$ was applied in combination with cationic polyelectrolytes Superfloc C-573 and Wisprofloc-P. Positive effects were achieved at polyelectrolyte dosage ≥0.5 g/m^3 and coagulant dose ≥10 g Fe(III)/m^3 . Figs. 5.27 and 5.28 represent an overview of the particle and turbidity removal efficiency results from this part of the research (1-4), compared to the full scale sedimentation (5) and filtration (6) processes.

The one-way ANOVA for different treatment options, including the full-scale WRK III sedimentation and filtration treatment (5 (SED) and 6 (FILTR) in Figs. 5.27 and 5.28), showed that treatment significantly affected both particle (P=0.00022) and turbidity (P=0.000001) removal efficiency. However, one- way ANOVA performed on only the DAF options (1 through 4 (DAF) in Figs. 5.27 and 5.28) yielded no statistically significant difference between the considered treatment options, either for particle (P=0.085) or for turbidity (P=0.686) removal efficiency.

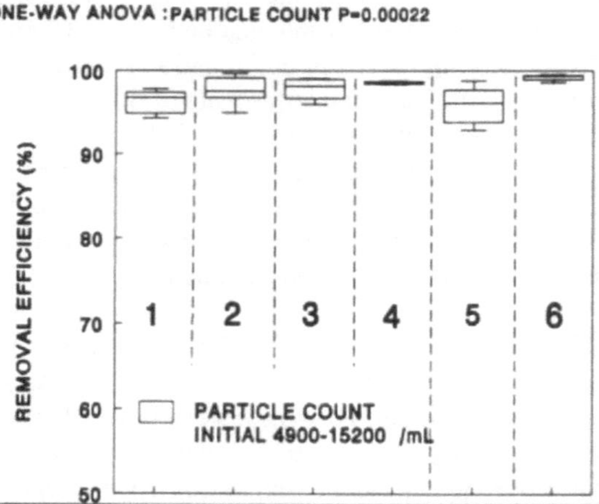

CONDITIONS	Fe(III) (g/m3)	KMnO4 (g/m3)	C-573 (g/m3)	W-P (g/m3)
1 (DAF)	10	-	-	-
2 (DAF)	10	1.0	-	-
3 (DAF)	15	1.0	0.3	-
4 (DAF)	15	1.0	0.5	-
5 (SED)	24	-	-	0.5
6 (FILTR)	24	-	-	0.5

Fig. 5.27 One-way ANOVA of the pilot plant (1-4 (DAF)) vs. full scale plant sedimentation (5 (SED) and filtration (6 (FILTR)) particle removal efficiency (of particles >2.75 μm).

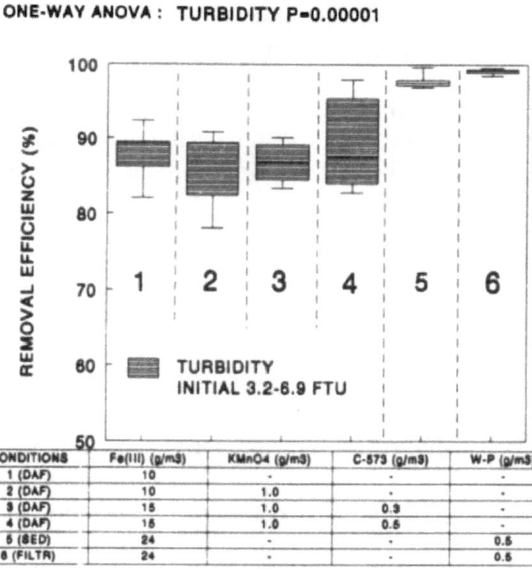

CONDITIONS	Fe(III) (g/m3)	KMnO4 (g/m3)	C-573 (g/m3)	W-P (g/m3)
1 (DAF)	10	-	-	-
2 (DAF)	10	1.0	-	-
3 (DAF)	15	1.0	0.3	-
4 (DAF)	15	1.0	0.5	-
5 (SED)	24	-	-	0.5
6 (FILTR)	24	-	-	0.5

Fig. 5.28 One way ANOVA of the pilot plant (1-4 (DAF)) vs. full scale plant sedimentation (5 (SED)) and filtration (6 (FILTR)) turbidity removal efficiency.

This suggests that the tendency that DAF improves particle removal efficiency after $KMnO_4$ conditioning, as well as the improvement after combined $KMnO_4$ conditioning and addition of cationic coagulant aid, were not statistically confirmed in the pilot plant operation. However, multiple comparison analysis (excluding the WRK III full-scale filtration) on particle removal efficiency yielded statistically significant differences only for treatment options 4 and 5 (the number of comparisons being 4, the comparisonwise error rate was P=0.05/4=0.0125, and since for treatment options 4 and 5 P=0.0093, thus <0.0125), approaching the comparisonwise error rate of 0.0125 in the cases of treatment options 1 and 4.

This suggests that the combined $KMnO_4$ and 0.5 g Superfloc C-573/m^3 polyelectrolyte treatment performed statistically significantly better in terms of particle removal efficiency than the full-scale sedimentation, and that the combined $KMnO_4$ conditioning and ≥0.5 g Superfloc C-573/m^3 polyelectrolyte treatment may be considered appropriate to significantly raise the DAF particle removal efficiency. The multiple comparison analysis on turbidity removal efficiency suggests that the full-scale sedimentation performed statistically significantly better than all the DAF options (P<0.0125), albeit at much higher coagulant dose. This is mainly due to the coagulation chemistry at the lower coagulant doses that are generally applied in the DAF process.

The $KMnO_4$ effect on coagulation and flocculation is considered a combined consequence of microorganism inactivation; modification of the algal cell wall and in-situ production of natural, algae derived coagulant aid; oxidation and subsequent adsorption of coagulation interfering organic matter; and accelerated flocculation because of increased particle concentration due to MnO_2 colloids (Chapter 4) [29, 37]. The main overall difference between DAF jar test experiments on model water and the pilot plant work, is that the latter provided consistently good effluent water quality, especially in terms of particles (algae) removal (94.9-99.7%), unlike the former which was characterised by variations in effluent quality and relatively low efficiency (27.0-88.2%). One of the possible causes of this phenomenon is the different NOM concentration and characteristics in the water used in the two experimental modes.

In case of the model water experiments (Chapter 3 and 4) the TOC concentration before spiking the water with the algae was in the range of 1.8-2.5 mg C/L, mostly comprised of DOC (>95%). This organic matter remained from the originally present organic matter in the (Biesbosch reservoirs) water after full treatment in the Kralingen water treatment plant (Rotterdam, the Netherlands). The treatment includes Fe (III) coagulation/flocculation, sedimentation, ozone application (3 mg O_3/L), rapid sand and granular activated carbon filtration. Since ozone generally reduces the molecular weight of NOM compounds and increases their functional group content, it is surmised that it also reduces the precipitating tendency of the remaining NOM molecules by the applied Fe (III) coagulant in the bench-scale experiments. Furthermore, the remaining NOM molecules after the full treatment likely lack functional groups that would enable sorption processes with the metal (coagulant) oxide surfaces [38]. After spiking the algae in this water, the TOC and DOC concentration rose to 2.3-3.0 mg C/L and 2.0-2.7 mg C/L, respectively. Thus, the algae and their EOM (as well as IOM in case of oxidants application) were probably the most significant portion of the organic matter exerting a coagulant demand and prone to react with the $KMnO_4$ in the bench-scale experiments. In these circumstances, $KMnO_4$ addition resulted in the creation of stress conditions for the algae and increased EOM and IOM concentrations, which coupled to other $KMnO_4$ induced mechanisms (Section 4.3.4.a) resulted in improved coagulation/flocculation and increased DAF efficiency. Since the impact of the conditioning process relied mostly on the quality of the cultured algae and their exudates (EOM and IOM composition, dependant on algae age), occasional changes of permanganate demand were recorded, as well as a relatively large standard deviation of the DAF efficiency.

On the other hand, the reservoir water used in the pilot plant experiments was characterised by a TOC concentration in the range of 5-15 g C/m^3, mostly comprised of DOC (90-95%). The relatively high DOC was comprised mostly of humic matter, which itself exerted a high coagulant demand. The importance of the OM content and composition in the presence of $KMnO_4$ conditioning, for the enhancement of coagulation, has been documented by others as well [39]. It is related to the documented existence of NOM in the form of fine matrices or meshes of filaments in waters of similar quality [27, 28]. These fine matrices can embed both algal EOM and IOM, and MnO_2 precipitates, and thus promote flocculation via a bridging mechanism. Thus, they are suggested as a factor which enhanced DAF pilot plant efficiency. It can also be surmised that the NOM fine matrices embedding the flocculated material form a gel-like structure which improves the particle-bubble attachment and DAF efficiency.

The particle count for size 0.3-5 μm showed that the effluent from the $KMnO_4$ + DAF + filtration scheme had a considerably higher particle count than that of the full-scale sedimentation + filtration scheme, especially in the colloidal size range <1 μm. It is suggested that the lower coagulant doses applied in the DAF (as compared to sedimentation), although efficient in destabilising particulate matter (including algae), could not provide complete sweep coagulation, especially with regard to the high DOC fraction of the NOM. Thus, it is the DOC rather than the particles that controls the coagulant dose [40]. Here, the high particle count in the <1 μm size range is compounded by the residual colloidal MnO_2 (of typical size 0.3-0.4 μm) and the difficult to remove organo-iron complexes (of typical size <1 μm). On the other hand, the considerably higher coagulant doses in the full-scale sedimentation (24 g Fe(III) + 0.5 g Wisprofloc-P/m^3, against 10 g Fe(III)/m^3 alone, or in combination with 1.0 g $KMnO_4$ /m^3 used in DAF) resulted in complete sweep coagulation and proved favorable for efficient sedimentation due to the very high particle concentration. This observation is supported by residual coagulant measurements that were considerably lower in case of the sedimentation.

This means that as measured against particle removal efficiency including particles in the 0.3-1 μm colloidal size range and the 1-150 μm algae representative size range, the best results were obtained by the sedimentation option (87.4%), followed by the pilot plant $KMnO_4$ + DAF (65.8%), and the pilot plant DAF without $KMnO_4$ conditioning (62.5%). Similar order of particle removal efficiency (0.3-150 μm) was noted for the different overall treatment options, although with smaller differences between each treatment option (92.7% for the sedimentation + filtration option, 88.8% for the pilot plant $KMnO_4$ + DAF + filtration option, and 86.1% for the DAF + filtration option). The particle removal efficiency for particles in the algal size range (>1 μm) of the DAF + filtration option achieved at higher coagulant (≥10 g Fe(III)/m^3) and coagulant aid (≥0.5 g Superfloc C-573/m^3) doses, alone or in combination with 1.0 g $KMnO_4$ /m^3 conditioning, was similar to that of the full-scale sedimentation + filtration option (24 g Fe(III) + 0.5 g Wisprofloc-P/m^3); however, this did not apply for the submicron particle size range. This suggests that due to the incomplete sweep coagulation achieved by the generally lower coagulant doses applied in the DAF, and the lower efficiency of the DAF unit in the removal of the colloidal particles in general [22, 41] (as compared to the favorable conditions created by the high coagulant doses in the sedimentation), the pilot filter in the DAF + filtration option was subjected to higher loads of colloidal matter, which will probably result in shorter filter runs [42, 43].

5.3.5 DAF recirculation ratio

Conservatively high recirculation ratio and saturator pressure proved to be of little influence on DAF efficiency in bench-scale model water experiments [41]. Similar results were obtained with the pilot plant (Fig. 5.29). Slightly better turbidity and particle removal efficiency was obtained at a recirculation ratio of 7% (86.3% and 95.6%, respectively), as compared to 5 % (84.0% and 93.9%, respectively) and 10% (84.6% and 95.6%, respectively) for a constant pressure of 500 kPa. The recirculation ratio R and saturator pressure determine the ratio between bubble and particle concentrations and their size, and have often been suggested to determine the DAF efficiency. However, measurements and calculations on the DAF single collector collision efficiency model [41] suggest that only relatively low bubble concentration is obtained with R=5% and saturator pressure of 500 kPa and saturator efficiency of 90% (N_b=25,600 bubbles/ml of 62 µm mean size) compared to the high particle concentration N_p=16,650 particles/ml after flocculation at 10 g Fe(III) + 0.3 g Superfloc C-573/m^3 (Fig. 5.23) which does not fully explain the overall high process efficiency. Most likely, it is again the NOM that accounts for the more efficient particle-bubble attachment in DAF.

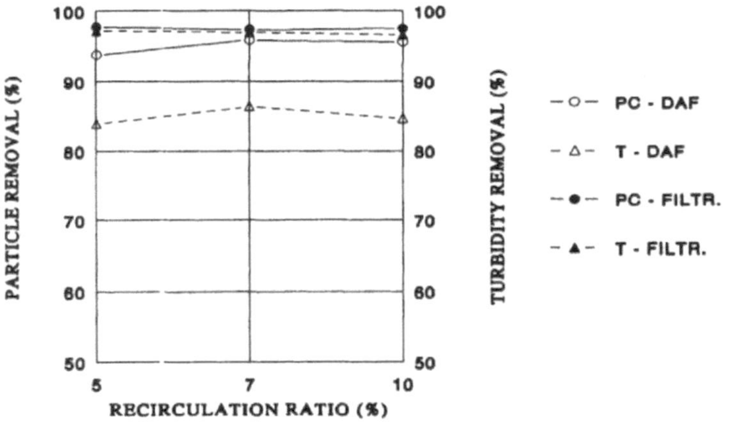

Fig. 5.29 Particle removal efficiency after DAF and after filtration, as a function of recirculation ratio R. Conditions : 7 g Fe(III)/m^3, pH 8, two-stage flocculation with t_{f1}=18.2 min and G_1=50 s^{-1} in the first stage, and t_{f2}=3.8 min and G_2=30 s^{-1} in the second stage, P=500 kPa, T=5.2-5.8 FTU, PC=11,040-11,670/mL.

Highest particle removal efficiency by filtration was obtained with a recirculation ratio of 5%, and efficiency slightly decreased with increasing recirculation ratio. This could be caused by the introduction of more bubbles in the DAF unit at increased recirculation ratios, increasing the chance that some of them are carried over with the DAF effluent and onto the filter. There, the bubbles may disturb the filter pore structure and contribute to the passage of particles through it.

5.3.6 Filtration efficiency and the MFI (Modified Fouling Index)

DAF effluent quality generally had a pronouncedly positive effect on the filter effluent quality, but the sedimentation + filtration treatment led to slightly higher overall removal efficiency. The turbidity and particle removal efficiency of the filtration step following DAF under optimal pilot plant conditions (7 g Fe(III)/m^3, two step flocculation t_{f1}=18.2 min, G_1=50 s^{-1}, t_{f2}=3.8 min, G_2=30 s^{-1}, R=7%, P=500 kPa, without coagulant aid) were 96.9-97.3% and 97.4-98.8% respectively, compared to the full-scale sedimentation + filtration average removal efficiency of 98.6 and 99.7%, respectively. The residual iron in the former case was 0.08-0.17 g Fe_{total}/m^3.

Recharge of the aquifer with treated water as practiced by WRK III leads to the use of the MFI as a major effluent quality parameter. Our attempts to link MFI, turbidity and residual iron for the full-scale plant proved unsuccessful (Fig. 5.6); it proved partly successful to establish a correlation link between filtrate particle count and turbidity [12]. The difference in particle removal efficiency between the pilot plant and the full-scale treatment options was even more pronounced with respect to the MFI. The full-scale filtrate MFI values during the 1995 research period were within the 5-26 s/L^2 range (mean value of 8.83 s/L^2), and for the 1996 research period within the 5-12 s/L^2 range (mean value of 7.2 s/L^2). Pilot scale filtrate MFI values for the 1995 research period were 46.6-97.8 s/L^2, the lower value being associated with an unusual low coagulation pH of 6.35. Longer term filter operation with the use of Wisprofloc-P coagulant aid resulted in drop of the MFI value from the initial 97.8 to 40.5 s/L^2 in six hours, however, accompanied by rapid filter clogging. In the case of a 7 g Fe(III)/m^3 dose the MFI value was 95±5 s/L^2 . These high values could be attributed predominantly to the relatively high DAF residual iron.

Further optimisation of the agglomeration phase by applying a combination of a higher coagulant dose (>7 g Fe(III)/m^3) and cationic polyelectrolyte Superfloc C-573 led to a reduction of the MFI value. A combination of 15 g Fe(III)/m^3 and 0.5 g Superfloc C-573/m^3 coagulant aid resulted in increased DAF efficiency (turbidity removal 93.6% and particle removal 98.6%, compared to efficiency after sedimentation of 96.7% and 97.7%, respectively). This had a positive impact on the filter efficiency, resulting in turbidity removal efficiency of 98.9% and particle removal efficiency of 99.8%; both results were better than those of the full-scale plant average. Both after DAF and filtration, the residual iron was lower than usual (0.97 and 0.08 g Fe(total)/m^3, respectively), suggesting that coagulation and flocculation were more efficient. The corresponding MFI value of the filtrate was 19.1 s/L^2, in the range of the full-scale plant MFI values for this part of the year (1995).

During the experiments (September-October, 1996), the pilot filter lowered the residual manganese below the 50 µg/L MAC (in the range of 35 µg/L), thus overcoming one of its potential limitations. However, the high particle and turbidity filter removal efficiency was not accompanied by a reduction of the MFI. The 1996 pilot plant MFI values were again >50 s/L^2 and were higher than the full-scale filtration 7.2 s/L^2 MFI mean value for this period of the year. Although the filter pore size of the MFI measurement apparatus was 0.45 µm, a large part of the colloidal MnO_2 and the organo-iron complexes was retained on the filter. This resulted in a characteristic brown colour of the deposit and high MFI values.

Unfortunately, the effect of ozone conditioning on the filtrate MFI value could not be assessed. However, based on model water experiments positive results were expected, and the MFI is expected to further decrease towards the prescribed 5 s/L^2 value. This effect should be the result of improved particle removal efficiency and possibly lower coagulant dose requirements (Chapter 4). On the other hand, $KMnO_4$ conditioning improved removal of particles >2.75 µm, however, the residual Mn and Fe, although below the prescribed limit of 50 µg/L and 200 µg/L respectively, still resulted in comparatively high MFI values >50 s/L^2.

5.4 CONCLUSIONS

Pilot plant investigations conducted with reservoir water that was characterised by a seasonal cyanobacteria bloom (of mainly *Microcystis aeruginosa*) indicated that the acceptable coagulant dose for DAF related to maximised algal removal was two to three times lower than the one applied in the full-scale sedimentation (7-12 g Fe(III)/m^3 compared to 20-24 g Fe(III) + 0.2-0.5 Wisprofloc-P/m^3). The DAF removal efficiency (before filtration) was somewhat lower in terms of turbidity removal and residual iron content, probably due to some degree of (partial) preferential complexation of the iron by the organic matter into difficult to remove organo-iron complexes (<1 µm), and hence a limited availability of coagulant for destabilisation. The incomplete coagulation occurring at the lower coagulant doses applied prior to DAF, especially with regard to the high NOM/DOC concentrations, and the generally lower efficiency of DAF in the removal of colloidal particles, are suggested responsible for the lower efficiency of the DAF + filtration, compared to the sedimentation + filtration treatment option.

However, the primary task of this study being particles (algae) removal (>1 µm), DAF was found very efficient and appropriate even at these relatively low coagulant doses. The mean DAF particle removal efficiency (>2.75 µm) for a coagulant dose of 7 g Fe(III)/m^3 was 94.5%, as compared to 97.3% for sedimentation with 20 g Fe(III) + 0.2 Wisprofloc-P/m^3. For a coagulant dose of 10 g Fe(III)/m^3 the DAF particle removal efficiency was higher than the one achieved by the sedimentation with a combination of 24 g Fe(III) + 0.5 Wisprofloc-P/m^3 (i.e. 96.3% and 95.9%).

The highest turbidity and particle removal efficiency was achieved at the lowest tested pH 6.35 (93.2 and 97.5%, respectively), suggesting the aiding effect of adsorption coagulation to the otherwise dominant sweep coagulation. Electrophoretic mobility measurements did not confirm complete particle charge neutralisation. The turbidity and particle removal efficiency decreased with pH increase (to pH 7 and 8), falling bellow the 90% value for pH 9, possibly related to precipitation of calcite particles.

Flocculation times longer than 15 minutes were found necessary, and removal efficiency was insensitive to the flocculation energy input (G value). The particle removal efficiency slightly decreased (from 96.6 to 94.5%) with increase of the G value from 10 to 70 s^{-1} (30 and 50 s^{-1} were also tested) while turbidity removal efficiency did not follow the same trend, reaching its 'optimum' at G=50 s^{-1} (89.9%). The results suggest that the heterodisperse curvilinear flocculation model for turbulent flocculation could be applicable; this model assumes that the G value has little effect compared to the rectilinear model.

No significant improvement of the DAF effluent quality was observed for the combination of the coagulant dose of 7 g Fe(III)/m^3 with cationic coagulant aid (organic or synthetic) in the dose range of 0.1-0.7 g/m^3. This coagulant dosing was accompanied by only a slight particle surface charge decrease compared to the charge neutralisation that occurred in the experiments with model water, possibly influenced by the high NOM content.

The positive effect of ozone and potassium permanganate conditioning on the removal of particles (algae) by DAF, were demonstrated on bench-scale experiments with model water for the first, and on bench-scale and pilot plant scale for the second. Due to technical limitations of the pilot plant the efficiency of the first could not be comprehensively tested in the pilot plant. Results showed that the relatively low ozone dosages (in the range of 0.2 g O_3/g TOC or 1.2 g O_3 /m^3) which significantly improved DAF process efficiency (by 40-50%) in the bench-scale experiments, resulted at pH 7 in bromate levels below the WHO provisional guideline value of 25 μg/L and approached the European guideline of 10 μg/L.

$KMnO_4$ conditioning at a dosage of 1 g/m^3 tended to raise DAF process efficiency, especially with respect to particle removal (97.5% mean particle removal of particles >2.75 μm, as compared to 96.3% in case of no $KMnO_4$ conditioning). The particle removal efficiency achieved by $KMnO_4$ conditioning and coagulant aid Superfloc C-573 use (dose ≥0.5 g/m^3) tended to further increase DAF efficiency typically by 5% (98.6% mean particle removal efficiency of particles >2.75 μm); this treatment option was statistically significantly more efficient than that of the full-scale sedimentation option. The residual Mn of the DAF process mode after filtration was below the 50 μg/L MAC (approximately 35 μg/L), while residual Fe_{total} was also below the 200 μg/L MAC.

The variation of the recirculation ratio (5, 7 or 10%) had a minor effect on DAF turbidity and particle removal efficiency; slightly better results were obtained for the recirculation ratio of 7 % (typically 1-2%). On the other hand, the filtration particle removal efficiency for the 5% recirculation ratio was slightly better (1-2%) than for the other values, possibly due to the reduced chance of filter pore structure disturbances from bubbles, which is related to the lower bubbles concentration at this recirculation ratio and decrease of the opportunity of their carry over onto the filters.

The pilot plant filter removal efficiency of particles and turbidity was slightly lower than that of the full-scale filters. This is related to the incomplete coagulation of the NOM and the lower efficiency of the DAF in the removal of the submicron particles, thus subjecting the filter in the DAF + filtration option to higher loads of colloidal matter. However, it should be taken into account that the full-scale plant results had been positively influenced by the fact that the filters were operated at half of their design capacity. The full-scale plant experienced difficulties in attaining the stringent MFI guideline of 5 s/L^2, especially during periods of algae blooms. Optimisation of the coagulation and DAF by increasing the coagulant dose to ≥15 g Fe(III)/m^3 in combination with ≥0.5 g Superfloc C-573 coagulant aid tended to outperform the sedimentation in terms of particle removal (>2.75 μm), whilst resulting in comparable turbidity removal. The MFI value accomplished in the filtrate was below 20 s/L^2. Further optimisation of the coagulation + DAF + filtration scheme, offers a good prospect for treating the algae laden WRK III reservoir water. Further research attention for ozone application in WRK III circumstances emerges from the possible improvement of the resultant MFI values due to expected ozone induced particle removal efficiency improvement, coupled to the expected improval of organic matter removal (including pesticides) by the GAC filtration. On the other hand, the count of particles in the <1 μm range suggested that MnO_2 and organo-iron complexes are the main contributors to the corresponding high MFI values (>50 s/L^2) in case of $KMnO_4$ conditioning, asking for further optimisation of the filtration step in order to maximise the positive effects of the permanganate induced increase of DAF removal efficiency of particles in the algal size range.

REFERENCES

1. Schippers J.C., 1989. *Vervuiling van Hyperfiltratiemembranen en verstopping van infiltratiepunten (Fouling of reverse osmosis membranes and clogging of artificial recharge wells)*. Ph.D. thesis, Keuringsinstituut voor Waterleidingartikelen - KIWA N.V., Rijswijk, the Netherlands.
2. Kooij van der D., 1992. Assimilable organic carbon as an indicator of bacterial regrowth. *J. AWWA*, **84**, 2.
3. Vlaški A. and van Breemen N., 1992. *A preliminary investigation of the membrane filtration index and characterisation of particles of water produced at WRK III treatment plant*. Internal communication, Delft University of Technology, the Netherlands.
4. Vlaški A., van Breemen A.N. and Alaerts G.J., 1996. The algae problem in the Netherlands from a water treatment perspective. *J. Water SRT-Aqua*, **45**, 4, pp.184-194.
5. Oskam G. and van Genderen J., 1995. Eutrophication and development of algae in surface water - a threat for the future ? Conference paper,. *Special subject 8, IWSA Congress*, Durban, South Africa.
6. Evans R., 1957. Review of experiences with microstrainer installations. *J. AWWA*, **49**, 5, pp. 541-552.
7. Janssens J.G., van Hoof F. and Dirckx J.J., 1987. Comparative study on preozonation and prechlorination applied in treatment of surface waters from impounding reservoirs. Proc. *AWWA Annual Conference*, Kansas City, MO., USA.
8. Jodlowski A., 1990. *Preoxidation effects on algae related to direct filtration water treatment*. Internal communication, Delft University of Technology, the Netherlands.
9. Langlais B., Reckhow D.A. and Brink D.R. (Eds.), 1991. *Ozone in Water Treatment, Applications and Engineering*. AWWA and Campagnie Generale des Eaux, Lewis Publishers, USA.
10. Jun Ma and Guibai Li, 1992. Laboratory and full scale plant studies of potassium permanganate oxidation as an aid in coagulation. Proc. *IAWQ/IWSA Joint Specialised Conference : Control of Organic Material by Coagulation and Floc Separation Processes*, Geneva, Switzerland.
11. Petruševski B., van Breemen A. N. and Alaerts G.J., 1993. Pretreatment in relation to direct filtration of impounded surface waters. Proc. *European Water Filtration Congress, KVIV*, Oostende, Belgium.
12. Petruševski, B., 1996. *Algae and particle removal in direct filtration of Biesbosch water*. Ph.D. thesis TU and IHE - Delft, published by A.A. Balkema, the Netherlands.
13. Raman, R.K., 1985. Controlling algae in water supply impoundments. *J. AWWA*, **77**, pp. 41-43.
14. Nederlands Normalisatie Instituut, 1982. NEN: 6460: Water - Determination of iron content by atomic absorption spectrometry (flame technique).
15. Nederlands Normalisatie Instituut, 1982. NEN: 6466: Water - Determination of manganese content by atomic absorption spectrometry (graphite furnace technique).
16. APHA, AWWA and WPCF, 1985. *Standard methods for examination of water and wastewater*. Am. Publ. Health Assoc., Washington D.C., USA.
17. Bruggink C., Rossum van W.J.M. and Smeenk J.G.M.M., 1995. Method for BrO_3^- determination in surface drinking water, based on ion chromatography with macroinjection, H_2O, **28**, 11, pp.343-347.

18. Rijk de, S.E., van der Graaf, J.H.J.M. and den Blanken, J.G., 1994. Bubble size in flotation thickening. *Wat. Res.*, **28**, 2, pp. 465-473.
19. Vlaški A., van Breemen A.N. and Alaerts G.J, 1996. The role of particle size and density determining dissolved air flotation and sedimentation efficiency. Proc. *4th IAWQ-IWSA International Conference The Role of Particle Characteristics in Separation Processes*, Jerusalem, Israel.
20. Vlaški A., van Breemen A.N. and Alaerts G.J., 1995. Optimisation of coagulation conditions for the removal of cyanobacteria by dissolved air flotation or sedimentation. *J.Water SRT-Aqua*, **45**, 5, pp. 253-261.
21. Janssens J.G., 1990. The application of dissolved air flotation in drinking water production, in particular for removing algae. Conference paper, *DVGW Wasserfachlichen Aussprachetagung*, Essen, Germany.
22. Edzwald J.K., 1994. Principles and application of dissolved air flotation. Keynote address, *IAWQ-IWSA-AWWA Joint Specialised Conference on Flotation Processes in Water and Sludge Treatment*, Orlando, USA.
23. Bernhardt H. and Clasen J., 1991. Flocculation of micro-organisms - *J.Water SRT-Aqua*, **40**, 2, pp. 76-87.
24. Malley J.P. and Edzwald J.K., 1991. Laboratory comparison of DAF with conventional treatment - *J. AWWA*, **83**, pp. 56-61.
25. Arnold R.S. and Harvey P., 1994. Recent applications of DAF pilot studies and full scale design. Proc. *IAWQ-IWSA-AWWA Joint Specialised Conference on Flotation Processes in Water and Sludge Treatment*, Orlando, USA.
26. Lawler D.L., 1993. Physical aspects of flocculation : from microscale to macroscale -*Wat. Sci. Tech.*, **27**, 10, pp. 165-180.
27. Fillela M., Buffle J. and Leppard G.G., 1992. Characterisation of submicron colloids in freshwaters : evidence for their bridging by organic structures. Proc. *IAWQ/IWSA Joint Specialised Conference on Control of Organic Material by Coagulation and Flocculation Processes*, pp. 71-80.
28. Buffle J. And Leppard G.G., 1995. Characterization of aquatic colloids and macromolecules. 1. Structure and behaviour of colloidal material. *Environmental Science & Technology*, **29**, 9, pp. 2169-2175.
29. Petruševski B., van Breemen A.N. and Alaerts G.J., 1995. Effect of permanganate pre-treatment and coagulation with dual coagulants on algae removal in direct filtration. Proc. *IAWQ/IWSA Workshop on Removal of Microorganisms From Water and Wastewater*, Amsterdam, the Netherlands.
30. Edzwald J.K. and Paralkar A., 1992. Algae, coagulation, and ozonation - *Chemical Water and wastewater treatment II*, Klute R. and Hahn H.(Eds.), Springer-Verlag, pp. 263-279.
31. Chang S.D. and Singer P.C., 1988. The impact of ozonation on removal of particles, TOC and THM precursors. Proc. *AWWA Annual Conference*, Orlando, Florida, USA.
32. Reckhow D.A., Singer P.C. and Trussel R.R., 1986. Ozone as a coagulant aid. Proc. *AWWA National Conference : Recent Advances and Research Needs*, USA.
33. Singer P.C., 1990. Assessing ozonation research needs in water treatment. *J. AWWA*, **82**, pp.78-88.
34. Sukenik A., Teltch B., Wachs A.W., Shelef G., Nir I. and Levanon D., 1987. Effect of oxidàtion on microalgal flocculation. *Wat. Res.*, **21**, 5, pp 533-539.
35. Vlaški A., van Breemen A.N. and Alaerts G.J., 1996. The application of ozone for improved algae flocculation and DAF removal. In prep.

36. Orlandini E., Siebel M.A., Graveland A. and Schippers J.C., 1994. Pesticide removal by combined ozonation and GAC filtration, *Water Supply*, **14**, pp. 99-108.
37. Vlaški A., van Breemen A.N. and Alaerts G.J., 1996. The application of $KMnO_4$ for improved algae flocculation and DAF removal. In prep.
38. Edwards M. and Benjamin M.M., 1992.Effect of preozonation on coagulant-NOM reactions. *JAWWA*, **84**, pp. 63-72.
39. Ma J., Graham N. and Li G., 1997. Effect of permanganate preoxidation in enhancing the coagulation of surface waters-laboratory case studies. *J.Water SRT-Aqua*, **46**, 1, pp. 1-10.
40. Edzwald J.K., 1993. Coagulation in drinking water treatment : particles, organics and coagulants. *Wat. Sci. Tech.*, **27** (11).
41. Vlaški A., van Breemen A.N. and Alaerts G.J., 1996. Evaluation and verification of the single collector collision efficiency dissolved air flotation (DAF) kinetic model. In prep.
42. Tobiason J. E., Johnson G.S., Westerhoff P.K. and Vigneswaran B., 1993. Particle and chemical effects on contact filtration performance. *Journal of Environmental Engineering*, **119**, 3, pp. 520-539.
43. Mackie R.I. and Bai R., 1993. The role of particle size distribution in the performance and modelling of filtration. *Wat. Sci. Tech.*, **27**, 10, pp. 19-34.
44. Petruševski B., van Breemen A. N. and Alaerts G.J., 1994. Optimisation of coagulation conditions for in-line direct filtration. Proc. *Workshop on Optimal Dosing of Coagulants and Flocculants*, Műlheim an der Ruhr, Germany.

Chapter 6

THE ROLE OF FLOC SIZE AND DENSITY IN DISSOLVED AIR FLOTATION AND SEDIMENTATION

-Parts of this chapter were published by A.Vlaški, A.N. van Breemen and G.J.Alaerts in (1997) *Water Science and Technology*, 36, 4, pp. 177-189.

ABSTRACT : Sedimentation and dissolved air flotation (DAF) were studied for particle removal, in particular of the single cells form of the cyanobacterium *Microcystis aeruginosa*. This cyanobacterium species is considered a suitable representative of algae and particles of the problematic size range (3-10 µm) in the assessment of removal efficiency. The agglomeration (coagulation/flocculation) phase was proven to determine the down-stream process efficiency. Relevant process parameters were studied on a bench-scale (using model water) and pilot plant scale (using reservoir water), including the influence of coagulant dose ($FeCl_3$), coagulation pH, flocculation time, energy input (G value), single stage versus tapered flocculation, and application of cationic polyelectrolyte as coagulant aid. The process efficiency was interpreted in function of the preceding agglomeration (coagulation/flocculation) phase and the obtained particle (floc) size distributions. The particle (floc) size - density relationship was addressed.

6.1 INTRODUCTION

The algal morphology and physiology have been reported to influence the treatment efficiency of algae laden water in various ways [1]. In principle, algae can be regarded as particles whose stability is due to : (i) electrostatic surface charge which is negative for pH 2.5-11.5, (ii) steric effects due to water bound on cell surface, to adsorbed macromolecules or extracellular organic matter (EOM) onto the cell [2, 3]; (iii) morphology and extremities, and (iv) motility [1, 4]. Compared to other suspended particles of inorganic nature and similar stability algae have the peculiarity that they are living matter comprised of numerous species with different morphological and physiological characteristics that may affect charge phenomena and steric behaviour, such as: size, shape, motility, cell wall and EOM composition, etc. For example, flagellated *Rhodomonas* spp. have been noted to escape from floc material because of their motility [4]. Therefore, no single treatment is equally effective for the removal of all species.

The single cell *Microcystis aeruginosa* size of 3-10 µm, its round shape and surface characteristics contribute to its persistent occurrence in filtrate water in many surface water treatment plants, even though all other quality criteria have been fulfilled [5]. It is considered to be among the most widespread algal species occurring in water reservoirs, as well as among the species that are most difficult to remove in water treatment (Chapter 1). This makes it a suitable reference particle to study efficiency of removal of algae or other classes of microorganisms of comparable size such as *Cryptosporidium parvum* and *Giardia lamblia*.

In the treatment line, the agglomeration (coagulation/flocculation) phase is typically the most critical to achieve overall efficient removal of algae (Chapters 3). It can be enhanced by the application of polyelectrolytes as coagulant aids and/or conditioning of the algae (e.g. by application of chemical oxidants such as ozone or $KMnO_4$) (Chapters 4 and 5). Physical pre-treatment (microstraining) can reduce a substantial share of the algal load, especially the larger size colonial and filamentous species. Optimisation of the subsequent solid-liquid separation process is essential for algal removal (Chapters 3, 4 and 5). The commonly applied separation processes are either plain or lamellar sedimentation, or dissolved air flotation (DAF), followed by rapid sand filtration. The destabilisation by the coagulation and the consequent flocculation of particles, as well as eventual polyelectrolyte and/or conditioner application, define the resulting particle (floc) size distribution. The particle charge, floc size, floc density, and the relation between floc size and floc density play decisive roles in the down-stream processes.

Previous particle size analysis research with relatively clean model water (polystyrene particles with diameter d=0.87 μm in distilled water [6], or kaolin particles with diameter d=1.88 μm [7] in tap water buffered with 100 mg $NaHCO_3$), indicated that the particle size distribution after coagulation/flocculation with Al(III) or Fe(III) salts in a dose range of 3-5 mg/L was most pronounced in the <100 μm size range. In the former case, floc growth continued until steady state (no change of floc dimensions) was achieved, as a function of the shear rate (G value) and coagulant concentration. Increased coagulant concentration at constant G value produced larger and more open floc structures. Higher coagulant concentrations at low G values resulted in increased floc strength, while increasing G for the higher coagulant doses resulted in increased amount of floc breakage (more significant than for the lower doses) into smaller and more compact floc structures [6]. Pilot plant DAF research with water from the Göta River in Gothenburg [8], Sweden defined the particle size range of 10 μm as the most significant and responsible for the high DAF efficiency after coagulation/flocculation with 5.3 mg Fe(III)/L, and with 5 or 20 min flocculation time at G=30-70 s^{-1}. This is in agreement with the conceptual DAF kinetic model of Edzwald et al. [3, 9, 10], while flocs of size in the range of 100 μm and larger are advised for sedimentation. On the other hand, Fukushi et al. [11] state that floc size in water treatment is in the range 10^1- 10^3 μm and that the production of larger flocs should be the target in a DAF based treatment, since large flocs have a much higher collision efficiency (proportional to the third power of the collision radius). Their arguments are based on water from a diluted kraft pulp black liquor with colour of 50 Units and to which 5 mg Al(III)/L coagulant were applied at pH 6.7. Under such conditions 80% of the particles was of size <500 μm.

The existing data diversity regarding the water treatment relevant particle (floc) size may partly arise from : (i) diversity of applied particle size determination techniques and absence of a standardised measuring procedure, (ii) susceptibility of some particles, especially flocs to structural changes due to the application of often inappropriate measurement techniques (e.g. the use of flow through counters in which high shear may disturb the aggregates structure), and (iii) different raw water quality characteristics (coagulation pH, temperature, initial particulate, colloidal and dissolved inorganic and organic matter concentration and composition, etc.), which can influence the obtained particle size distribution.

Organic or inorganic polyelectrolyte coagulant aids of cationic, non-ionic or anionic nature have been applied in sedimentation in numerous cases, especially at low initial particle concentrations. On the other hand, experiences with the use of polyelectrolytes as coagulant aids in DAF are controversial with respect to obtained benefits [12, 13]. In the Netherlands only one non-ionic organic, and one anionic organic (starch based) polyelectrolytes are in use, although previous research showed that synthetic cationic polyelectrolytes can be more efficient [14]. Their impact on the process efficiency depends on their charge and molecular weight characteristics, and is asserted through effects on the surface charge of the algae, as well as the obtained floc size distribution [15] (Chapters 3, 4 and 5).

The application of ozone and $KMnO_4$ as algae conditioners has been associated with spontaneous microflocculation, even before the addition of coagulant [16, 17, 18, 19]. However, ozone has also been found to cause an opposite effect resulting in an increase of the smaller size particle fraction [20, 21], as well as decrease of overall particle volume and mass and lower filter loading [22]. Both ozone and $KMnO_4$ have been associated with the creation of stress conditions for the algae, resulting in EOM and IOM release, which then act as coagulant aids and promote spontaneous microflocculation [16], (Chapters 4 and 5).

The situation with respect to dominant particle (floc) sizes in treatment becomes more complicated if the submicron particle size range is considered, since it is most often determining the coagulation/flocculation efficiency (Chapter 5). In addition to the spontaneous microflocculation effect, ozone induces the formation of colloidal matter from dissolved organic matter and enhances its subsequent removal [16, 17], however, affecting the coagulation and flocculation process and the resulting particle size distribution. On the other hand, an important aspect of permanganate conditioning is the effluents quality is terms of residual Mn. This pertains to the dissolved permanganate, as well as to the colloidal and (presumably) filterable MnO_2 manganese. Strictly avoiding overdosing and permanganate reduction by Na_2SO_3 are two possible strategies to prevent dissolved Mn residuals [18, 19] (Chapter 4). Filtration is able to reduce the most substantial part of the colloidal Mn fraction [18, 19]. On the other hand, MnO_2 manganese and organo-iron complexes were suggested to be partly responsible for the significantly higher DAF effluent turbidity and high filter effluent MFI, compared to the sedimentation based treatment (Chapter 5). Nevertheless, it was also reported that the combined action of permanganate conditioning, and iron and cationic polyelectrolyte coagulation resulted in improved DAF efficiency and low effluent Fe and Mn residuals.

The present study addresses critical aspects of the agglomeration phase in advance of the DAF and sedimentation separation processes. It intends to quantify particle size frequency and particle volume distributions before and after the coagulation/flocculation step, and after the down-stream DAF or sedimentation. This is done under different process conditions, including varied coagulant dose, coagulation pH, flocculation velocity gradient and time, tapered versus single stage flocculation, as well as for different treatment options, including the application of Fe(III) coagulant alone or in combination with cationic polyelectrolyte coagulant aid, and ozone or $KMnO_4$ conditioning. This information would be used to interpret overall treatment performance, and identify the key factors in the process kinetics.

6.2 MATERIALS AND METHODS

6.2.1 Laboratory experiments with model water

The laboratory experimental set-up consisted of a batch-wise jar test apparatus with incorporated DAF facilities (Chapter 3 and 4). For both the sedimentation and the DAF modes of operation, the standard experimental procedure included coagulation at pH 8, temperature 20°C and a G value of 1,000 s^{-1} for 30 s. Other coagulation pH conditions were also tested for DAF (pH 6, 7, 8, and 9); the desired coagulation pH was achieved by adding HCl or NaOH, in amounts determined previously by titration. Periodical check-ups showed that the pH remained stable throughout the experiments. The flocculation G and flocculation time t were varied (G_f=10, 30, 50, 70, 100 and 120 s^{-1} and t_f=5, 10, 15, 25, 30 and 35 min). The coagulant ($FeCl_3 \cdot 6H_2O$) doses were 0 - 15 mg Fe(III)/L, while the cationic polyelectrolyte (Superfloc C-573) doses were in the range of 0 - 1.5 mg Superfloc C-573/L for DAF and 0 - 3 mg Superfloc C-573/L for sedimentation. The rationale behind the dose range of polyelectrolyte was to approximately cover the range which corresponds to one tenth of the optimal Fe coagulant dose. The sedimentation time t_s was varied (t_s=10, 20, 30, 45, 60 and 90 min), while for DAF different recirculation ratios

R (R=5, 7 and 10%) and saturator pressures P (P=500, 600 and 700 kPa) were tested.

Ozone was used as an algae conditioner in the context of improving down-stream DAF efficiency. For ozone production, the Trailigaz LABO LO, France, ozone generator was used (Chapter 5, Fig. 5.12). Ozone was produced under standard ozone generator conditions of 220 V, 0.6 A, and 0.6-0.7 bar pressure. The ozone dosage (transferred ozone) was derived from the ozonation time which was varied from 1-4.5 min (Chapter 4). The pH of the raw (model) water subject to ozonation was previously set to 7.5 by HCl addition. A 5.5 L ozone resistant glass jar was used as a batch ozone reactor. The time gap between ozonation and subsequent coagulation and flocculation was kept at 2 min. The pH change after ozonation was 7.5±0.2 and it served as coagulation pH without additional corrections. Ozone was applied in combination with the previously determined optimal coagulant dose for a concentration of ≈10,000 cells/mL of *M. aeruginosa* (3 mg Fe(III)/L). The cationic polyelectrolyte Superfloc C-573 was tested as a sole coagulant (0.3-1.0 mg/L), or as a coagulant aid (0.1-1.0 mg/L) to Fe(III) coagulant, combined with a low ozone dose of 0.2 mg O_3/mg C (0.58 mg O_3/L).

For permanganate conditioning experiments a 0.0057 M stock solution was prepared at weekly intervals. The applied permanganate dose ranged from 0.1-2.0 mg $KMnO_4$/L, while the pH was previously set at 8. The permanganate was applied at a G value of 400 s^{-1} for a period of 1 min, followed by 30 min of slow mixing at a G value of 10 s^{-1}. It was combined with a previously optimised Fe(III) coagulant dose at coagulation pH 8.

The investigated cyanobacterium species was a semi-continuously laboratory cultured, single-cells form of *M. aeruginosa*. The model water originated from the Biesbosch storage reservoirs, the Netherlands, and was spiked with the cyanobacteria to achieve an initial concentration of ≈ 10,000 or 20,000 ±cells/mL. The experiments did not include a final filtration step.

6.2.2 Pilot plant experiments

The pilot plant research was conducted on the premises of the Princess Juliana treatment plant (WRK III) situated on the shores of the IJssel Lake in Andijk, the Netherlands (Chapter 5). The raw water used in the experimental work was reservoir water previously subject to microstraining over a 35 μm mesh. The period of the research (late summer, early autumn) coincided with a bloom of *M. aeruginosa*, a common species in the IJssel Lake in this period of the year. The DAF pilot plant investigations were carried out with a commercial (Purac, Sweden) pilot plant installation (Q_{max} =13 m^3/h). The unit itself comprised a 1 m^3 stainless steel tank with four needle valves located at the bottom of the inlet for the recycled water stream introduction. Coagulation occurred in a Kenics-type in-line mixer (Purac, Sweden); one, two or three-stage flocculation could be applied with variable speed flocculation paddles. An automated system regulated the pumps, the saturation and the recycled flow. The experiments did not include a final filtration step.

$Fe_2(SO_4)_3 \cdot 9H_2O$ was used as coagulant in a dosage range of 0-20 g Fe(III)/m^3. pH in coagulation was set at pH 8 using $Ca(OH)_2$, as in the full-scale WRK III process. The synthetic cationic polyelectrolyte Superfloc C-573 was tested as a coagulant aid in the concentration range of 0.1-0.7 g/m^3. The Kenics static mixer was used for the rapid mix with a calculated G value of ≈3,500 s^{-1} for a flow of 4.6 m^3/h. This flow was applied in the DAF pilot plant in order to avoid carry-over of bubbles into its effluent and onto the filters, a common occurrence in case of higher flows. Different flocculation time, energy input (G value) and flocculation sequence (one-, two- and three-stage flocculation) were tested; the values were partly dictated by the technical provisions of the pilot plant. The recirculation ratio R in the DAF unit was set at R=5, 7 and 10%, while the saturator pressure P was 500 kPa. The water temperature over the experimental period decreased from 16.7°C to 10.1°C in 1995, and from 16.4 °C to 12.9 °C in 1996.

Because of disproportion between the pilot plant flow (Q=4.6 m^3/h) and the continuous-flow ozone reactor (Q_{max}=1.14 m^3/h), only a limited number of pilot plant ozone conditioning experiments was carried out, assessing primarily the issue of bromate formation (Chapter 4). Potassium permanganate (0.0021M working solution) was dosed immediately before pH correction and coagulant addition, allowing for very short contact times (<1 s) before coagulation and flocculation.

Model water experiments were intended to simulate relatively "clean" water circumstances and provide conditions in which the assessment of the removal of the spiked algae would be affected to a limited degree by other water quality parameters (i.e. the presence of larger quantities of other suspended, colloidal and dissolved matter of inorganic or organic origin). Thus, the algae were the most significant factor determining the optimal process conditions and the achieved efficiency. These experiments were used to optimise DAF and sedimentation process conditions on bench-scale and gain deeper insight into the governing process mechanisms, which was to serve as basis for the DAF pilot plant optimisation. On the other hand, the WRK III reservoir (Chapter 5) provided for the "real" water conditions in which the naturally occurring cyanobacteria bloom of *M. aeruginosa* was coupled with other water quality factors of importance for process optimisation and of influence on process efficiency (i.e. temperature, NOM concentration and composition, inorganic colloidal and particulate matter, etc.). These results were than compared to the results achieved by the full scale WRK III sedimentation based treatment. Particle size and density relations emerging from both experimental modes are used to highlight important process phenomena.

6.2.3 Analytical techniques

Process evaluation was based on turbidity (Sigrist L-65, Switzerland), particle (algae) count measurements of particles in the size range 2.75-150 μm (HIAC-Royco PC-320, USA) and residual coagulant concentration (Fe_{total}). These measurements were compounded by inverted microscope count (M40-Wild Leitz, Switzerland), and electrophoretic mobility (Tom Lindström AB-Repar). Other particle analysis techniques included particle count of particles in the size range of 0.3-5 μm (Particle Measuring Systems Inc./Liquid Batch Sampler LBS-100, and the Microlaser Particle Spectrometer, Boulder, Co., USA) and computer image analysis (Mini-Magiscan, IAS 25/IV25 Joyce-Loebl Ltd., UK). The use of the former technique was limited to a number of $KMnO_4$ conditioning experiments and the assessment of the colloidal MnO_2 and Fe residuals involvement in the overall DAF effluent quality.

The computer image analysis results are presented in the form of particle size frequency and volume distributions. The frequency and volume distributions represent the number and the volume of recorded particle size range. For this purpose the recorded particles and/or floc material were approximated to spheres and presented in size ranges of 10 μm, from 0-200 μm. Particles >200 μm were also recorded (by image analysis and visually), in particular after flocculation; their frequency increase was associated with higher coagulant doses, lower flocculation G values and longer flocculation times, as well as with the raw water quality : larger flocs were observed in the reservoir water than in the model water experiments. However, large flocs represented only a small percentage of the overall particle concentration, especially for the "clean" model water circumstances (by concentration <2.5% of particles (floc) were of size >100 μm for a dose range of 0-10 mg Fe(III)/L, pH 8, G=70 s^{-1} and t=10 min). The removal of large particles by the solid-liquid separation process (DAF or sedimentation) was complete, while the smaller particles proved to be more problematic in this respect. Thus, the adopted size limit in the discussion of the results is 200 μm, while the particle size of 50 μm, which is roughly the size limit of visually observable aggregates, was tentatively adopted as the size dividing the smaller from the larger particle size range.

The image analyses were performed on fresh samples, as well as on photographs of samples. Photo analysis was preferred over fresh sample analysis, since it enabled registration of representative situations at different spots along the treatment line, without the potential danger of disturbing the process or the quality of the sample [23]. For this purpose, a specially devised flat photo cell was connected at different locations of the jar test apparatus and the pilot plant unit, and samples were photographed before and after flocculation, and after particle removal. Special care was taken to avoid break-up of particle agglomerates (flocs) during the passage of the sample from the jar test unit to the photo cell, by allowing low flow velocities adjusted through a system of valves. In case of DAF, the released air bubbles were also photographed and analysed. Professional high resolution black-and-white film AGFA 25 or Kodak TMAX 100 was used for the purpose, while forced development was applied for film processing. 30-100 fields, or more than 500 particles of each film shot were processed and analysed, using a ccd-camera (604*288 pixels) mounted to a Nikon Optiphot microscope (at 40x magnification). The minimum detectable size under these circumstances was 1.9 μm. The software package Genias25 developed by Joyce-Loebl (UK) was used for data processing.

The computer image analysis is subject to a measurement error which varies with the size of the recorded particle and the applied magnification. Previous work [23] showed that the smaller the particle and the magnification, the larger the measuring error. This is due to the fact that the particle area used for calculations of the particle diameter is measured as the full area of all the computer screen pixels over which the particle image is spread, even if the pixels at the edge of the image are not fully covered. In the case of 40x magnification (pixel size of 1.9x1.9 μm), we showed that for an 8 μm size *M. aeruginosa* cell the calculated cell diameter was 25-50% larger than the actual cell dimension. This explains the particle frequency peak of the raw (model) water which occurred at the 10-20 μm, instead of at the expected 0-10 μm range (Fig. 6.2 and 6.3). For larger particles (flocs>10 μm) the measurement error led to up to 10% higher values than the actual sizes. To verify the reproducibility of the applied analytical technique, multiple shots of the same sample were analysed regularly.

Particle size distributions departed strongly from Gaussian shapes. Statistical comparison of data on particle size frequency distributions from experiments that were performed under identical conditions was carried out using the chi-square test. Testing for the hypothesis that the obtained particle size distributions were identical resulted in a probability level of p=0.999, confirming highly comparable particle size distributions and hence high experimental reproducibility.

6.3 RESULTS AND DISCUSSION

6.3.1 Coagulant dose

The settling characteristics of a flocculent suspension depend on floc size, floc density, water temperature [24] and on initial particle number concentration [25]. The influence of the coagulant dose on the overall performance of the sedimentation based process is presented in Fig. 6.1 in the form of particle volume distribution at initial algal concentration of 10,000 cells/mL and for coagulant dose of 6 mg Fe(III)/L and 15 mg Fe(III)/L. The flocculation step for the 6 mg Fe(III)/L coagulant dose resulted in the production of a substantial volume of flocs in the larger particle size range (>50 μm); the 15 mg Fe(III)/L dose contributed even more significantly to this particle size range, whilst maintaining the generation of a large volume composed of the smaller particles (<50 μm).

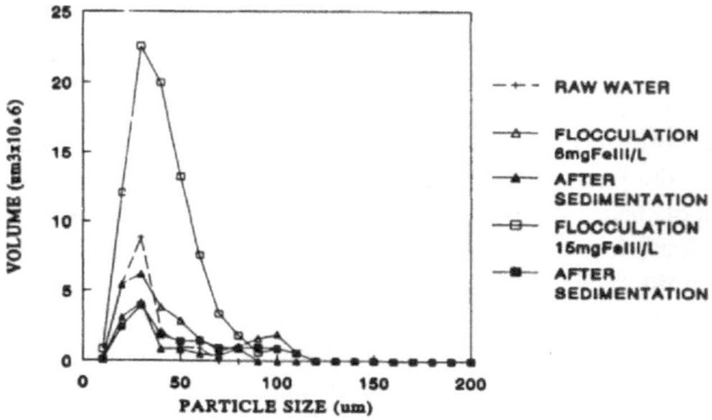

Fig. 6.1 Particle volume distribution of raw water, and before and after sedimentation at coagulant dose of 6 versus 15 mg Fe(III)/L (bench-scale), pH 8, flocculation G_f=30 s^{-1} and t_f=30 min, and sedimentation t_s=60 min.

The optimal coagulant dose for sedimentation in case of an algal concentration of 10,000 cells/mL was 10 mg Fe(III)/L, three times higher than that for DAF (3 mg Fe(III)/L) (Chapter 3). Fig. 6.2 depicts particle (floc) size frequency distributions after flocculation and after DAF at the algal concentration of 20,000 cells/mL for the different coagulant doses of 1 (a), 3 (b), 5 (c) and 10 mg Fe(III)/L (d). The intermediate dose of 5 mg Fe(III)/L was most efficient (also when expressed as turbidity and particle count measurements), suggesting an approximately proportional stoichiometry between optimal coagulant dose and initial particle concentration [1, 26] (Chapter 4). The dose of 1 mg Fe(III)/L was inadequate for sufficient particle destabilisation and significant floc formation, which would shift the particle (floc) size distribution and result in a particle size distribution favouring DAF. The increase in the frequency in the smaller particle size range (<50 μm) due to iron flocs precipitation [8] and the small effect on the larger particle size range (>50 μm) resulted in poor DAF performance. Thus, the sweep coagulation conditions at this low coagulant dose were not fully appropriate resulting in inefficient flocculation and particle removal by DAF, coupled to high residual coagulant concentration (only ≈60% of the Fe coagulant dose was removed). HIAC particle count measurements of acidified effluent showed that the count at the higher coagulant doses can consist up to 15% of the residual coagulant.

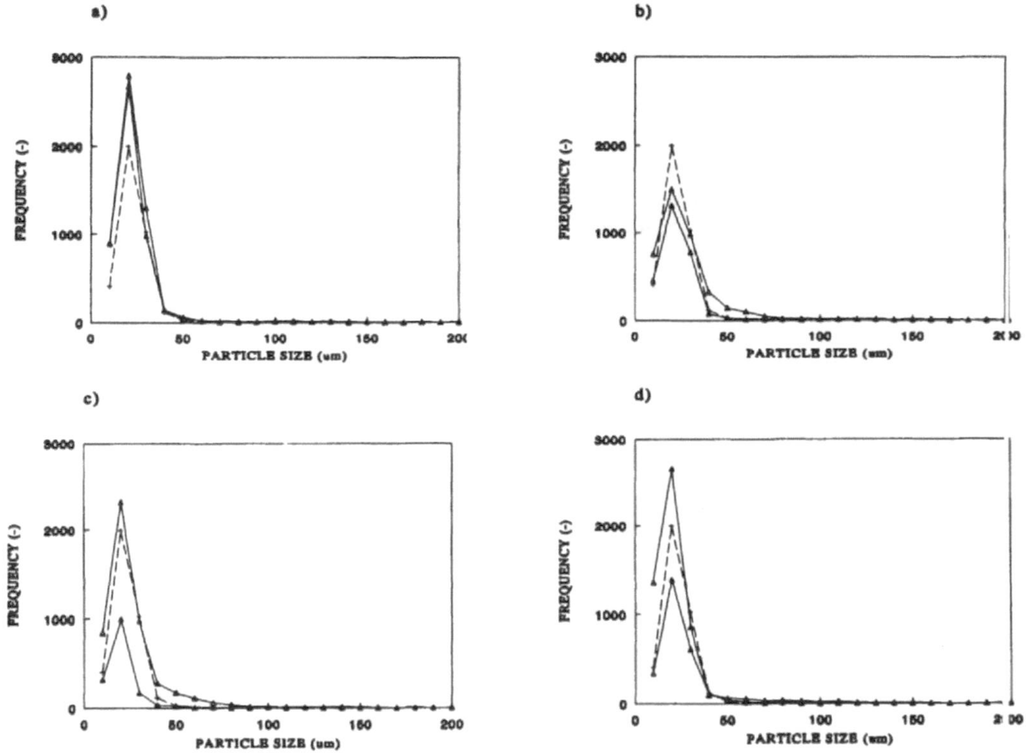

Fig. 6.2 Particle size frequency distribution before and after DAF at different coagulant dose (bench scale): (a) 1 mg Fe(III)/L, (b) 3 mg Fe(III)/L, (c) 5 mg Fe(III)/L and (d) 10 mg Fe(III)/L. Conditions : pH 8, G_f=70 s^{-1}, t_f=10 min, R=5%, P=5 bar. Notation : + raw water, △ water after flocculation, and ▲ after DAF.

Raising the dose to 3 mg Fe(III)/L enhanced the formation of particles in the larger size range (>50 μm) and resulted in improved DAF removal efficiency. At the most efficient particle removal dose of 5 mg Fe(III)/L, both the frequencies of particles in the smaller and the larger particle size ranges were increased by flocculation. The corresponding particle volume distribution (not presented) of the flocculated water at this coagulant dose, showed that the recorded increase of particle frequency in the smaller size range, corresponded to a decrease in the particle volume for this range (both relative to the raw water). This may be ascribed to a change in the nature of the particles of size <50 μm before and after flocculation : iron micro-flocs taking the place of the original algae which were flocculated and shifted to the larger size range. The 10 mg Fe(III)/L dose contributed to the <50 μm size range in a similar way as the 1 mg Fe(III)/L dose, but nevertheless resulted in better removal efficiency due to the flocculation of algae into flocs of size >50 μm. In this set of bench-scale experiments with model water the particle fraction of particle size >100 μm contributed only up to 1% of the total particle amount, but corresponding high particle volumes (40-50% of total particle volume after flocculation). In the pilot plant case, particles of larger size (up to 200 μm and larger) were present (Figs. 6.5 and 6.6), however, these were completely removed by the DAF treatment (data not shown).

6.3.2 Coagulation pH

The coagulation pH influences the extent and the rate of hydrolysis of metal coagulants, and it also determines the composition and surface charge of polymeric metal hydrolysis species [14]. The pH simultaneously controls the stoichiometry and kinetics of other chemical reactions, and influences the surface charge of algae and other particulate matter. The effect of coagulation pH on particle agglomeration and on overall removal efficiency after DAF is presented in Fig. 6.3.

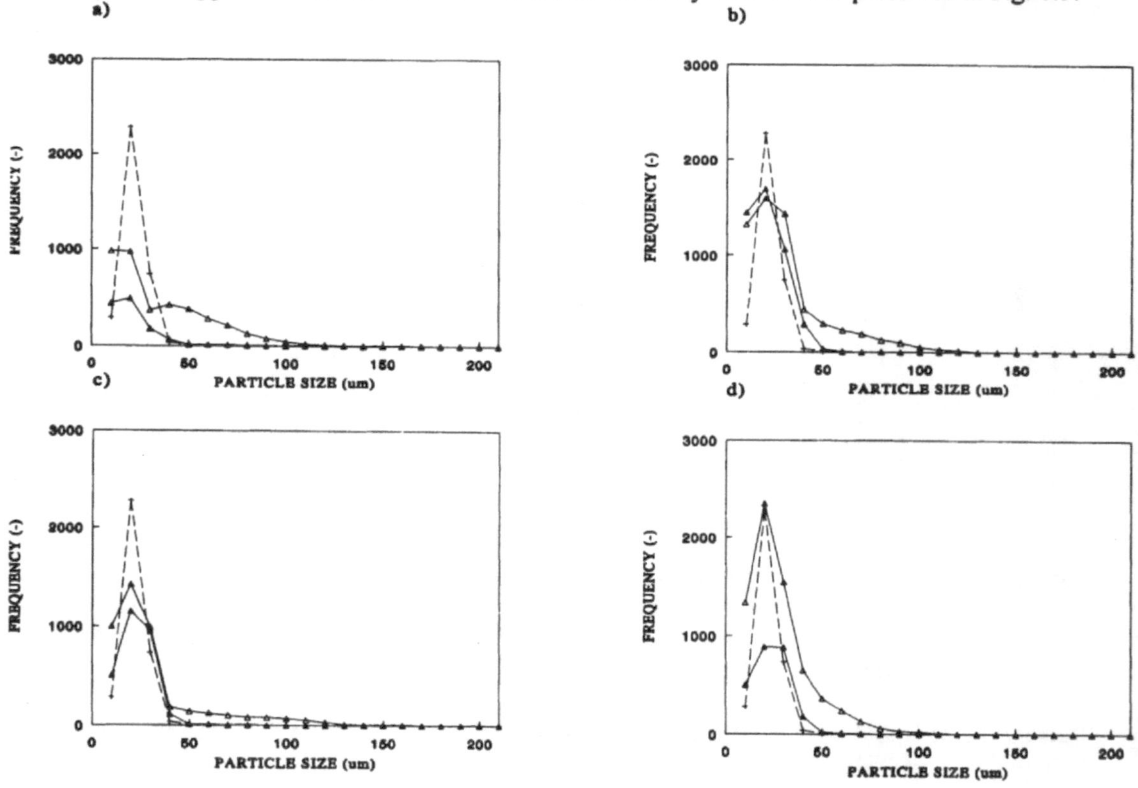

Fig. 6.3 Particle size frequency distributions before and after DAF at different coagulation pH (bench-scale) : (a) pH 6, (b) pH 7, (c) pH 8 and (d) pH 9. Other conditions : 10,000 cells/mL, 10 mg Fe(III)/L, G_f=70 s^{-1}, t_f =10 min, R=5%, P=500 kPa. Notation : + raw water, ▵ water after flocculation, and ▴ after DAF.

Image analysis data after flocculation show a more pronounced reduction in the smaller particle size range, and an increase in the larger size range at pH 6 (Fig. 6.3.a), compared to pH 7 and 8 (Figs. 6.3.b and 6.3.c). In all cases, at the low and the neutral pH, the coagulant dose tended to produce more large size aggregates which were efficiently removed by DAF. At the neutral pH of 7-8, particle agglomeration was likely to have been achieved solely through sweep coagulation [27] (Chapter 3). The surface charge of the destabilised colloidal and floc material remained negative. However, at pH 6 adsorption coagulation and charge neutralisation (verified by EM measurements, not presented) substantially contributed to particles destabilisation beside the dominant sweep coagulation. This almost doubled particle removal efficiency, suggesting that adsorption coagulation plays an important role in algae flocculation. The frequency increase at pH 9 (Fig. 6.3.d) of both the smaller and larger particles, may be ascribed to precipitation of $CaCO_3$ under alkaline conditions. Electrophoretic mobility measurements (data not shown) confirmed that algal charge neutralisation occurred at the low pH conditions (pH 6), which is in accordance with other research [15, 28]. The turbidity data and particle count measurements correlated with the image analysis results (data not shown). The charge neutralisation mechanism is also suggested to cause the higher removal efficiency of particles in the smaller particle size range (<50 µm) at pH 6, as the adsorption efficiency between neutralised particles and negatively charged air bubbles was higher [29].

6.3.3 Flocculation conditions

The agglomeration process which has already begun in the coagulation phase with the fast precipitation of metal coagulant species, continues during the flocculation phase under slow mixing. The optimal slow mixing intensity and period are a function of the down-stream process. These two parameters further influence the particle (floc) size distribution and density, and thus the down-stream process efficiency.

Floc size and flocculation energy input (G) were found by image analysis to be inversely proportional (Fig. 6.4).

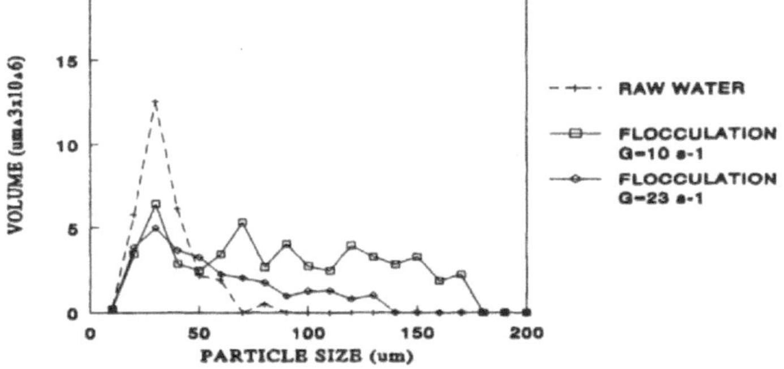

Fig.6.4 Particle volume distribution for different flocculation Gs (bench-scale). Conditions : pH 8, 3 mg Fe(III)/L, t_f=10 min.

Jar test experiments at a low G value of 10 s^{-1} resulted in a floc volume distribution which led to highest DAF removal efficiency (Chapter 3), although flocs were of weaker structure compared to those generated at higher G values (as in the case of G=23 s^{-1}). Therefore, the statement [12, 15] that DAF requires high flocculation G in order to produce smaller, stronger and shear resistant flocs, may require adjustment. On the other hand, floc volume distributions obtained at flocculation G values of 10-30 s^{-1} favoured sedimentation (Fig. 6.1). It is therefore suggested that the floc density plays an equally important role as the floc size, yielding slightly better results at G=30 s^{-1}, in combination with flocculation time ≥30 min (Chapter 3).

Image analysis data of reservoir water flocculated in pilot plant circumstances with G=10 and 50 s^{-1}, are presented in Fig. 6.5. In this case, the inverse relation between floc size and flocculation energy input was less pronounced; an increase of flocculation G seemed necessary, possibly to provide more contact opportunities for the particulate matter in the <50 μm size range. The reduction of the frequency (data not shown) of recorded particles in the range <50 μm increased with increasing G. The extent of the reduction is less obvious on the presented cumulative particle volume graph (Fig. 6.5) due to the significantly smaller share of the small particles (<50 μm) in the volume (about 10% of the total volume after flocculation). The flocculation G=50 s^{-1} value produced the best turbidity removal (89.9%), while the particle removal was comparable for all G values (≈95%), slightly better at G=10^{-1} (96.6%).

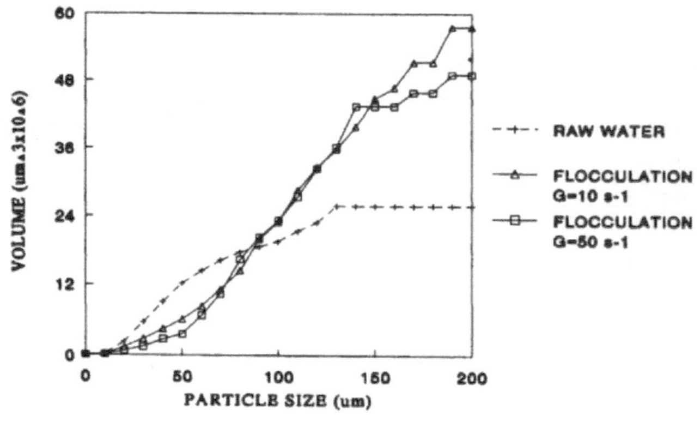

Fig. 6.5 Cumulative particle size volume distribution for different flocculation G values (pilot plant). Conditions : pH 8, 10 g Fe(III) + 0.3 g Superfloc C-573/m^3, t_f=20 min, R=7%, P=500 kPa.

Similar to the jar test experiments with model water, the pilot plant experiments showed that tapered flocculation did not significantly improve overall DAF performance (Chapter 5). Particle volume distribution after each of the two stages of the applied tapered flocculation in pilot plant experiments showed that there was an increase in the volume of particles in the larger size range, on account of a decreasing volume of the smaller particles, as G decreased in the second flocculation step (Fig. 6.6). However, overall tapered flocculation had little impact on the final DAF effluent quality compared to that of the single-stage flocculation DAF at similar flocculation G and flocculation times. It is therefore suggested that in the case of DAF total flocculation time is more important than the tapering effect [9] (Chapter 5).

6.3.4 Polyelectrolyte as coagulant aid

The use of organic or inorganic coagulant aids of cationic, non-ionic or anionic nature has been known to sedimentation in numerous cases, especially at low initial particle concentrations. On the other hand, experiences with the use of polyelectrolytes as coagulant aids in DAF are controversial with respect to obtained benefits [12, 13]. In the Netherlands only one non-ionic organic, and one anionic organic (starch based) polyelectrolytes are approved, although previous research showed that synthetic cationic polymers can be more efficient [14]. Relatively low cationic polyelectrolyte (Superfloc C-573) doses of approximately one tenth of the optimal coagulant dose can substantially improve product water quality (Chapter 3). Particle removal efficiency (as measured by HIAC) for DAF of model water rose from 71% to 94% (at a dose of 0.5 mg Superfloc C-573/L), and for sedimentation from 88% to 99.5% (at a dose of 1.0 mg Superfloc C-573/L. This was accompanied by charge neutralisation of the particulate matter at a dose somewhat higher than the determined optimal one (Fig. 6.8), similar to what was found by other research [30]. Particle volume distributions of the flocculated model water with and without the addition of coagulant aid (Fig. 6.7) showed that agglomeration improved in the former case. It resulted in a more pronounced larger particle size fraction, on account of a reduction of the smaller particles fraction. Furthermore, an increase of the polyelectrolyte dose resulted in a decrease of the volume associated with the larger particle size fraction. This suggests that denser floc structures were formed at the higher polyelectrolyte dose; 0.5 mg Superfloc C-573/L favoured DAF, whereas the 1.0 mg/L favoured sedimentation.

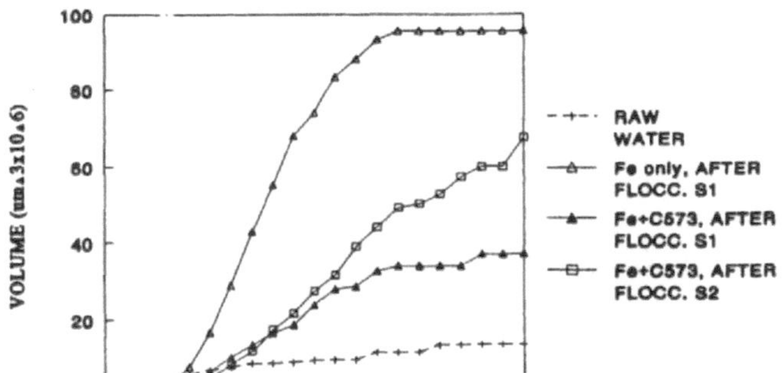

Fig. 6.6 Cumulative particle volume distribution for tapered flocculation (pilot plant), 1st stage (S1) : G_1 =50 s^{-1} , t_{f1} =18 min, 2nd stage (S2) : G_2 =30 s^{-1} , t_{f2} =4 min, pH 8, 15 g Fe(III) + 0.5 g Superfloc C-573/m^3.

Image analysis of reservoir water after flocculation confirmed these observations (Fig. 6.6). The concentration of particles in the 30-150 µm size range increases in case of application of 15 g Fe(III)/m^3 only. This situation changes in case of combined application of 15 g Fe(III)/m^3 and 0.5 g Superfloc C-573/m^3 coagulant aid, resulting in denser floc structures, with slight increase of particles or flocs in the >150 µm range. In the case of the DAF pilot plant investigations, the positive impact of the cationic polyelectrolyte was less pronounced than in the laboratory jar test experiments with model water. This improvement was accompanied by a slight surface charge decrease (from -1.29 to -0.88 μS/V/cm at coagulation pH 6.35, and from -0.84 to -0.78 μS/V/cm with 0.5 g/m³ cationic coagulant aid at coagulation pH 8, and with 7 g Fe(III)/m³ in both cases), as compared to the complete charge neutralisation which occurred in the model water experiments (Fig. 6.8). As discussed in Chapter 5, this suggests the importance of the NOM concentration and composition for the coagulation process.

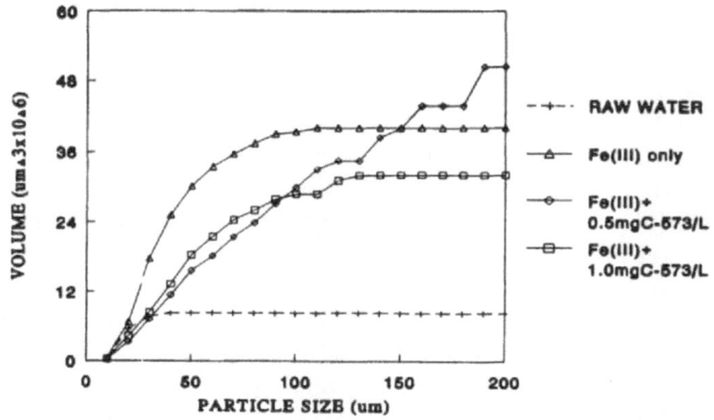

Fig. 6.7 Cumulative particle volume distribution for flocculation without and with polyelectrolyte Superfloc C-573 coagulant aid (bench-scale). Conditions : pH 8, 10 mg Fe(III)/L + Superfloc C-573, G_f=30 s^{-1}, t_f=30 min.

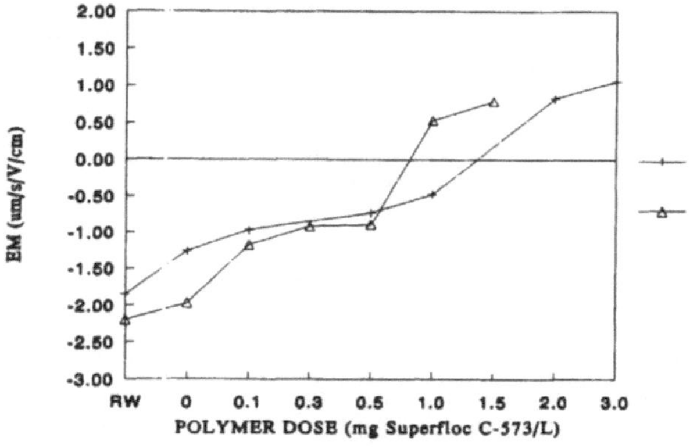

Fig. 6.8 Electrophoretic mobility (EM) as a function of polyelectrolyte Superfloc C-573 coagulant aid dose for bench scale DAF (+) (3 mg Fe(III)/L, G_f=70 s^{-1}, t_f=10 min), and sedimentation (▵) (10 mg Fe(III)/L, G_f=30 s^{-1}, t_f=30 min) with model water.

6.3.5 Ozone conditioning

The spontaneous microflocculation induced by ozone conditioning was discussed in Chapter 4 (Figs. 4.14 and 4.18). Here, the influence of different treatment options on the particle (volume) distribution is discussed : Fe(III) coagulant only treatment (Fig. 6.9), combined ozone and Fe(III) coagulant (Fig. 6.10), combined ozone, Fe(III) and Superfloc C-573 coagulant aid (6.11), and finally ozone and Superfloc C-573 as sole coagulant (Fig. 6.12).

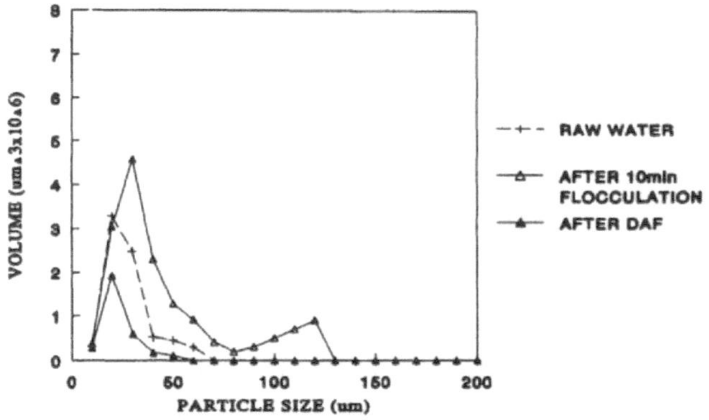

Fig. 6.9 Particle volume distribution after different treatment stages in the case of treatment with Fe(III) coagulant only (bench-scale). Conditions : 3 mg Fe(III)/L, pH 7.5, G_f=10 s^{-1}, t_f=10 min, R=7%, P=500 kPa.

Fig. 6.9. indicates the efficient removal of the flocculated material and the absence of larger size particles in the flotate, however, the residuals in the lower particle size range were still considerable. The impact of the relatively low ozone dose of 0.2 mg O_3/mg C on the particle volume distribution can be observed by comparing the raw water particle distribution in Fig. 6.9 and the ozonated water particle distribution in Fig. 6.10; the latter was the 'raw water' that was further subjected to coagulation and flocculation. The impact of the 3 mg Fe(III)/L coagulant dose resulted in this case in more efficient particle coagulation and subsequent flocculation, as expressed in the more pronounced larger particle (floc) size fraction of the flocculated water. The EOM and IOM release are suggested to have eventually resulted in improved DAF particle removal efficiency, including that of the lower particle size ranges.

Similar to our previous experiences with coagulant aid (Chapter 4), the combined ozone, Fe(III) and Superfloc C-573 cationic polyelectrolyte treatment (Fig. 6.11) led to a denser floc structure. This phenomenon benefitted DAF efficiency, resulting in further improvement of the particle removal compared to the two previously discussed options. Particle count (HIAC) and turbidity measurements support these findings. On the other hand, the combination of ozone conditioning and 1 mg Superfloc C-573/L as sole coagulant did not yield different results from the ozonated sample prior to other treatment, suggesting that the relatively high polyelectrolyte dose resulted in comparatively denser floc formation. The combination of the 1 mg Superfloc C-573 with the 0.2 mg O_3/L resulted in comparably good DAF particle (HIAC) and turbidity removal efficiency.

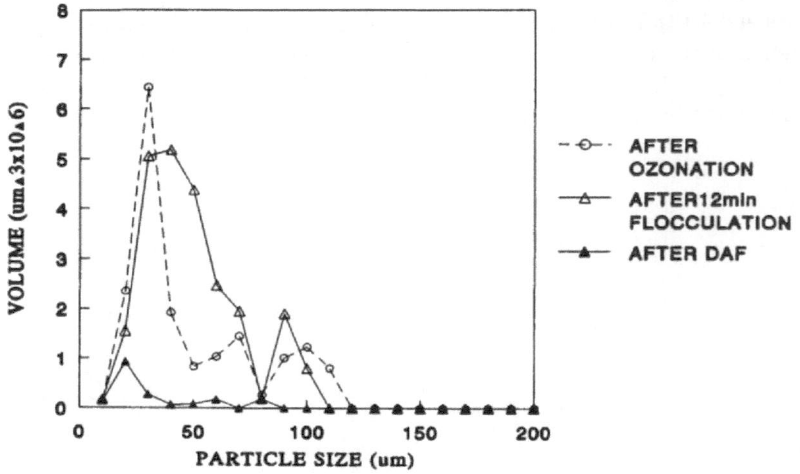

Fig. 6.10 Particle volume distribution after different treatment stages in the case of combined ozone conditioning with 0.2 mg O_3 /mg TOC and 3 mg Fe(III)/L coagulant treatment (bench-scale). Conditions: ozonation pH 7.5, coagulation pH 7.5 ±0.2, G_f=10 s^{-1}, t_f=10 min, R=7%, P=500 kPa.

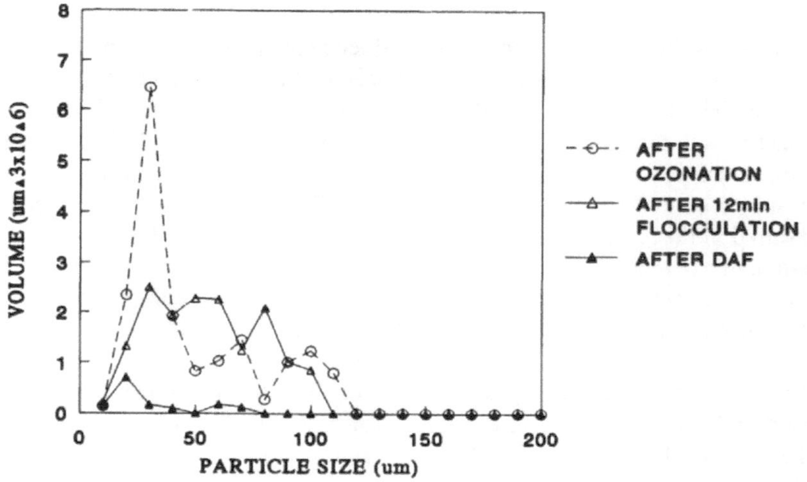

Fig. 6.11 Particle volume distribution after different treatment stages in the case of combined ozone conditioning with 0.2 mg O_3 /mg TOC + 3 mg Fe(III)/L + 0.5 mg Superfloc C-573/L coagulant aid treatment (bench-scale). Conditions: ozonation pH 7.5, coagulation pH 7.5 ±0.2, G_f=10 s^{-1}, t_f=10 min, R=7%, P=500 kPa.

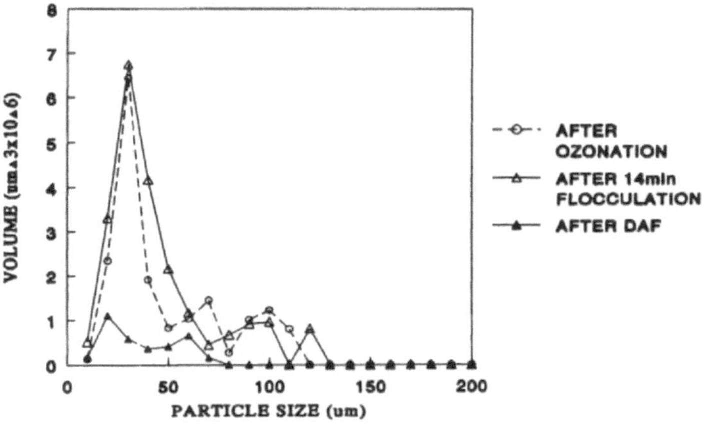

Fig. 6.12 Particle volume distribution after different treatment stages in the case of combined ozone conditioning with 0.2 mg O_3 /mg TOC + 1.0 mg Superfloc C-573/L coagulant treatment (bench-scale). Conditions: ozonation pH 7.5, coagulation pH 7.5 ±0.2, G_f=10 s^{-1}, t_f=10 min, R=7%, P=500 kPa.

6.3.6 $KMnO_4$ conditioning

As discussed in Chapter 4, permanganate conditioning also resulted in spontaneous microflocculation, initiated again by phenomena related to EOM and IOM release, even before the application of the iron coagulant. The subsequent application of 5 mg Fe(III)/L resulted in further increase of particle volume both in the larger and the lower particle size fractions (Fig. 6.13). DAF experiments with model water resulted in statistically insignificant improvement of the particle removal efficiency due to permanganate conditioning (mean particle removal of 49% compared to 40% without permanganate conditioning). The overall relatively low particle removal efficiency in the case of permanganate conditioning (Fig. 4.28) was also observed in the experiment described in Fig. 6.13, especially in the lower particle size fraction. In Section 4.3.4.a a hypothesis was offered to explain the unexpectedly poor efficiency of DAF in case of model water after permanganate conditioning, relating it to TOC/DOC concentration and composition.

Fig. 6.14. describes the particle size distributions (0.3-1μm) after the DAF and filtration in the pilot plant experiments, and for comparison purposes after the full-scale sedimentation and filtration (WRK III treatment plant in Andijk). The full-scale sedimentation (E) and filtration (F) plant was operated at a coagulant dose of 24 g Fe(III)/m^3 and 0.5 g Wisprofloc-P/m^3 cationic coagulant aid. The pilot plant results relate to DAF (A) and filtration (B) at coagulant doses of 15 g Fe(III)/m^3 and 0.5 g Superfloc C-573/m^3 . The combination of conditioning with 0.96 g $KMnO_4$/m ^{3} and coagulation with 15 g Fe(III)/m^3 + 0.5 g Superfloc C-573/m^3 is also presented for DAF (C) and filtration (D). The distributions show up to 100% more particles of the colloidal size <0.5 μm to remain after pilot plant DAF (A and C, C with $KMnO_4$ conditioning) than after full-scale sedimentation (E).

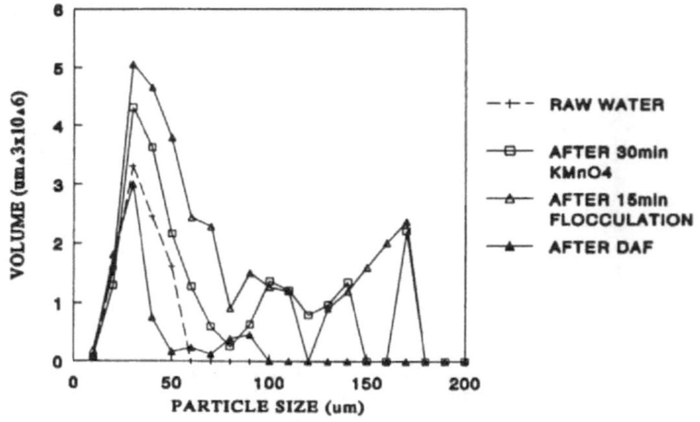

Fig. 6.13 Particle volume distribution after different treatment stages in the case of combined permanganate conditioning with 0.7 mg $KMnO_4$/L and 5 mg Fe(III)/L coagulant treatment (bench-scale). Conditions : $KMnO_4$ contact time 30 min, pH 8, G_f=70 s^{-1}, t_f=15 min, R=7%, P=500 kPa.

	A	B	C	D	E	F
TREATMENT	DAF	DAF+FILTR.	DAF	DAF+FILTR.	WRK-III (SED)	WRK-III (FILTR)
KMnO4(g/m3)	-	-	0.96	0.96	-	-
Fe(g/m3)	15	15	15	15	24	24
C-573(g/m3)	0.5	0.5	0.5	0.5	-	-
W-P (g/m3)	-	-	-	-	0.5	0.5

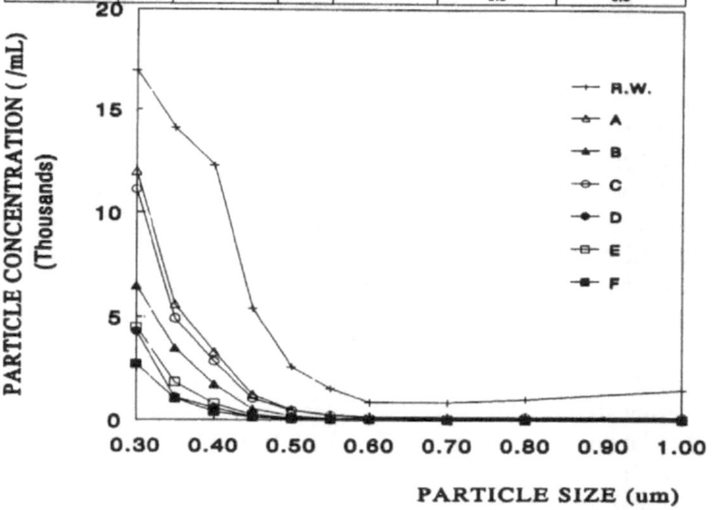

Fig. 6.14 Laser particle count in the colloidal size range (<1μm) for the DAF + filtration (pilot plant) in the case of Fe(III) coagulant and Superfloc C-573 coagulant aid treatment, and combined $KMnO_4$ conditioning, Fe(III) coagulant and Superfloc C-573 coagulant aid treatment, versus the full-scale WRK III plant with sedimentation + filtration (pilot plant MFI = 56.6 s/L^2).

At the same time, Fe (in case of A) and Fe and Mn residuals (in case of C) (data not shown) were substantially higher than after sedimentation (E) which applied very high metal coagulant doses. This suggests that the lower metal coagulant dose used for DAF fails to remove the organo-Fe complexes and MnO_2 colloids well, as does the higher coagulant dose used for sedimentation. The rising slope of the distribution curves towards the smaller size particle region suggests possible higher involvement of colloidal particles <0.3 μm (lower size instrument limitation) in the resultant turbidity. The particle distribution after filtration in the case of combined Fe + polyelectrolyte coagulation (B) suggests that the residual Fe and the likely residual polyelectrolyte contributed to the relatively high effluent MFI. The permanganate conditioning tended to increase the DAF particle removal efficiency both for particles >2.75 μm (Fig. 5.27) and for submicron particles (C versus A). The same applies for the submicron particle size range of DAF + filtration (D versus B), which approaches that of the full-scale sedimentation + filtration submicron particle removal efficiency (E). Nevertheless, the total particle count in the size range of 0.3-1 μm is still by more than 30% higher for DAF + filtration than for sedimentation + filtration. The higher DAF + filtration effluent particle counts in the size range <0.3 μm, suggest that these particles were the source of the observed higher MFI as compared to those of the sedimentation + filtration treatment.

Finally, the sedimentation + filtration efficiency relied on high metal coagulant doses which can periodically reach up to 45 g Fe(III)/m^3 combined with Wisprofloc-P as coagulant aid (up to 0.5 g Wisprofloc-P/m^3). These high coagulant doses resulted in high overall floc volume concentration, more efficient co-precipitation of DOC, and possibly more efficient colloid destabilisation, resulting in efficient removal of initial particles, as well as lower coagulant residuals and MFI values.

6.3.7 Bubble size distribution

Bubbles influence DAF kinetics through their size distribution and their (negative) surface charge characteristics. Our theoretical model analysis has shown that for a given particle size distribution, process efficiency can be influenced if the bubble size distribution can be modified [31] (or see Chapter 7).

Variation of the saturator pressure P and recirculation ratio R resulted in different bubble size distribution (Fig. 6.15). The mean bubble size was 48-61 μm. The increase of R resulted in a higher bubble concentration and a 'wider' bubble size distribution (with a higher standard deviation) with 2-3 times more bubbles of size <40 μm and 30% fewer bubbles > 60 μm. An increase of P at constant R (data presented in Chapter 7) also shifted the bubble size distribution towards smaller bubble sizes that theoretically are more efficient, and reduced the standard deviation of the population. However, the overall DAF efficiency theoretically depends on the combination of the bubble and particle (floc) size distributions (Chapter 7). Our experimental work with model water showed that the combination of these bubble size distributions with different particle size distributions resulted in insignificant difference of DAF efficiency (Chapters 3, 5 and 7). It is concluded that under our typical conditions the high bubble concentration relative to the particle concentration (>1.9 μm) is an important determinant for DAF efficiency. Consequently, combinations of low to medium R and P (e.g., R=5-7% and P=500 kPa) suffice for obtaining high DAF efficiencies (see also Chapter 7).

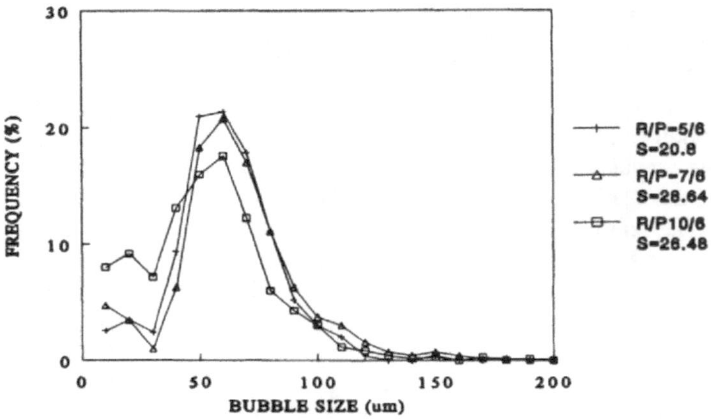

Fig. 6.15 Bubble size distributions for a saturator pressure P=600 kPa and recirculation ratios of R=5, 7 and 10 % (bench-scale DAF jar test unit with model water, T=20°C, S=standard deviation).

6.3.8 Floc size - density relationship

The floc size - density relationship has been investigated among others by Boller and Kavanaugh [32]. In their attempt to model the headloss development in a granular media filter bed, they assumed that particle deposition in filters occurs in a manner similar to the phenomena occurring during the flocculation process. Flocculation is the process of aggregation of suspended and colloidal particles in water into larger agglomerates called flocs which are of a loose structure and entrap water as well. The mass of a floc m_{fl} [M] containing N particulates of volume V_p [L^3] and density ρ_s [M/L^3] is given by :

$$m_{fl} = N \cdot V_p \cdot \rho_s + V_w \cdot \rho_w \tag{1}$$

V_w [L^3] being water volume in a floc and ρ_w [M/L^3] being the density of water. The ratio between the floc mass m_{fl} and the floc volume V_{fl} [L^3] given by $V_{fl} = V_w + N\ V_p$ expresses the suspended floc density (effective density) ρ_{fl} as :

$$\rho_{fl} = m_{fl} / V_{fl} = (N \cdot V_p \cdot \rho_s + V_w \cdot \rho_w) / (V_w + N \cdot V_p) \tag{2}$$

By rearranging equation (2), the relative density of a floc (p) can be expressed as a ratio of the product of the aggregated number of primary particles (N) and their volume (V_p), and the obtained floc volume (V_{fl}) :

$$p = (\rho_{fl} - \rho_w) / (\rho_s - \rho_w) = N \cdot V_p / V_{fl} \tag{3}$$

Boller and Kavanaugh also summarised the results of work of different authors [32] who assessed the floc size-density function and expressed it in a model equation relating the ratio between floc volume (V_{fl}) and primary particles volume (V_p) as a function of the aggregated number of particles (N) :

$$V_{fl} / V_p = a \cdot N^b \qquad (4)$$

in which a and b are empirical dimensionless equation coefficients. Thus, referring to equation (3) and (4), the relative density of a floc can be expressed as :

$$p = 1/a \cdot N^{(1-b)} \qquad (5)$$

The coefficients a and b have been theoretically derived by a number of authors whose assumptions are based on the aggregation of mono-disperse primary particles. The coefficient values differ slightly depending on the definition of floc volume and size. Assuming that the particle size distribution of the raw water sample containing the single cells *M. aeruginosa* may be treated as fairly mono-disperse, and approximating the floc volume with that of a sphere with the same silhouette surface area as the projected area of the floc as measured by computer image analysis, the model coefficients were chosen as : a=1 and b=1.305, based on work of Medalia (as discussed by Boller and Kavanaugh [32]).

Using volume distribution data obtained by image analysis of water before and after flocculation under different process conditions and eqs. (3) and (4) the flocs relative density (p) can be calculated for an *M. aeruginosa* flocculated suspension. Further use of equation (3) can yield the suspended floc density ρ_{fl}, which is significant for kinetic model calculations. The relative floc density calculated from the jar test results with model water at different coagulation (coagulant and polyelectrolyte doses) and flocculation conditions (G value) are presented in Fig. 6.16.

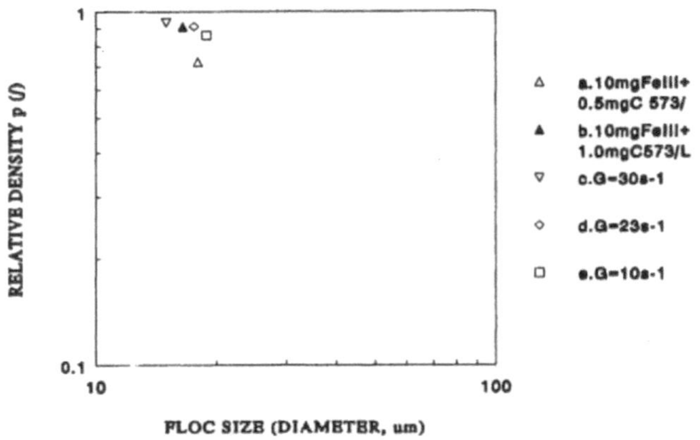

Fig. 6.16 Relative floc density p (eqs.3 and 4) versus (mean) floc size (from computer image analysis) for different coagulation and flocculation conditions (bench-scale jar test experiments with model water). Conditions : a. and b. - pH 8, G_f=30 s^{-1}, t_f=30 min, and c., d., and e. - pH 8, 3 mg Fe(III)/L, t_f=10 min.

The polyelectrolyte coagulant aid dose of 0.5 mg Superfloc C-573/L had resulted in larger flocs with lower density p=0.72, while the dose of 1.0 mg Superfloc C-573/L had produced smaller and denser flocs of p=0.91 (see also Fig. 6.7). Similarly, an increase of the flocculation G from 10 to 23 s^{-1} (and 30 s^{-1}) had resulted in a decrease of the mean floc size and the consecutive increase of relative density p from 0.86 to 0.91 and 0.93 respectively (Fig. 6.4). The results confirm an inverse dependence of relative density to floc size, as well as inverse dependence of floc size and flocculation G.

Calculations results for the jar tests with model water regarding the effect of ozone conditioning and different treatment combinations are presented in Fig. 6.17. As observed in this figure (compare data for raw water and ozonated water in Figs. 6.9 and 6.10) ozone induced microflocculation produced flocs of lower density (p=0.81) and larger size. This effect of ozone conditioning on the reduction of flocs relative density was even more pronounced than the one caused by the 3 mg Fe(III)/L coagulant dose (p=0.91). Higher reduction of floc density and increase of floc size was caused by the combination of ozone conditioning and iron coagulant (p=0.74, see also Fig. 6.10), while the inclusion of Superfloc C-573 in the ozone + Fe treatment scheme resulted in the opposite trend of producing denser and smaller flocs (p=0.78, see also Fig. 6.11). The last treatment scheme led to the highest DAF particle removal efficiency. Finally, the ozone + Superfloc C-573 coagulant combination resulted in floc density and size characteristics similar to the iron only case (p=0.88), however accompanied by comparably higher DAF efficiency (see also Section 4.3.3.c). This suggests that although the particle size-density ratio plays an important role in determining the DAF process efficiency, efficient particle destabilisation and surface charge related effects play a more determining role.

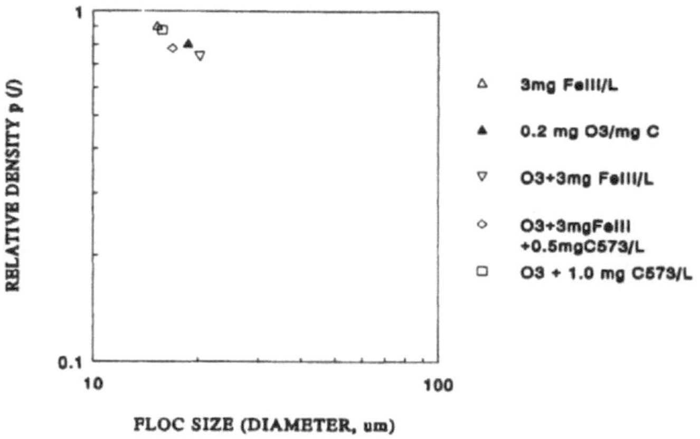

Fig. 6.17 Relative floc density p (eqs. 3 and 4) for different treatment combinations including ozone conditioning (0.2 mg O_3/mg TOC), and a reference situation with only Fe(III) treatment (bench-scale). Conditions : ozonation pH 7.5, coagulation pH 7.5±0.2, G_f=10 s^{-1}, t_f=10 min.

Calculation results for jar tests with model water regarding the effect of $KMnO_4$ conditioning are presented in Fig. 6.18. Similar to the case of ozone conditioning, the permanganate induced microflocculation resulted in larger and lower-density flocs (p=0.89 after 1 min contact time, before coagulant addition, see also Fig. 6.13), while the effect of longer contact time was not so pronounced (p=0.85 for 30 min contact time before coagulant addition). The iron coagulant further reduced the floc density and increased the floc size, as observed after 10 minutes flocculation (p=0.74).

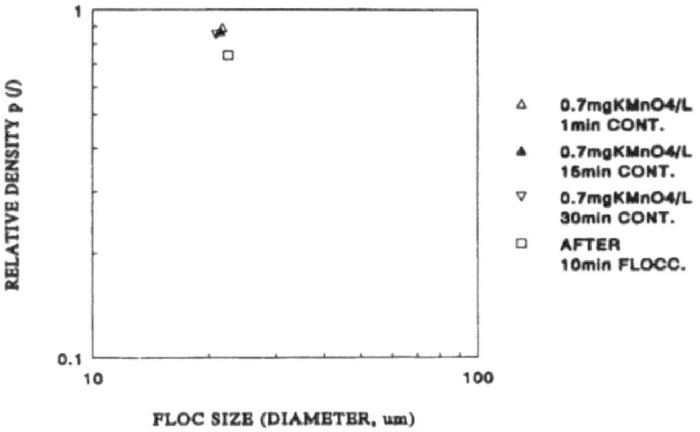

Fig. 6.18 Relative floc density p (eqs. 3 and 4) in case of $KMnO_4$ conditioning (bench-scale), after different contact time. Conditions : pH 8, 5 mg Fe(III)/L, G_f=70 s^{-1}, t_f=15 min.

6.3.9 Considerations for modelling

Sedimentation is generally regarded as a process which is favoured by low flocculation G resulting in larger flocs with supposedly good settling characteristics [35]. Our bench-scale analysis showed that better results were achieved at higher coagulant dose than applied in DAF (2-3 times higher) and resultant increase of the initial particle concentration after flocculation (Chapter 3). A flocculation G of 30 s^{-1} resulted in smaller and denser flocs than at G=10 s^{-1}, with better settling characteristics [35]. Assuming constant water temperature and viscosity μ $[ML^{-1}T^{-1}]$ and laminar flow conditions (Re<1), the sedimentation efficiency is proportional to settling velocity v_s $[LT^{-1}]$ which is defined by the floc size d_{fl} [L] and density ρ_{fl} $[ML^{-3}]$:

$$v_s = g\, d_{fl}^2\, (\rho_{fl} - \rho_w) / 18\, \mu \tag{3}$$

Attempts of modelling of DAF are primarily focussed on the reaction zone (region where the saturated recycle flow is introduced) or the separation zone. DAF modelling is discussed in more detail in Chapter 7. According to the single collector collision efficiency DAF model [10, 36], the efficiency in the reaction zone is expressed as dN_{fl}/dt (reduction of number of primary particles - flocs with time), and depends on floc and bubble sizes (d_{fl} [L] and d_b [L]), their concentration (N_{fl} $[L^{-3}]$ and Φ_b $[L^3/L^3]$), the single collector collision efficiency coefficient η_T, and the particle-bubble attachment efficiency α_{pb} :

$$dN_{fl} / dt = - (3/2)\, (\alpha_{pb}\, \eta_T)\, (\Phi_b\, v_b\, N_{fl}) / d_b \tag{4}$$

Particle (floc)-bubble interception is considered the most relevant mechanism in DAF; it depends on the floc and bubble sizes (d_{fl} and d_b) and is incorporated in the η_T. This term also covers particle/floc-bubble collision by Brownian diffusion, settling and drag. Efficient particle removal by DAF generally requires smaller flocs than sedimentation, which should also be of preferably lower density [35]. Particles should be larger than 1 μm, preferably of few tens of micrometers size, in combination with bubbles in the size range of 10-100 μm and a concentration that is one to two orders of magnitude higher [10, 36]. Our experiments showed that α_{pb} is significantly affected by surface charge, which also affects DAF via the floc concentration (N_{fl}) and its size distribution (d_{fl} and η_T), as in the case of coagulation pH 6 versus pH 7, 8 and 9 given in Figs. 6.3.a through 6.3.d, or when cationic polyelectrolyte was used as coagulant aid versus Fe coagulant treatment given in Figs. 6.6 and 6.7.

Assuming laminar flow conditions, the efficiency of the separation zone as expressed by the floc-bubble agglomerate rising velocity (v_{flb}) is defined by the floc-bubble agglomerate size (d_{flb}) and density (ρ_{flb}) :

$$v_{flb} = g\, d_{flb}^2\, (\rho_w - \rho_{flb}) / 18\, \mu \qquad (5)$$

$$v_{flb} > v_{os} / (1-m) \qquad (6)$$

d_{flb} and ρ_{flb} depend on the floc size - density ratio and concentrations, and the size and number of bubbles comprising the floc-bubble agglomerate. Equation (6) describes a prerequisite for efficient DAF, with v_{os} the DAF overflow rate and m the fraction of DAF tank dead space [35].

In the case of DAF experiments with model water, similar particle removal efficiency was obtained both at the low G value of 10 s^{-1} (mean particle removal of 61.4%) and that of G=70 s^{-1} (mean particle removal of 61.7%) although the resultant floc size - density ratios were different. Increase of the flocculation time further increase particle removal efficiency in case of the G value of 10 s^{-1} (mean particle removal of 67.2%), while it decreased the removal efficiency for the G value of 70 s^{-1} (mean particle removal of 54.0%) The equations for the reaction and separation zone suggest that DAF efficiency should be comparable in both cases (considering the measured mean floc size of d_{fl}=15-20 μm, and the mean bubble size of d_b=40 μm). The v_{flb} in both cases was approximately 3 m/h, implying efficient floc removal in the available 4 minutes of flotation during the DAF jar tests. The larger and weaker flocs are susceptible to the shear effect of the recycled flow. Generally low flocculation G and resulting weak flocs are considered a drawback for efficient DAF, but this may be compensated by the production of flocs that are larger (thus causing higher α_{pb} due to interception) and less denser (and thus easier to be lifted to the surface).

Although the relative density equation and the adopted coefficient values apply to monodisperse suspensions such as was the case with our model water, similar calculations were carried out for the heterodisperse reservoir water in the pilot plant experiments. At a coagulant dose of 10 g Fe(III)/m^3 and 0.3 g Superfloc C-573/m^3 best turbidity removal after DAF was obtained at G=50 s^{-1} (89.9%), typically 1-2% higher than at G=10 and 30 s^{-1} (Fig. 5.18). Image analysis data (Fig. 6.5) showed more efficient flocculation for this G value, expressed as highest reduction of the concentration of primary particles (thus lowest number of particles after flocculation N_{fl}) and formation of larger size flocs with lowest density p=0.72. In the case of tapered flocculation (Fig.

6.6) the second flocculation stage increased the volume of the larger particle size range while reducing the volume of smaller particles, and resulted in a decrease of relative floc density (from p=0.71 to 0.61). It is suggested that the relatively high coagulant doses (>10 g Fe(III)/m^3) in combination with cationic polyelectrolyte coagulant aid resulted in a high α_{pb}. In general, process efficiency tended to increase with increasing floc size (based on image analysis data mean floc size rose up to d_{fl}=40 μm) and decreasing relative floc density (as low as p=0.61); this is in accordance with the DAF kinetic model.

Referring to eqs. (3) and (4) it is suggested that for a constant raw water quality (thus constant ρ_w and μ, as well as colloidal and particulate matter concentration and composition) and varied coagulation/flocculation conditions (coagulant dose, pH, flocculation G and flocculation time), it may be possible to define the settling efficiency expressed as v_s (3), or the floc-bubble agglomerate rising velocity v_{flb} (4) in the form of a surface area (v_s or v_{flb} on z-axis) in function of the resulting variations of floc or floc-bubble agglomerate size (d_{fl} or d_{flb} on x axis) and floc or floc-bubble agglomerate density (ρ_{fl} or ρ_{flb} on y axis). This may possibly be achieved by the application of the statistical response surface methodology [33, 34]. This technology aims at overcoming the weaknesses of the classical one-variable-at-a-time strategy, such as e.g. assessing sedimentation or flotation efficiency (expressed as v_s or v_{flb}) as a function of floc and floc-bubble agglomerate size (d_{fl} or d_{flb}) or density (ρ_{fl} or ρ_{flb}) separately. It consists of a group of techniques which can be used in empirical studies of relationships between one or more responses, such as in our case sedimentation or flotation efficiency (expressed as v_s or v_{flb}), and a number of simultaneous input variables, such as in our case floc or floc-bubble agglomerate size (d_{fl} or d_{flb}) and density (ρ_{fl} or ρ_{flb}). This prospect seems achievable with the latest developments in computer image analysis technology.

6.4 CONCLUSIONS

Particle counting techniques (e.g., HIAC or similar) are being recognised as a parameter of filter effluent quality. This technique can be applied to a limited degree to sedimentation and DAF effluents, since the residual coagulant may significantly influence the count. An additional draw-back may be the possible floc break-up due to high shear forces during the passage of the sample through the recording chamber, which discredits it for characterisation of flocculated water. The computer image analysis technique, on the other hand, allows more qualitative and quantitative characterisation of the coagulation/flocculation process (with a size limitation of 1.9 μm). Characterisation of particles in the size range <1 μm will substantially improve the qualitative and quantitative assessment of the coagulation/flocculation process.

To achieve efficient sedimentation it appeared necessary to increase the frequency of the larger particle size fraction (>50 μm), and to increase the overall particles concentration. This is generally achieved by higher coagulant doses than required for DAF. For efficient DAF on the other hand, increase of the larger particle size (>50 μm) on account of the disappearance of the smaller size fraction (<50 μm) was necessary. This was more pronounced at coagulation pH<IEP (pH 6, compared to 7, 8 and 9), while at the same time, the smaller particle size fraction removal was improved by the achieved particle charge neutralisation. The latter phenomenon is related to the reduction of the particle - bubble repulsion (related to the negative charge of air bubbles) and resulting increase of the α_{pb}.

In the model water experiments low flocculation G= 10 s^{-1} contributed significantly to the larger particle size fraction which resulted in more efficient DAF, than at higher G values. On the other hand, low to medium flocculation energy input of G=10-30 s^{-1} resulted in more efficient sedimentation, the latter value resulting in a denser floc structure and higher sedimentation efficiency (relative floc density of p=0.93 versus 0.86).

Particle charge neutralisation achieved by cationic coagulant aid application raised the α_{pb}, as well as the sedimentation efficiency; a 0.5 mg /L dose resulted in floc material favoured by DAF, and a 1.0 mg/L dose favoured by sedimentation. The cationic polyelectrolyte shifted the particle (floc) size distribution towards the higher size range; a higher polyelectrolyte dose resulted in denser floc structures (relative floc density of p=0.91 versus 0.72, for 1.0 and 0.5 mg Superfloc C-573/L).

Ozone and $KMnO_4$ induced flocculation improvements that were mostly related to EOM and IOM release, resulting in spontaneous microflocculation and visibly affecting the particle size-density relationship before the addition of coagulant. A 0.2 mg O_3/mg TOC without the application of coagulant resulted in relative floc density p of 0.81, while the 0.7 mg $KMnO_4$ dose resulted in a relative floc density of 0.89 after 1 min contact time, further decreasing to 0.85 after 30 min contact time. Different treatment combinations using ozone, iron and polyelectrolyte coagulant were tested. The production of flocs in the larger particle size range, created by ozone conditioning and iron coagulant (0.2 mg O_3/mg TOC + 3 mg Fe(III)/L), resulted in relative floc density of p=0.74 and in improved DAF efficiency, as compared to the case of 3 mg Fe(III)/L and the resultant p=0.91. Although inclusion of the Superfloc C-573 coagulant aid resulted in an opposite trend in terms of the relative floc density (p=0.78), the DAF process efficiency still increased suggesting that particle (algae) destabilisation and surface charge related effects determine it. The application of the polyelectrolyte Superfloc C-573 as a sole coagulant for ozone conditioned model water experiments demanded larger doses to achieve efficient particle destabilisation, which resulted in dense floc material (p=0.88) and relatively good DAF efficiency, mainly related to the efficient particle surface charge effects.

Bubble size distribution was affected by the recirculation flow conditions. Higher saturation pressure resulted in a slight decrease of the mean bubble size to theoretically more efficient sizes (from approximately 60 to approximately 50 μm), as well as in a narrower distribution. Higher recirculation ratios resulted in an increase of the bubbles concentration and of the distributions width. The overall high bubble concentration, and the generally high bubble concentration to particle concentration ratio is suggested to be an important factor determining DAF efficiency.

Reservoir water pilot plant experiments showed a comparable effect on the particle volume distribution in the larger size range for all tested G values (G=10, 30 and 50 s^{-1}), and similar particle removal efficiency. However, an increase of the G created better contact opportunities for the colloidal and particulate matter, reducing the number of smaller particles (<50 μm) and turbidity more pronouncedly (p=0.72). A lower velocity gradient in a second flocculation stage (G reduced from 50 to 30 s^{-1}) increased the volume of larger size particles at the expense of smaller ones and reduced the relative floc density from p=0.71 to 0.61.

The kinetics and particle removal efficiency for DAF and sedimentation were found to be defined by the particle agglomeration (coagulation/flocculation). The increase of the larger particle (floc) size fraction resulted in a relative floc density decrease and a consecutive increase in DAF efficiency. While for sedimentation the floc size - density relation directly affects the efficiency, bubble size and bubble concentration are an additional factor of influence for DAF. Particle and bubble characterisation can further allow calculation of the floc size - density ratio (p) and the absolute floc density (ρ_{fl}), important parameters in DAF and sedimentation model equations.

REFERENCES

1. Bernhardt H. and Clasen J., 1991. Flocculation of micro-organisms. *J. Water SRT-Aqua*, **40**, 2, pp. 76-87.
2. Ives K.J., 1950. The significance of surface electric charge on algae in water purification. *J. Biochem. Microbiol. Technol. Engineer.* **1**, pp. 37-47.
3. Edzwald J.K., 1993. Algae, bubbles, coagulants and dissolved air flotation. *Wat. Sci. Tech.*, **27**, 10, 67-81.
4. Petruševski B., Vlaški A., van Breemen A. N. and Alaerts, G.J., 1993. Influence of algal species and cultivation conditions on algal removal in direct filtration. *Wat. Sci. Tech.*, **27**, pp. 211-220.
5. Vlaški A., van Breemen A.N. and Alaerts G.J., 1996. The algae problem in the Netherlands from a water treatment perspective. *J. Water SRT-Aqua*, **45**, **4**, **pp. 184-194.**
6. Spicer P.T. and Pratsinis S.E., 1996. Shear-induced flocculation : the evolution of floc structure and the shape of the size distribution at steady state. *Wat Res.*, **30**, 5, pp. 1049-1056.
7. Hanson A.T. and Cleasby J.L., 1990. The effect of temperature on turbulent flocculation: fluid dynamics and chemistry. *JAWWA*, **82**, pp. 56-73.
8. Valade M.T., Edzwald J.K., Tobiason J.E., Dahlquist J., Hedberg T. and Amato T. 1996. Particle removal by flotation and filtration : pretreatment effects. *JAWWA*, **88**, pp. 35-47.
9 Edzwald J.K., 1994. Principles and application of dissolved air flotation. Opening address of the *Joint IAWQ-IWSA Specialised Conference on Flotation Processes in Water and Sludge Treatment*, Orlando, USA.
10. Malley J.P. and Edzwald J.K., 1991. Concepts for dissolved-air flotation treatment of drinking waters - *J. Water SRT-Aqua*, **40**, 1, pp. 7-17
11. Fukushi K., Tambo N. And Matsui Y., 1994. A kinetic model for dissolved-air flotation in water and wastewater treatment. Proc. *Joint IAWQ-IWSA Specialised Conference on Flotation Processes in Water and Sludge Treatment*, Orlando, USA
12. Janssens J.G., 1990. The application of dissolved air flotation in drinking water production, in particular for removing algae. *DVGW Wasserfachlichen Aussprachetagung* Essen, Germany.
13. Van Puffelen J., 1993. Flotation - State of the art, *Cursus moderne drinkwaterzuiveringstechnieken (Course Modern Water Treatment Technologies)*, Technical University-Delft, Delft, the Netherlands.
14. Petruševski B., van Breemen A. N. and Alaerts G.J., 1994. Optimisation of coagulation conditions for in-line direct filtration. Proc. *Workshop on Optimal Dosing of Coagulants and Flocculants*, Műlheim an der Ruhr, Germany.
15. Edzwald J.K. and Wingler B.J., 1990. Chemical and physical aspects of DAF for the removal of algae. *J. Water SRT-Aqua*, **39**, pp. 24-35.

16. Jun Ma and Guibai Li, 1992. Laboratory and full scale plant studies of potassium permanganate oxidation as an aid in coagulation. Proc. *IAWQ/IWSA Joint Specialised Conference : Control of Organic Material by Coagulation and Floc Separation Processes*, Geneva, Switzerland.
17. Reckhow D.A., Singer P.C. and Trussel R.R., 1986. Ozone as a coagulant aid - Proc. *AWWA National Conference : Recent Advances and Research Needs*, USA.
18. Petruševski B., van Breemen A. N. and Alaerts G.J., 1993. Pretreatment in relation to direct filtration of impounded surface waters. Proc. *European Water Filtration Congress, KVIV*, Oostende, Belgium.
19. Petruševski B., van Breemen A.N. and Alaerts G.J., 1995. Effect of permanganate pre-treatment and coagulation with dual coagulants on particle and algae removal in direct filtration. Proc. *IAWQ-IWSA Workshop on Removal of Microorganisms from Water and Wastewater*, Amsterdam, the Netherlands.
20. Langlais B., Reckhow D.A. and Brink D.R., editors, 1991. *Ozone in Water Treatment, Applications and Engineering*. AWWA and Campagnie Generale des Eaux, Lewis Publishers.
21. Petruševski, B., 1996. *Algae and particle removal in direct filtration of Biesbosch water*. Ph.D. thesis TU and IHE - Delft, published by A.A. Balkema, the Netherlands
22. Edwards M., Benjamin M.M. and Tobiason J.E., 1994. Effect of ozonation on NOM using polymer alone and polymer/metal salt mixtures. *JAWWA*, **86**, pp. 105-116.
23. Rijk de, S.E., van der Graaf, J.H.J.M. and den Blanken, J.G. (1994). Bubble size in flotation thickening. *Wat. Res.*, **28**, 2, pp. 465-473.
24. Bhargava D.S. and Rajagopal K., 1992. An integrated expression for settling velocity of particles in water. *Wat. Res.*, **26**, 7, pp. 1005-1008.
25. Patry G.G. and Takacs I., 1992. Settling of flocculent suspensions in secondary clarifiers. *Wat. Res.*, **26**, 4, pp. 473-479.
26. Vlaški A., van Breemen A.N. and Alaerts G.J., 1995. Optimisation of Coagulation Conditions for the Removal of Cyanobacteria by Dissolved Air Flotation or Sedimentation, *J.Water SRT-Aqua*, **45**, 5, pp. 253-261.
27. Alaerts G.J. and Van Haute A., 1981. Stability of colloid types and optimal dosing in water flocculation. *Physicochemical Methods for Water and Wastewater Treatment*, Pawlowski, L. (eds.), Elsevier Publ., New York/Amsterdam.
28. Bernhardt H. and Clasen J., 1992. Studies on the removal of planktonic algae by flocculation and filtration. Proc. *8th ASPAC-IWSA Regional Water Supply Conference*, Kuala Lumpur, Malaysia, October 26-30.
29. Malley J.P., 1994. The use of selective and direct DAF for removal of particulate contaminants in drinking water treatment. Conference paper, *IAWQ-IWSA Joint Specialised Conference on Flotation Processes in Water and Sludge Treatment*, Orlando, USA.
30. Edzwald J.K. and Paralkar A., 1992. *Algae, coagulation and ozonation - chemical water and wastewater treatment II*. Klute, R. and Hahn, H. (eds.), Springer-Verlag.
31. Vlaški A., van Breemen A.N. and Alaerts G.J., 1996. Evaluation and verification of the single collector collision efficiency dissolved air flotation (DAF) kinetic model. In prep., IHE-Delft, the Netherlands.
32. Boller M. A. and Kavanaugh M. C., 1995. Particle characteristics and headloss increase in granular media filtration. *Wat. Res.*, **29**, 4, pp. 1139-1149.

33. Box G.E.P., 1954. The exploration and exploitation of response surfaces : some general considerations and examples. *Biometrics*, **10**, pp. 16.

34. Hill W.J. and Hunter W.G., 1966. A review of response surface methodology : A literature survey. *Technometrics*, **8**, pp. 571.

35. Gregory J., 1993. The role of colloid interactions in solid-liquid separation. *Wat. Sci. Tech.* **27**, 10, pp. 1-17.

36. Schers G.J. and van Dijk J.C., 1992. Flotatie, de theorie en de praktijk (Flotation, the theory and the practice). *H_2O*, **25**, 11, pp. 282-290.

Chapter 7

DISSOLVED AIR FLOTATION (DAF) KINETIC MODELLING - A TOOL FOR IMPROVED PROCESS DESIGN AND OPERATION

ABSTRACT : This study discusses and assesses the kinetic models of dissolved air flotation (DAF), in particular the single collector collision efficiency model. Experimental data were fitted into the model equations and results discussed. Bubble and particle size distributions obtained by computer image analysis were used for determination of the process parameters: mean particle and bubble size of raw water, after coagulation/flocculation and after DAF, as well as the respective particle (and bubble) concentration and density. The particle-bubble collision efficiency coefficient (α_{pb}) was used to describe process efficiency. In bench-scale experiments with model water with algae as reference particles, previously assumed α_{pb} values of 0.5 were accomplished under optimal process conditions. This value, however, was achieved easier in DAF pilot plant experiments with reservoir water. The model appeared to be more sensitive to particle size than bubble size, while analysis showed that it is unnecessary to apply high saturator pressures (>500 kPa) and recirculation ratios (>7%), suggesting cost savings. Coagulation/flocculation was found to be the most critical factor for DAF overall efficie.ncy, particularly in terms of its effects on the particle (floc) size distribution and surface charge characteristics. Coagulation depends on water characteristics such as pH, temperature, and the nature and concentration of organic matter. Hydrodynamic conditions (flow pattern, flocculation G and t) determine the flocculation. Both chemical and physical factors determine particle-bubble attachment in the DAF reactor.

NOMENCLATURE

i. Reay and Ratcliff dispersed air flotation model (1973):

E	=collection efficiency (-)
E_1	=collision efficiency (-)
E_2	=attachment efficiency (-)

ii. Single collector collision efficiency DAF model (Edzwald and Malley, 1990):

R	=recycled flow (%)
α_{pb}	=particle-bubble attachment efficiency (-)
η_T	=total single collector (bubble) efficiency (-)
η_D	=single collector efficiency for Brownian diffusion (-)
η_I	=single collector efficiency for interception (-)
η_s	=single collector efficiency for sedimentation (-)
η_{in}	=single collector efficiency for inertia (-)
k	=Boltzmann's constant ($1.38 * 10^{-23}$ J/°K)
Re	=Reynolds number (-)
T	=absolute temperature (°K)
T°	=temperature (°C)
d_p	=particle diameter (m)
d_b	=bubble diameter (m)
ρ_p	=particle density (kg/m^3)
ρ_w	=water density (kg/m^3)

μ =viscosity (Ns/m^2)
ν =kinematic viscosity (m^2/s)
g =gravitational constant of acceleration (9.81 m/s^2)
N_p =concentration (number) of particles during DAF treatment (/m^3)
t =residence time in the contact zone (s)
N_p' =particle number concentration in the influent (/m^3)
N_p'' =particle number concentration in the effluent (/m^3)
ϕ_b =bubble volume concentration (m^3/m^3)
v_b =bubble rise velocity (relative velocity of bubble with respect to fluid) (m/s)
v_{pb} =particle-bubble agglomerate rising velocity (m/s)
v_{os} =overflow rate in the DAF separation zone (m/s)
m =fraction of dead space in the DAF separation zone (-)

iii. Population balance DAF model (Fukushi et al., 1994):

$_o$ =denotes original or initial value
μ =water viscosity (10^{-4}·Ns/cm^2)
α =particle-bubble collision-attachment factor (-) (α_o=0.3-0.4)
β =$(1/15)^{1/2}$ constant (-)
ε =mean energy dissipation rate (W/cm^3)
t =elapsed time of mixing (within the DAF reaction zone) (s)
d_b =bubble size (cm)
d_f =floc size (cm)
d_{fb} =floc-bubble agglomerate size (cm)
n_b =bubble concentration at a given moment (/cm^3)
n_f =floc concentration (/cm^3)
F =dimensionless floc size (-)
N_b =dimensionless free bubble concentration ($N_b=n_b/n_{bo}$) (-)
i_f =average number of bubbles on an F-size floc (-)
m_f =maximum number of F-size flocs with i bubbles (-)
θ =universal mixing time taking into account the decrease of free bubbles as attachment progresses
T =normalised mixing time
ρ_a =air density (g/cm^3)
k =constant of drag force (k=16 for bubbles and 45 for flocs) (-)
a =constant of the floc density function (g/cm^3)
k_p =constant of the floc density function (-)

7.1 INTRODUCTION

7.1.1 Scope of the study

Dissolved air flotation (DAF) is currently considered as one of the best available drinking water treatment technologies to remove low-density particles like algae [1, 2, 3, 4]. Initially used in the paper industry for recovery of fibres in the 1920's, DAF entered potable water treatment in the mid 1960's. The design and the construction of many DAF plants in the 1970's was based on these experiences, particularly in Scandinavia and South Africa [5]. Attempts were made to analyse the process parameters and model the process in order to provide better insight into the DAF process and optimise it [1, 2, 6, 7, 8, 9, 10, 11]. In this context, knowledge and experience related to dispersed air flotation (bubbles being generated by electrolysis, or by forcing air through a porous plate or through spargers) modelling has often been considered [12, 13]. In particular the single collector collision efficiency DAF kinetic model proposed by Edzwald and Wingler [1] and Malley and Edzwald [6] provided the framework for the research in this study.

This chapter aims at discussing and assessing the basic concepts of existing DAF kinetic models, and in particular addresses the single collector collision efficiency model. The role of pre-treatment and flotation tank parameters is assessed based on experiments and theoretical work. As the agglomeration (coagulation/flocculation) phase is the critical step to achieve down-stream treatment efficiency, irrespective of the solid-liquid separation process applied (DAF, sedimentation or filtration) [14, 15, 16, 17, 18] (Chapters 3, 4, 5), this study focuses on the chemical and physical aspects of the coagulation and flocculation. These data were also used as source data incorporated into model equations and calculations. Finally, some practical implications for full scale DAF plant installations are given and discussed. Particle agglomeration and removal performance was evaluated after each step using particle and bubble size distribution.

7.1.2 Physical principles and aspects of DAF

DAF is based on the formation of air bubbles which collect colloidal and suspended impurities during rise to the water surface, from where they are periodically skimmed off. In a DAF unit typically 4-10% of treated water flow is recycled, pressurised at 400-800 kPa and saturated with air in a so-called saturation vessel (in fact only 70-90 % of equilibrium saturation is achieved [6, 8, 19]). The degree of saturation is defined by the Boyle's gas law and thus depends on the pressure in the saturator unit, the water temperature, and the concentrations of nitrogen and oxygen in the air and in the water. A high salts or impurity concentration may decrease the saturation concentration. The degree of saturation depends on the size of the air/water contact area, the saturation coefficient (the difference between the equilibrium and the actual gas concentration in the water) and the gas transfer coefficient (which depends on the gas diffusion coefficient, which itself depends on temperature, viscosity and water composition) [19]. Thus, saturation of water by air may be a slow and incomplete process.

The flotation tank consists of two zones (Fig. 7.1): a reaction or contact zone and a separation or clarification zone. The recycled saturated flow is subjected to a pressure drop when introduced into the DAF tank in the reaction zone via needle valves or injection nozzles, merging with the main stream to be treated. Upon release of the saturated recycle flow, micro-bubbles are formed in the 10-100 μm (typically 40 μm) size range. Theories about the formation of the micro-bubbles include (1) the spontaneous precipitation of nitrogen and oxygen molecules due to their natural affinity, (2) nucleation in which gas molecules diffuse towards gas adsorbed on impurities or/and on walls, and most importantly (3) cavitation. In the cavitation theory due to the pressure drop in the DAF unit, explosion of an air bubble is followed by implosion and breaking-up of the bubble in tens of small, stable micro-bubbles [20]. Research [21] has shown that only 60-90% of the bubbles can be characterised as micro-bubbles, i.e. smaller than the arbitrary limit of 150 μm, due to bubble coalescence. The coalescence is suggested to occur primarily in the air release zone close to the nozzles, and to depend on turbulence and bubble size. The micro-bubbles fraction percentage varies and depends on the configuration of the conduit between the saturator vessel and the nozzles, the nozzle type and orifice size, nozzle configuration (existence of and distance to an impinging surface), etc. As a consequence, large bubbles and bubble agglomerates are formed which have a larger rising velocity and may disturb the flotation process. Furthermore, large bubbles are less efficient in flotation because they have a comparatively small surface area, and deprive the suspension of a substantial amount of air. On the other hand, oxygen deficiency in the treated water may be an additional drawback resulting in the redissolution of the oxygen from the air bubbles in the water and reducing the precipitated air available for bubble formation [21]. The air precipitation efficiency (amount of dissolved air ending-up in bubbles) within the DAF tank varies from 67-94% and is affected by the saturator's pressure and efficiency, the diameter of the nozzle orifice and the presence of an impinging surface near the recycled stream re-introduction location [22].

The formed 'cloud' of air bubbles, or often referred to as 'white water', rises to the water surface whilst colliding with colloidal and suspended impurities and collecting them. Efficient particle-bubble attachment requires destabilised floc material [1, 2, 6]. The bubble volume concentration is typically 5 - 10 L/m^3 while the corresponding number concentration is 1.5 - 3 * 10^{11} /m^3 [7]. A raw water particles concentration of 10 g/m^3 and average size of 100 μm results in a particle number concentration of 1.9 * 10^7 /m^3. The much larger bubble concentration compared to the average particle concentration is suggested to be an essential factor to determine DAF performance.

7.1.3 Particle-bubble interactions

The attachment of particles to bubbles occurs in the reaction zone (Fig. 7.1) and was conceptualised by Kitchener and Gochin [8]. They distinguished three bubble-floc aggregation mechanisms : (1) entrapment of bubbles within a condensing network of flocs, (2) bubble growth from nuclei within flocs, and (3) adhesion of flocs upon collision with bubbles. They recognised the last one as the principal DAF mechanism occurring under normal drinking water treatment circumstances; mechanism (1) typically occurs under high-rate coagulation conditions but not in very concentrated suspensions and (2) when the saturated recycle water flow is injected above settled flocs. Based on mechanism (2) they proposed a modification of the DAF process : a direct flotation process in which the flocculation process immediately follows the addition of the saturated recycle flow and coagulant, allowing for nucleation and growth of air bubbles within the growing flocs. They also pointed out the importance of organic matter impurities in natural waters (generally > 1 mg C/L), which act as air bubble collectors, resulting in inadequate particle-bubble adhesion in case of very clean water.

The purpose of the separation zone which is preferably divided from the reaction zone by an inclined baffle (Fig. 7.1), is to provide quiescent conditions for the rise of the bubble-particle agglomerates to the surface. The driving force for flotation is the ratio $(\rho_w - \rho_{pb})/\rho_w$, the particle-bubble agglomerate density depending on the number of attached bubbles per particle (floc). The higher this number, the higher is the agglomerate's rising velocity; up to one single layer of air bubbles around a particle is suggested to be physically possible [7]. This suggests that in order to achieve higher rising velocities of agglomerates (and higher process loading), the production of larger flocs with a larger surface area and the possibility for attachment of more bubbles is advisable.

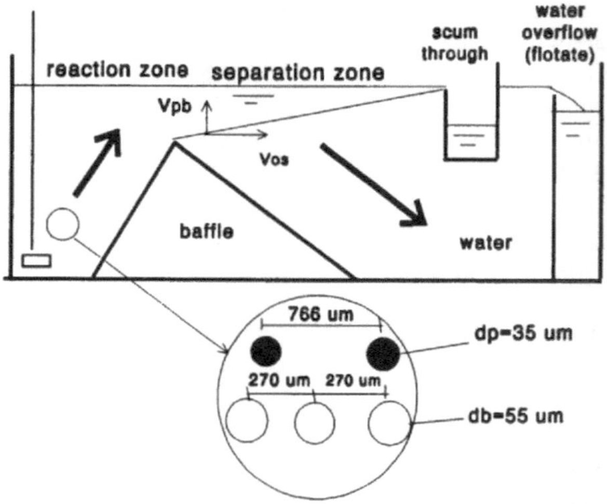

Fig. 7.1 Flow scheme of a DAF unit. Insert : typical sizes of bubble and particle; mean sizes and distances measured/calculated by computer image analysis for a flocculation suspension with 10 g Fe(III)/m^3, 0.3 g Superfloc C-573/m^3, $G_f=10\ s^{-1}$, $t_f=20$ min, R=7%, P=500 kPa.

Reay and Ratcliff [12] modelled dispersed air flotation and recognised that the particle-bubble contacts occur either by collision or by Brownian diffusion, depending on the particle size. Their hydrodynamic analysis showed that in the collision regime (bubble-floc aggregation mechanism (3), as defined by Kitchener and Gochin [8]) there are great kinetic advantages in reducing the bubble size below 100 μm, which is typical for dissolved air flotation [1, 6]. They further suggested that in a concentrated suspension the collection of particles on air bubbles occurs in the wake (down-stream) of the rising bubble, until this side is eventually completely covered; this phenomenon occurs due to the concentration of surface active ions in a spherical cap at the back side of the bubble (relative to its motion). These ions cause an electrical surface charge which according to Tambo et al. [23] is highly negative (150 mV) at neutral pH; the IEP (isoelectrical point) is reported to be pH 4.5 [24]. High-speed cine-camera analysis [9] and high-speed video-endoscopy [25] proved that flocs are attached as a tail to the rising air bubbles, erratically oscillating over its down-stream part of the surface as water streams past. This characterizes the bubble as a carrier of one or more particles (depending on the particle size and density), rather than viewing the particle as a site which is completely covered by bubbles. Furthermore, bubble collision does not necessarily result in bubble growth due to charge related high repulsion, except in the vicinity of the nozzles where high turbulence may lead to coalescence. Bubble-particle attachment occurs typically at positively charged sites on the particle (floc) surface. The use of surface active agents, although not allowed in drinking water treatment, may also improve the efficiency of DAF by making the flocculated material more hydrophobic and amenable for easier particle-bubble attachment and flotation [8].

7.1.4 DAF kinetics modelling

a. Dispersed air flotation modelling

The incorporation of the influence of water chemistry into kinetic models is complex. Modelling attempts have been concentrated on the hydrodynamics and the physical aspects of the process. Flint and Howarth [13] concluded that particle-bubble collision in dispersed air flotation depends on the balance of viscous, inertial and gravitational forces acting on the particle and the form of the streamlines around the bubbles. Considering the two-dimensional motion of a small spherical particle relative to a spherical bubble rising in an infinite pool of liquid, they developed a set of motion equations. According to these, the collision efficiency (see below) is independent of the flow conditions for very fine particles (Stokes or potential flow), while for large particles the collision efficiency depends strongly on inertial forces. This implies that: (1) collision efficiency of fine particles increases with the decrease of bubble size, (2) collision efficiency of large particles increases with increase of the bubble size. Similarly, high bubble concentrations cause straightening of the liquid streamlines along the rising bubbles, and can significantly increase the collision efficiency, suggesting why increased bubble concentrations are beneficial for flotation efficiency.

Reay and Ratcliff [12] defined three collection efficiency regimes as a function of the particle size (for a constant bubble size) as diffusion, transition and collision regime. For the collision regime they introduced the term collection efficiency (E) as the product of the fraction of particles in the bubble's rising trajectory which actually collide with the bubble (collision efficiency E_1) and the fraction of the collided particles which actually stick to it (attachment efficiency E_2). By making a number of assumptions, including Stokes flow around the air bubble (experimentally validated for bubble sizes of up to 0.1 mm), they defined the collision efficiency $E_1 = f(r_p/r_b, \rho_p/\rho_w)$. They concluded that the E is proportional to bubble concentration and independent of bubble size over the entire range of particle sizes; the bubble size should be preferably as small as possible for a given air rate, in order to maximise their concentration. Furthermore, for particles in the collision regime E increases with particle size, while for sub-micron particles collection is mainly by diffusion. Their experimental results (determination of the removal efficiency of glass microspheres $d_p = 1$-20 μm by hydrogen bubbles of size 20-100 μm, generated by electrolysis) further suggest that hydrodynamic retardation (tendency to decrease particle collisions due to the water layer between particle and bubble which prevents actual contact because of inertia) should be neglected for the examined particle and bubble sizes.

b. The single collector collision efficiency DAF reaction zone model

The conceptual model proposed for the DAF reaction zone by Edzwald and Wingler [1] and Malley and Edzwald [6] is a continuation of the work of Reay and Ratcliff [12]. It is based on a number of theoretical assumptions, most importantly that in drinking water treatment circumstances particle-bubble aggregation relies most significantly on their collision, and that laminar flow conditions exist around the bubbles. Consequently, it neglects the other two mechanisms of particle-bubble aggregation proposed by Kitchener and Gochin [8], the entrapment of bubbles within a condensing network of flocs, and the bubble growth from nuclei within flocs. It is a batch kinetic model that must be incorporated into a continuous flow system [26]. The conceptual model uses the single collector efficiency concept in order to describe particle transport to the bubble surface. This concept was used for modelling the removal of particles by deep bed filtration by Yao et al. [27]. It considers the rising air bubbles 'cloud' as a rising filter bed, in which the equivalent measure of collector volume for DAF is ϕ_b the bubble volume concentration, while for filtration it is (1-*f*) (*f* : filter-bed porosity). The removal efficiency by a single bubble collector is expressed by:

$$R = \alpha_{pb}\eta_T(100\%) \tag{1}$$

in which the single collector collision efficiency term η_T (equivalent to the E_1 term used by Reay and Ratcliff [12]) is a sum of the expressions for the single collector efficiency for Brownian diffusion (η_D), interception (η_I), sedimentation (η_S), and inertia (η_{IN}), while the α_{pb} (equivalent to the E_2 term used by Reay and Ratcliff [12]) represents the particle - bubble attachment efficiency (fraction of collisions resulting in particle - bubble agglomerates) [6].

$$\eta_T = \eta_D + \eta_I + \eta_S + \eta_{IN} \tag{2}$$

$$\eta_D = 0.9\ (k\ T/\mu\ d_p d_b v_b)^{2/3} \tag{3}$$

$$\eta_I = 3/2\ (d_p/d_b)^2 \tag{4}$$

$$\eta_S = (\rho_p - \rho_w)\ g\ d_p^2/(18\ \mu\ v_b) \tag{5}$$

$$\eta_{IN} = (g\ \rho_p \rho_w\ d_b d_p^2)/(324\ \mu^2) \tag{6}$$

The inertia term (η_{IN}) is significant for particles and bubbles larger than 100 μm, and therefore not of importance for DAF, but it may be relevant for e.g. dispersed flotation where such particles and bubbles are encountered. Similar to filtration, the minimal value for total single collector efficiency is achieved for particles with the size of approximately 1 μm (Fig. 7.2). However, η_T for flotation is an

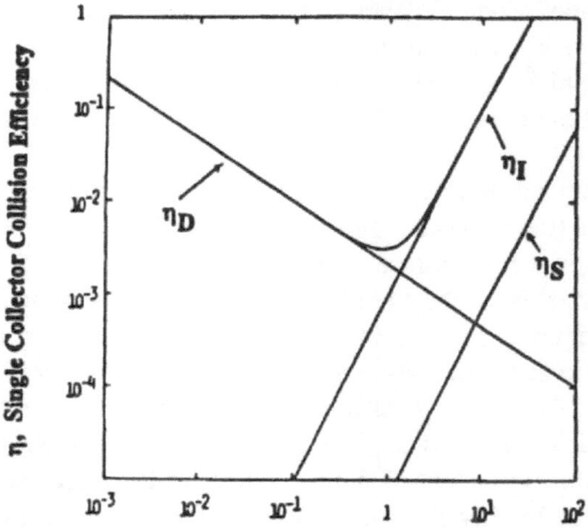

Fig. 7.2 Single collector efficiency vs. particle size for ρ_p=1010 g/m^3, d_b=40 μm and 25 °C in DAF (from Edzwald [26]).

order of magnitude higher than for filtration. A typical bubble size of 40 μm and an assumed particle size of 10 μm result in an η_T value of 0.09, while an increase of the particle size to 100 μm results in an η_T value of 1.0, compared to η_T values of 0.000 and 0.028, respectively, for rapid sand filtration at 10 m/h and sand grain size of 0.8 mm [7]. An extension of the single bubble removal efficiency equation for a system of bubbles characterised by a total bubble number concentration N_b, represents the batch rate particle removal equation :

$$dN_p/dt = -(3/2)\,(\alpha_{pb}\,\eta_T)\,(\phi_b\,v_b\,N_p)\,/\,d_b \qquad (7)$$

The same equation can be expressed as particle removal rate per flotation tank depth :

$$dN_p/dH = -(3/2)\,(\alpha_{pb}\,\eta_T)\,(\phi_b\,N_p)\,/\,d_b \qquad (8)$$

The right-hand side of the equation represents the operational parameters that affect flotation performance. It considers the influence of pre-treatment (effect of coagulation/flocculation on α_{pb}, η_T [d_p] and N_p) and of the flotation facility design (effects of saturator pressure, recycle flow and injection on ϕ_b and η_T [d_b]). The most important variables that determine DAF efficiency are the particle and bubble size and their concentrations.

All the parameters involved in the model equation can be measured or calculated based on these measurements, except α_{pb} which is in concept identical to α for particle-particle attachment in flocculation and particle-medium attachment in filtration. The α_{pb} value is affected by electrostatic particle-bubble interactions and by the hydrophilic nature of organic particles : most particles, as well as bubbles, carry an electro-negative charge and repel each other even if the two entities have neared each other [2, 23, 28, 29]; to attach, particles and bubbles in each others pathway have to overcome resistance caused by water adsorbed on the hydrophilic particle surface. An additional stability factor are the steric effects of algae due to adsorbed macromolecules or extra-cellular organic matter (EOM) and protruding cell features [2, 14]. Nakamura et al. [30] experimentally determined the diameter of a *Microcystis aeruginosa* cell to be 4.8 μm, while the diameter of the cell including its surrounding slimy layer amounted to 7.2 μm, although the dry weight of the slimy layer itself contributed only to 1.3 % of the fresh mass [31]. Other algal steric effects may be caused by morphology and physiology: spines, flagella, motility, etc [14, 32]. Favourable conditions for particle-bubble attachment require therefore coagulation conditions that reduce particle charge and produce more hydrophobic particles. It has been suggested that even without the addition of surface active agents, hydrophobic particles are formed in treatment [8, 33]. Namely, reservoir water generally contains organic carbon levels in excess of 1 mg/L and since most natural compounds are surface active, their eventual adsorption on precipitating iron or aluminium hydroxides results in the formation of hydrophobic compounds [8].

In addition, although DAF requires particle size above 1 μm, it has been suggested that particles should be not too large since such flocs presumably require more attached bubbles to produce particle-bubble agglomerates of a density lower than that of water. A floc of a 50 μm size and 1.01 g/cm^3 particle density would get lifted by a single air bubble at a velocity of 2.1 m/h, but this is suggested impossible for a floc of 200 μm size. Typically, flocs are recommended to be < 100 μm and preferably 10-30 μm [26]. However, the increase of floc size results in decrease of relative floc density (see Chapter 6) which may compensate for the larger size, and still make the larger flocs recommendable for DAF.

Schers and van Dijk [7] extended the single collector collision efficiency model [1, 6] by introducing the inertia term η_{IN} (Equation 6). As with the sedimentation term η_S, the inertia term starts determining the collision efficiency of particles of approximately 1,000 μm size or larger. Their calculations at 10°C show that for a bubble size of 40 μm interception determines the collision efficiency for particle sizes > 5 μm, η_T reaching a value of 1 if the particle size is > 32 μm. Considering the usual floc size of 100-1,000 μm, they state that interception is the key mechanism involved in particle-bubble interactions in the reaction zone. They further modified Stokes' equation for appropriate flow conditions (1<Re<50) and expressed the particle-bubble agglomerate rising velocity in the separation zone by the following equation:

$$v_{pb} = g^{0.8} [(\rho_w - \rho_{pb})/\rho_w]^{0.8} d_{pb}^{1.4} / 10 \nu^{0.6} \quad (9)$$

Removal of the particle-bubble agglomerate would occur under plug-flow conditions, if its rising velocity v_{pb} is larger than the overflow rate (v_{os}) for the separation zone, considering the fraction of dead space m within the separation zone:

$$v_{pb} > v_{os} / (1 - m) \quad (10)$$

c. The population balance DAF model

Fukushi et al. [9] agree about the floc size in water treatment being generally 10-1,000 μm, but state that the large particle sizes are accompanied by higher collision efficiencies proportional to the third power of the collision radius. Furthermore, larger flocs have presumably lower density which provides for better flotation. They introduced a collision-attachment factor α which represents the coverage of the floc surface by precipitated cationic coagulant and the number of attached bubbles (calculated initial value at zero air bubbles attached α_o=0.3-0.4). Importantly, they approached particle-bubble collisions as a heterogeneous flocculation process, considering that the bulk fluid flow is turbulent in both, the reaction and the separation zone. They considered that the transport between particles and bubbles occurs due to differences in fluid velocities arising from turbulent mixing and energy dissipation in the DAF reaction zone and proposed a population balance model (simplified as "flocculation" of bubbles and flocs in the reaction zone). They created a series of kinetic equations and converted them into normalised dimensionless equations, which were solved by means of Laplace transformation. The solution is given by the following equation which expresses the time variation of the average number of bubbles i_f attached on F-size flocs:

$$i_f = m_f [1 - \exp\{-(1 + F)^3 / \theta\, m_f\}] \quad (11)$$

F = d_f/d_b - dimensionless floc size,
m_f = $\alpha_o F^2$ - maximum number of F-size flocs with i bubbles,
θ = $d\theta/dT = N_b$ - universal mixing time taking into account the decrease of free bubbles as bubble attachment progresses ($N_b = n_b/n_{bo}$), and
T = $(3/2)\, \pi\beta\, (\varepsilon_o/\mu)^{1/2}\, n_{bo}\, d_b^3\, \alpha_o\, t$ - normalised mixing time.

By introducing the change of floc density which occurs due to bubble attachment into Stokes' equation and considering the drag constant k (combination of k values for bubbles and flocs), they developed a particle-bubble agglomerate rising velocity equation:

$$v_{pb} = 4\,g\,[\,i\,(\rho_w - \rho_a) - a\,(d_f/1)^{-kp}\,(d_f/d_b)^3\,]\,d_{fb}\,/\,3\,\mu\,k\,i\,(d_f/d_b)^3 \tag{12}$$

$$d_{fb} = (d_f^3 + i\,d_b^3)^{1/3} \tag{13}$$

$$k = 16\,i + 45\,(d_f/d_b)^2\,/\,i + (d_f/d_b)^2 \tag{14}$$

They measured and calculated the cumulative floc number (concentration of F-size flocs with i number of bubbles). By expressing the cumulative floc number in function of the measured and calculated rise velocity of the agglomerate, they obtained a very good fit, thus verifying the validity of their modelling approach.

d. Research considerations and rationale

The two modelling approaches of Fukushi et al. (1994) [11] and Edzwald et al. [1, 6, 26] differ in the definition of the bubble-particle collision rate: the first consider that the transport between particles and bubbles occurs due to differences in fluid velocities arising from turbulent mixing and assumes coalescence (equation (13)), while the second consider an equivalent collision rate term which defines bubble-particle interception under laminar flow conditions in the close vicinity of the air bubble. Edzwald [26] concluded that the bulk liquid flow in the DAF contact (and separation) zone is correctly considered by Fukushi et al. as turbulent. Schers and van Dijk [7] found that velocities of 20-30 m/s exist close to the nozzles during introduction of the recycle flow creating turbulent (potential) flow conditions around the bubbles. It would be advisable to apply the recycle flow in a separate space within the reaction zone, which would be connected to the contact zone with a network of free outflow openings. An impinging surface which would accept and dampen the turbulence effect of the recycle flow is another option which has the additional benefit of forming a higher percentage of micro-bubbles (< 150 μm) [21]. However, if one considers the transport of particles in the close vicinity of the bubble itself, and assumes the flocculated suspension mean floc and bubble size below roughly 100 μm, then the assumed laminar flow condition in the bubble's vicinity is accurate. Both models, however, consider the independence of bubbles and particles and particle-bubble collision as a prerequisite for their agglomeration, while neglecting the two other bubble-particle aggregation mechanisms of entrapment of bubbles within a condensing network of flocs, and bubble growth from nuclei within flocs [8].

Our calculations based on the research of Flint and Howarth [13] show that for a bubble size of e.g. 60 μm, a floc size of up to 160 μm, and turbulent (potential) flow around the bubble, the collision efficiency is indeed independent of the particle weight and drag; this would be applicable under Stokes' flow conditions for particles of up to 500 μm. Thus, particle size can determine the collision efficiency factor η_T.

Comparison of the models, especially those of Edzwald et al. [1, 6] and Fukushi et al. [9], has raised some controversy. Fukushi et al. [9] confirmed the validity of their approach by model water experiments with synthetic clay suspensions and coloured water (diluted kraft pulp black liquor), by applying a mathematical procedure of which they present scarce data and information. Recently, Edzwald [10] considered the actual floc size and concentration after flocculation for validation of his model. Schers and Van Dijk [7] report on DAF removal of total iron from tests with water from the Andelse Meuse (the Netherlands) as a surrogate for particle removal. However, iron removal also depends largely on the process chemistry and the raw water quality; iron coagulant may form difficult to remove colloidal organo-iron complexes and may lead to biased results. Based on our model water and pilot plant experimental work which features flocculated water particle size distributions with mean particle size generally in the range of 20-40 μm, we assumed that the single collector collision efficiency model may well be applicable to practical treatment conditions. Thus, we opted to incorporate our experimentally obtained data into the model equations (equations 1 through 8).

Similar to the collision-attachment factor α in Fukushi's model (estimated α_o=0.3-0.4) [9], the particle-bubble attachment efficiency α_{pb} has to be assumed. Based mainly on bubble to surface water particle concentration ratios, it has been commonly taken as or close to 0.5 [2, 7], implying that every second particle-bubble collision results in a successful attachment. However, the actual particle concentration in the flocculated water particle remained subject of assumptions. Lately, calculated α_{pb} values based on curvilinear trajectory analysis and considering the hydrodynamic and inter-particle (van der Waals) forces, were 0.001-0.02, several orders of magnitude lower. These calculations of Han et al. [11] were based on a different approach, however, and lack experimental verification.

The objective of the research was to calculate the α_{pb} value by incorporating the experimentally measured, or calculated process parameters into the single collector collision efficiency model equations. Importantly, particle size distributions after the flocculation process served as starting point for the calculations. This was done both, for bench scale experiments with model water and for pilot plant experiments with reservoir water. The motive was to investigate and assess the influence of the different raw water characteristics on the α_{pb} value. Different process treatment options and conditions were assessed in the context of their impact on the α_{pb} value, including : coagulant dose, coagulation pH, flocculation conditions, the application of cationic polyelectrolyte coagulant aid, ozone or $KMnO_4$ conditioning. The objective was to describe the influence of different process conditions on α_{pb} and to develop a deeper insight into the process kinetics, to enable us to test the hypothesis that an α_{pb} value of 0.5 was representative of field DAF circumstances.

Furthermore, the efficiency prediction of the model was investigated, for assumed conditions typical for full scale treatment, bench scale experiments with model water and pilot plant experiments with reservoir water. The motive was to investigate the impact of the particle-bubble attachment efficiency α_{pb} on the DAF efficiency. The sensitivity of the model was tested with respect to the critical particle (floc) and bubble size. The options of influencing DAF process efficiency by varying the particle (coagulation and flocculation conditions) and the bubble (recirculation ratio and saturator pressure) size distributions were assessed.

7.1.5 DAF hydrodynamic modelling

The hydraulic conditions within the DAF tank also determine the process efficiency. Small differences in the tank design may result in large efficiency differences [34].The already relatively high flow rates applied in Dutch and UK circumstances (minimum flotation time of 6-9 minutes) [19] are being further optimised [7, 35]. The success of such high flotation rates depends largely on two factors: the water temperature, with lower temperature resulting in higher water viscosity and interference with the floc formation process; and the flow conditions i.e. the tank geometry and configuration. The second factor is amenable to optimisation regarding the creation of turbulence to blend the recycled flow with the main water stream, and sufficient floc-bubble contact opportunities. As for the separation zone, once floc-bubble aggregates have been formed, the imperative is to avoid short circuiting and provide stable flow conditions. Construction details like inlet baffle position, angle and height, tank depth and length to width ratio, subnatant withdrawal, floating sludge removal, etc., can significantly affect the DAF process efficiency [36].

Increasingly, DAF tank flow conditions are simulated by two-dimensional, one- or two-phase (depending whether mass and momentum conservation equations are solved for the water, or the water and the air phase, respectively) computational fluid dynamics (CFD) [33] as a complementary tool to pilot plant testing. Two dimensional, two phase CFD simulations [33] have shown that the flow patterns, both, in the DAF reaction and in the separation zone are different when the air is included in comparison with the flow pattern occurring if only water flow is simulated (one phase CFD). In the former case, buoyancy of the bubbles dominates the flow in the reaction zone resulting in a more uniform velocity distribution than for the one-phase flow. The recycled stream must possess a high enough momentum to penetrate the main flow and efficiently mix the air, in order to avoid air losses. Loading rates higher than the bubble rise velocity may result in undesired flocs and air carry-over to the filters. CFD can also be used to determine the existence of undesired three dimensional flows and optimise tank geometry. It can further determine non-uniform bulk and recycle flow and distribution, nozzle or valve blockages [37, 38], suggesting the existence of a large operational window for improvement in many DAF plants.

Although the nominal capacity of the pilot plant used in this study was 13 m^3/h, based on CFD simulations conducted by the manufacturer [39], the pilot plant was operated at $Q < 8$ m^3/h, in order to avoid carry-over of bubbles and flocculated material onto the down-stream filters.

7.2 MATERIALS AND METHODS

7.2.1 Laboratory experiments with model water

The laboratory experimental set-up consisted of a batch-wise jar test apparatus with incorporated DAF facilities (Chapter 3 and 4). The standard experimental temperature was 20°C and coagulation G was 1,000 s^{-1} for 30 s at pH 8. Other coagulation pH conditions were also tested for DAF (pH 4, 6, 7, 8, and 9); the desired coagulation pH was achieved by adding HCl or NaOH, in amounts determined previously by titration. Periodical check-ups showed that the pH remained stable throughout the experiments. The flocculation G and time were varied (G_f=10, 30, 50, 70, 100 and 120 s^{-1} and t_f=5, 10, 15, 25, 30 and 35 min). The coagulant ($FeCl_3 \cdot 6\,H_2O$) doses were 0 - 15 mg Fe(III)/L, while the cationic polyelectrolyte Superfloc C-573 was tested as a coagulant aid in a doses range of 0 - 1.5 mg Superfloc C-573/L. The rationale behind the dose range of polyelectrolyte was to approximately cover the range which corresponds to one tenth of the accompanying optimal Fe coagulant dose. Different recirculation ratios R (R=5, 7 and 10%) and saturator pressures P (P=500, 600 and 700 kPa) were tested. The flotation time was 5 minutes, of which during the first minute (after introduction of the recycle flow) continuous slow mixing ($G < 10\ s^{-1}$, or lowest possible mixing intensity with the available apparatus) was applied, in order to simulate the DAF reaction zone. During the remaining four minutes no mixing was applied in order to simulate the separation zone conditions.

Ozone was used as an algae conditioner in the context of improving down-stream DAF efficiency. For ozone production, the Trailigaz LABO LO, France, ozone generator was used (Chapter 5, Fig. 5.12). Ozone was produced under standard ozone generator conditions of 220 V, 0.6 A, and 0.6-0.7 bar pressure. The ozone dosage (transferred ozone) was derived from the ozonation time which was varied from 1-4.5 min (Chapter 4), corresponding to ozone doses of 0.48-2.16 mg O_3/L, or 0.2-0.9 mg O_3 /mg TOC [40]. The time gap between ozonation and subsequent coagulation and flocculation was kept at 2 min. The pH of the raw (model) water subject to ozonation was previously set to 7.5 by HCl addition. A 5.5 L ozone resistant glass jar was used as a batch ozone reactor. The pH change after ozonation was 7.5 ± 0.2 and it served in ozonation experiments as coagulation pH without additional corrections. Ozone was applied in combination with the previously determined optimal coagulant dose of 3 mg Fe(III)/L for a concentration of ≈10,000 cells/mL of *M. aeruginosa*. The cationic polyelectrolyte Superfloc C-573 was tested as a sole coagulant (0.3-1.0 mg/L), or as a coagulant aid (0.1-1.0 mg/L) to Fe(III) coagulant, combined with a low 0.2 mg O_3/mg C ozone dose.

For permanganate conditioning experiments a 0.0057 M stock solution was prepared at weekly intervals. The applied permanganate dose was 0.7 mg $KMnO_4$/L, while the pH was previously set at 8. The permanganate was applied at a G value of 400 s^{-1} for a period of 1 min, followed by 30 min of slow mixing at a G value of 10 s^{-1}. It was combined with a previously optimised Fe(III) coagulant dose at coagulation pH 8.

Model water was prepared by spiking water originating from the Biesbosch storage reservoirs (Biesbosch Water Storage Company, the Netherlands) with the semi-continuously laboratory cultured, single cells form of the cyanobacterium *M. aeruginosa*, to a standard initial concentration of ≈ 10,000 or 20,000 ±cells/mL. In order to avoid the introduction of larger clumps of algae and algae debris into the model water, upon spiking the algae suspension was subject to microstraining over a 35 μm mesh.

7.2.2 Pilot plant experiments

The pilot plant research was conducted on the premises of the Princess Juliana treatment plant (WRK III) situated on the shores of the IJssel Lake in Andijk, the Netherlands (Chapter 5). The raw water used in the experimental work was reservoir water previously subject to microstraining over a 35 μm mesh. The period of the research (late summer, early autumn) coincided with a bloom of *M. aeruginosa*, a common species in the IJssel Lake in this period of the year. The DAF pilot plant investigations were carried out with a commercial (Purac, Sweden) pilot plant installation (Q_{max} =13 m^3/h). The unit itself comprised a 1 m^3 stainless steel tank with four needle valves located at the bottom of the inlet for the recycled water stream introduction. Coagulation occurred in a Kenics-type in-line mixer (Purac, Sweden); one, two or three-stage flocculation could be applied with variable speed flocculation paddles. An automated system regulated the pumps, the saturation and the recycled flow. The experiments did not include a final filtration step.

$Fe_2(SO_4)_3 \cdot 9\ H_2O$ was used as coagulant in a dosage range of 0-20 g Fe(III)/m^3. pH in coagulation was set at pH 8 using $Ca(OH)_2$, as in the full-scale WRK III process. The synthetic cationic polyelectrolyte Superfloc C-573 was tested as a coagulant aid to Fe(III) coagulant in the concentration range of 0.1-0.7 g/m^3. The Kenics static mixer was used for the rapid mix with a calculated G value of ≈3,500 s^{-1} for a flow of 4.6 m^3/h. This flow was applied in the DAF pilot plant in order to avoid carry-over of bubbles into its effluent and onto the filters [39], a common occurrence in case of higher flows. Different flocculation time, energy input (G value) and flocculation sequence (one-, two- and three-stage flocculation) were tested; the values were partly dictated by the technical provisions of the pilot plant. The recirculation ratio R in the DAF unit was set at R=5, 7 and 10%, while the saturator pressure P was 500 kPa. The water temperature over the experimental period decreased from 16.7°C to 10.1 °C in 1995, and from 16.4° C to 12.9°C in 1996.

Only a limited number of pilot plant ozone conditioning experiments was carried out, because of flow limitation of the continuous-flow ozone reactor (Q_{max}=1.14 m^3/h, versus the applied Q=4.6 m^3/h for the pilot plant) (Chapter 5). Potassium permanganate (0.0021M working solution) was dosed immediately before pH correction and coagulant addition, allowing for very short contact times (< 1 s) before coagulation and flocculation (Chapter 5).

Model water experiments were intended to simulate relatively "clean" water circumstances and provide conditions in which the assessment of the removal of the spiked algae would not be significantly interfered by other water quality parameters (i.e. the presence of larger quantities of other suspended, colloidal and dissolved matter of inorganic or organic origin). On the other hand, the WRK III reservoir (Chapter 5) provided for the "real" water conditions in which the naturally occurring cyanobacteria bloom of *M. aeruginosa* was coupled with other water quality factors of importance for process optimisation and influence on process efficiency.

7.2.3 Analytical techniques

Process evaluation was based on turbidity (Sigrist L-65, Switzerland) and particle (algae) count measurements of particles in the size range 2.75-150 μm (HIAC-Royco PC-320, USA). Residual coagulant concentration (Fe_{total}) and residual manganese (Mn_{total} and $Mn_{dissolved}$) were measured by atomic absorption spectrometry at 248.3 nm for iron (NEN 6460) [41] and at 279.5 for manganese (NEN 6466) [42]. Other particle analysis techniques included computer image analysis (Mini-Magiscan, IAS 25/IV25 Joyce-Loebl Ltd., UK). These measurements were compounded by inverted microscope count (M40-Wild Leitz, Switzerland) and electrophoretic mobility (Tom Lindström AB-Repar) measurements.

The computer image analysis results are presented in the form of particle size frequency and volume distributions. The frequency and volume distributions represent the number and the volume of recorded particle size range. For this purpose the recorded particles and/or floc material were approximated to spheres and presented in size ranges of 10 μm, from 0-200 μm. Particles >200 μm were also recorded (by image analysis and visually), in particular after flocculation. However, they represented only a small percentage of the overall particle concentration, especially for the "clean" model water circumstances (<2.5% of particles (floc) were of size >100 μm under conditions of <10 mg Fe(III)/L, pH 8, G_f=70 s^{-1} and t_f=10 min). The removal of these large particles (flocs) by DAF was complete and of limited significance for further discussion. The adopted size limit of 200 μm was found more relevant for the assessment of the overall DAF efficiency, since it included a more significant portion of the overall particle (floc) concentration, of size which is more problematic for DAF removal than the larger particles (flocs). The particle size of 50 μm, which is roughly the size limit of visually observable aggregates, was tentatively adopted as the size dividing the small from the large particle size range.

The image analyses were performed on fresh samples, as well as on photographs of samples. Photo analysis was preferred over fresh sample analysis, since it enabled registration of representative situations at different spots along the treatment line, without the potential danger of disturbing the process or the quality of the sample [43]. For this purpose, a specially devised flat photo cell was connected at different locations of the jar test apparatus and the pilot plant unit, and samples were photographed before and after flocculation, and after particle removal. Special care was taken to avoid break-up of particle agglomerates (flocs) during the passage of the sample from the jar-test unit to the photo cell, by allowing low flow velocities adjusted through a system of valves. In case of DAF, the released air bubbles were also photographed and analysed. Measuring bubble size distributions under different recirculation ratio and saturator pressure conditions proved to be difficult, partly because bubbles tended to attach to the walls of the glass photo cell and appear on the following photos. To avoid this, the photo cell was thoroughly flushed after each snap shot with demi water in order to mechanically detach the bubbles. Professional high resolution black-and-white film AGFA 25 or Kodak TMAX 100 was used for the purpose, while forced development was applied for film processing. 30-100 fields, or more than 500 particles of each film shot were processed and analysed, using a ccd-camera (604*288 pixels) mounted to a Nikon Optiphot (Japan) microscope (at 40x magnification). The minimum detectable size under these circumstances was 1.9 μm. The software package Genias25 developed by Joyce-Loebl (UK) was used for data processing.

The computer image analysis is subject to a measurement error which varies with the size of the recorded particle and the applied magnification. Previous work [43] showed that the smaller the particle and the magnification, the larger the measuring error. This is due to the fact that the particle area used for calculations of the particle diameter is measured as the full area of all the computer screen pixels over which the particle image is spread, even if the pixels at the edge of the image are not fully covered. In the case of 40x magnification (pixel size of 1.9x1.9 μm), we found that for an 8 μm size *M. aeruginosa* cell the calculated cell diameter was 25-50% larger than the actual cell dimension. This explains the particle frequency peak of the raw (model) water which occurred at the 10-20 μm, instead of at the expected 0-10 μm range (Fig. 7.3). For larger particles (flocs>10 μm) the measurement error led to up to 10% higher values than the actual sizes. The same applies for the bubble size measurements, since bubbles were mostly of size >10 μm. To verify the reproducibility of the applied analytical technique, multiple shots of the same sample were analysed regularly.

Particle size distributions departed strongly from Gaussian shapes. Statistical comparison of data on particle size frequency distributions from experiments that were performed under identical conditions was carried out using the chi-square test. Testing for the hypothesis that the obtained particle size distributions were identical resulted in a probability level of p=0.999, confirming highly comparable particle size distributions and hence high experimental reproducibility.

7.3 RESULTS AND DISCUSSION

7.3.1 α_{pb} calculation for different process conditions

Under the assumption that all particles have the same residence time in the reaction zone (ideal up-flow plug flow conditions), the particle removal model equation (7) proposed by Edzwald et al. [1, 6, 26] can be solved for α_{pb} as follows:

$$\alpha_{pb} = -(2/3)\ (d_b / t\ \eta_T\ \phi_b\ v_b)\ \ln\left[1 - (N_p' - N_p'')/N_p'\right] \tag{15}$$

The influent, the flocculated and the effluent particle water concentration and size were measured by image size analysis. For both particles and bubbles, size and volume frequency were obtained. An example of these results is given in Figs. 7.3. and 7.4. The image analysis allows to quantify and qualify the influent before and after flocculation.

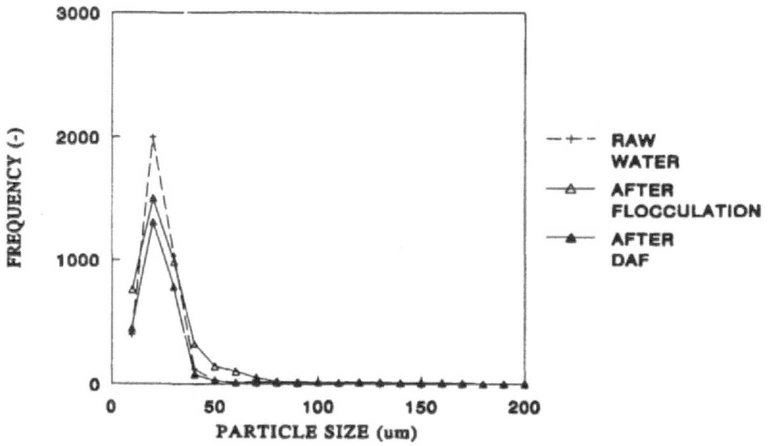

Fig. 7.3 Particle size frequency distribution for a bench-scale model water experiment. Conditions : pH 8, 3 mg Fe(III)/L, G_f=70 s^{-1}, t_f=10 min, R=5%, P=5 bar.

Similar to Fukushi et al. [11], we considered the calculated floc density obtained by applying the relative floc density concept, as discussed by Boller and Kavanaugh [44] (Chapter 6). In order to calculate the parameter values for equation (15), equations (1) through (6) were incorporated into a spreadsheet programme and the α_{pb} value was calculated for a range of process conditions. Table 7.1 in Appendix 7.1 contains the calculated α_{pb} values and the corresponding process conditions. In the following discussion two particle removal efficiency terms will be used, one relating to particle concentration after flocculation and one for overall removal efficiency relating to influent particle concentration. These terms do not necessarily follow the same pattern and are determined by the coagulation/flocculation conditions (discussed in Chapter 6). Thus, in some cases high particle removal after flocculation may not correspond with high overall removal efficiency, partly because flocculation increased the particle concentration. In some cases effluent quality was deteriorated relative to the influent (raw water) quality, although the particle (floc) removal efficiency was positive.

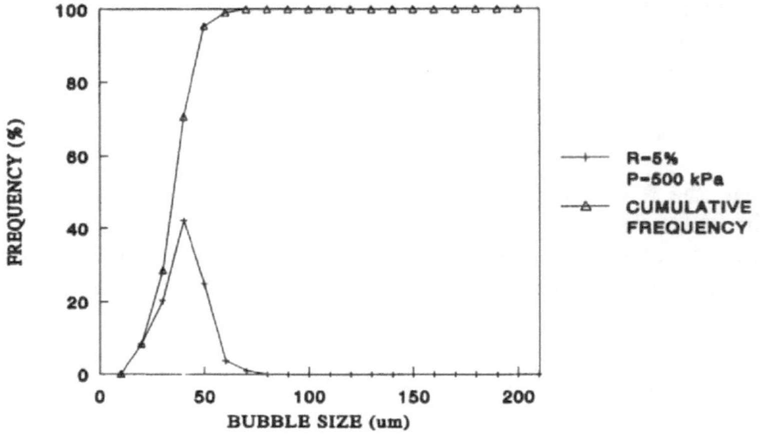

Fig. 7.4 Bubble size frequency (percentual) distribution for bench-scale jar test conditions, T=20°C, R=5%, P=5 bar.

a. Bench-scale experiments with model water

Figs. 7.3 and 7.4 represent image analysis results from a bench-scale experiment with model water. These results were used in the further calculations of the α_{pb} value. In the raw water a frequency peak ascribed to the algae was observed. A distinct increase in the frequency of particles in the larger particle size region (>50 μm) was observed for the flocculated sample (at pH 8 and with sweep coagulation as the destabilisation mechanism [45]). The flotate was always devoid of the larger particles, implying that the bigger flocs had been efficiently removed while the smaller particle sizes (<50 μm) were less affected by the DAF. This suggests that the η_T value, and especially its η_I fraction depend on particle size. Additional microscopic inspection of flocculated water samples showed that the formed floc size was ≤200 μm, and that the largest flocs of ≥300 μm were obtained for a dose of 15 mg Fe(III)/L. The concentration of large flocs was significantly lower than of flocs in the 50-100 μm range. A bubble size frequency distribution was observed with a bubble size mode of 40 μm and a standard deviation of s=9.39, in accordance with similar research [43].

α_{pb} for these circumstances (Figs. 7.3 and 7.4) was found to equal 0.11. The bubbles- to-particles concentration ratio N_b/N_p was 5.3, which is not in accordance with commonly suggested 10-100 value in surface water treatment [1, 6, 7]. This implies that for every particle present there were five bubble available for lifting it to the water surface, however, only every tenth particle-bubble collision resulted in their attachment. One of the important requirements for efficient DAF, i.e. a high N_b/N_p ratio, was not fulfilled. This, however, cannot fully explain the relatively low particle (floc) removal efficiency of 30% (23% overall efficiency). At a dose of 5 mg Fe(III)/L DAF performed significantly better under similar N_b/N_p ratio conditions (68% particle (floc), and 54% overall removal, and α_{pb}=0.6). This suggests that the coagulation destabilisation mechanisms play a primary role in determining DAF process efficiency.

Variations in the jar test coagulation/flocculation conditions resulted in different degrees of particle destabilisation and different particle size distributions after flocculation (Chapter 6). This also affected the α_{pb} value which could fluctuate from 0.04 to 0.75. The higher range would correspond to optimal DAF coagulation and flocculation (Table 7.1, A), as well as to experiments in which coagulation was enhanced either by cationic polyelectrolyte as coagulant aid (Table 7.1, C and D), or in the case of coagulation at pH 6 (Table 7.1, B).

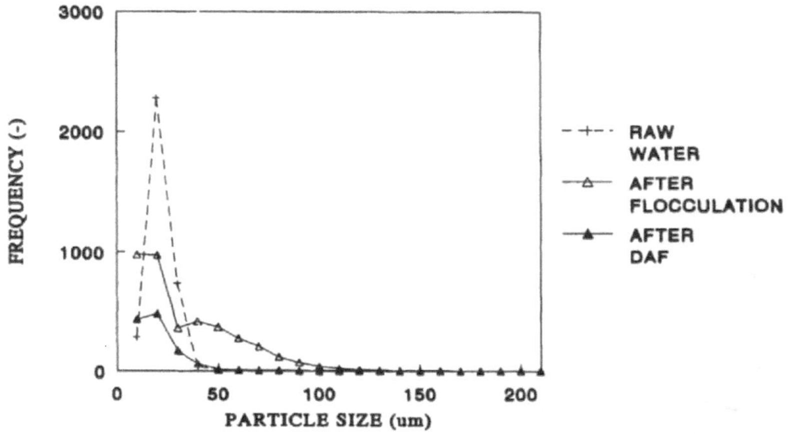

Fig. 7.5 Particle size frequency for a bench-scale experiment with model water. Conditions : pH 6, 10 mg Fe(III)/L, G_f=70 s^{-1}, t_f=10 min, R=5%, P=5 bar.

Under the two latter circumstances, the governing sweep coagulation was significantly aided by particle adsorption coagulation as was verified by electrophoretic mobility measurements. Furthermore, the presence of cationic polyelectrolyte and positively charged iron-hydroxo species (at pH 6) overcame the negative charge of the air bubbles and led to removal of the neutralised smaller flocs ($<50\ \mu m$) more efficiently than in the case of coagulation pH 8 (Fig. 6.3.c., Chapter 6). The charge neutralisation led to production of floc material of larger size at the expense of the smaller particles, as well as to more efficient floc-bubble agglomeration, both in the smaller and larger particle size ranges (Fig. 7.5).

The particle (floc) removal efficiency in case of coagulation pH 6 was 69% (overall 60%) compared to 36% (overall 30%) in case of identical process conditions but with coagulation pH 8. In the case of 0.5 mg Superfloc C-573 coagulant aid in combination with 3 mg Fe(III)/L the particle (floc) removal efficiency was 56% (overall 65%), compared to 51% (overall 54%) for the 3 mg Fe(III)/L only. Therefore, it is suggested that a high(er) α_{pb} does not necessarily indicate high(er) removal efficiency, since it is related to the quality and quantity of the flocculated particles (floc stability, concentration and size characteristics). A typical example supporting this observation is the higher coagulant aid dose (Table 7.1, D - 1.0 mg Superfloc C-573/L) characterised by an α_{pb} of 0.33, which resulted in particle (floc) removal efficiency of 37% (overall 74%). It was lower than the α_{pb} =0.75 for the 0.5 mg Superfloc C-573/L coagulant aid dose (Table 7.1, C), although the particle (floc) removal for the lower polyelectrolyte dose was higher (56% as compared to 37%), while the overall removal efficiency was lower (65% as compared to 74%).

The flocculation G value of 50 s^{-1} performed better than the 100 s^{-1} value, as the latter produced a detrimental effect on floc structure and composition (floc concentration was twice as high as raw water particle concentration, and flocs were of smaller size). This was accompanied by lower α_{pb} for G=100 s^{-1} at both pH 7 and 9 (α_{pb}=0.04 and 0.08, respectively) compared to α_{pb}=0.2 and 0.12, respectively at G=50 s^{-1}. The low α_{pb} values corresponded to low particle (floc) removal efficiency (59% and 44% for pH 7 and 9 at G=50 s^{-1}, and 16% and 33% for pH 7 and 9 at G=100 s^{-1}).

The use of ozone as a conditioner, especially when applied in combination with Fe (III) coagulant and cationic Superfloc C-573 coagulant aid, increased significantly DAF process efficiency (Chapter 4). This combined treatment resulted in more favourable floc characteristics (Fig. 7.6) (52% particle (floc) and 70% overall removal efficiency) than when Fe (III) was used as the sole coagulant (40% particle (floc) and 32% overall particle removal efficiency), and in similar characteristics as those of the combination of ozone conditioning and Fe coagulation (55% particle (floc) and 67% overall removal efficiency). The α_{pb} was highest for the first case at 0.44 (Table 7.1, F), compared to α_{pb} =0.38 and 0.34 for the other two cases, respectively (Table 7.1, G and E). The relatively low DAF efficiency in the case of model water $KMnO_4$ experiments (Chapter 4) was confirmed with an α_{pb} =0.19 (Table 7.1, H).

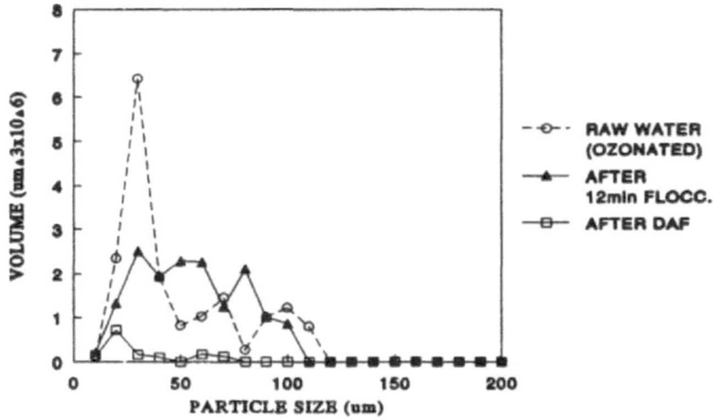

Fig. 7.6 Particle volume distribution for a bench-scale experiment with model water. Conditions : ozonation pH 7.5, 0.2 mg O_3/mg TOC, coagulation pH 7.5 $\pm$0.2, 3 mg Fe(III)/L, 0.5 mg Superfloc C-573/L coagulant aid, G_f=10 s^{-1}, t_f=10 min, R=7%, P=5 bar.

b. Pilot plant experiments with reservoir water

Selected pilot plant results for different conditions are presented in Table 7.1 and Fig. 7.7.

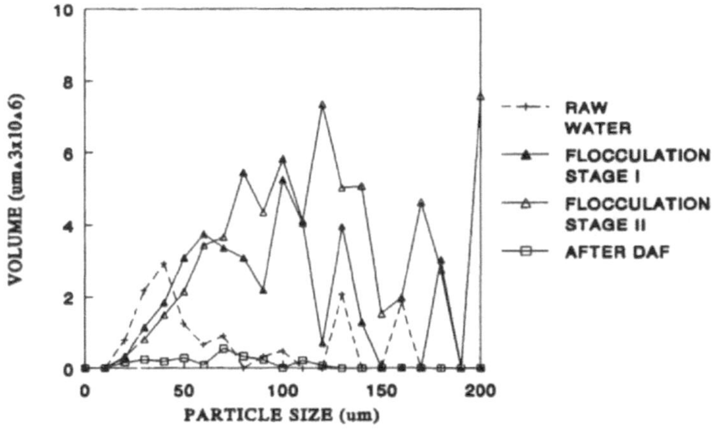

Fig. 7.7 Particle volume distribution for pilot plant experiments with reservoir water. Conditions : pH 8, 15 g Fe(III)/m^3, 0.5 g Superfloc C-573/m^3, two stage flocculation, stage I :G_1=50 s^{-1}, t_{f1}=18.2 min, and stage II :G_2=30 s^{-1}, t_{f2}=3.8 min, R=5%, P=500 kPa.

The higher DAF efficiency observed in the pilot plant experiments as compared to the bench-scale jar tests with model water was discussed in Chapter 5. In the pilot plant high coagulant doses ≥10 g Fe(III)/m^3 were required for efficient coagulation and flocculation of the abundant inorganic and organic matter. The raw water NOM composition and concentration (TOC was in the range of 5-15 g C/m^3, mainly related to humic matter) were suggested to be a major reason for the results. On the other hand, the obtained mean bubble size of d_b=62 μm and the relatively low water temperature of 10.1°C contributed to the relatively low bubble concentration and the low N_b/N_p ratio. The fact that this low ratio still resulted in relatively high DAF process efficiency compared to the bench-scale model water experiments (i.e.≈ 90% compared to ≈50%) suggests that other factors affect DAF efficiency, possibly NOM [8]. Consequently, the α_{pb} values were relatively high for the combined 10 g Fe(III)/m^3 and 0.3 g Superfloc C-573/m^3 coagulant aid treatment (at G=10 s^1) resulting in α_{pb}=1.79 (though values of α_{pb} >1 have no physical meaning) (Table 7.1, I); maximum recorded particle (floc) and overall removal efficiency were 83 and 88% respectively. The lower α_{pb} values for the 15 g Fe(III)/m^3 coagulant dose (Table 7.1, K and L) were accompanied by more significantly reduced particle concentration after flocculation, and larger mean floc size increasing the η_T value. The cationic coagulant aid had a positive effect, although not as distinct as in the case of model water experiments (Chapter 3, 4 and 5). Although the particle charge was not completely neutralised, the efficiency improved with increasing dose. Compared to the cases when Fe(III) was used as sole coagulant, the use of polyelectrolyte shifted the particle size distribution further towards the larger- particle size range (>50 μm), and produced lighter floc material more amenable to flotation. Higher polyelectrolyte doses had the same effect, however, they resulted in denser floc structures (Figs. 6.6 and 6.7, Chapter 6).

7.3.2 Application of the model

a. DAF kinetic model efficiency prediction

The calculated α_{pb} values for the bench-scale jar tests under optimal and enhanced coagulation conditions were in the range of the value of 0.5 that was adopted in the earlier theoretical evaluations of the single collector efficiency DAF model [2, 7]. This value was achieved easier under the pilot plant conditions. Here DAF efficiency was predicted as a function of the bubble size using the model (Fig. 7.8). Three cases were simulated : (1) assuming process conditions similar to Dutch DAF plant circumstances [7], d_p=30 μm and α_{pb}=0.5, (2) under the model water jar test circumstances of Figs. 7.3 and 7.4 (3 mg Fe(III)/l, pH 8, T=20°C, G_f=70 s^{-1}, t_f=15 min, R=5%, P=500 kPa) and with α_{pb} as determined from experimental data and eq. (15), and (3) under the pilot plant circumstances as discussed in Fig. 7.7 (15 g Fe(III)/m^3, 0.5 g Superfloc C-573/m^3, pH 8, G_1=50 s^{-1}, t_{f1}=18.2 min, G_2=30 s^{-1}, t_{f2}=3.8 min, R=7%, P=500 kPa) and with α_{pb} as determined from experimental data and eq. (15).

Assumed conditions [7]:		Model water jar tests :		Pilot plant :	
t	=1.6 min	t	=1 min	t	=1 min
d_p	=30 μm	d_p	=24 μm	d_p	=37 μm
d_b	=40 μm	d_b	=35 μm	d_b	=62 μm
ϕ_b	=5*10^{-3} m^3/m^3	ϕ_b	=2.8*10^{-3} m^3/m^3	ϕ_b	=3.2*10^{-3} m^3/m^3
T	=10 °C	T	=20 °C	T	=10 °C
α_{pb}	=0.5	α_{pb}	=0.11	α_{pb}	=0.28

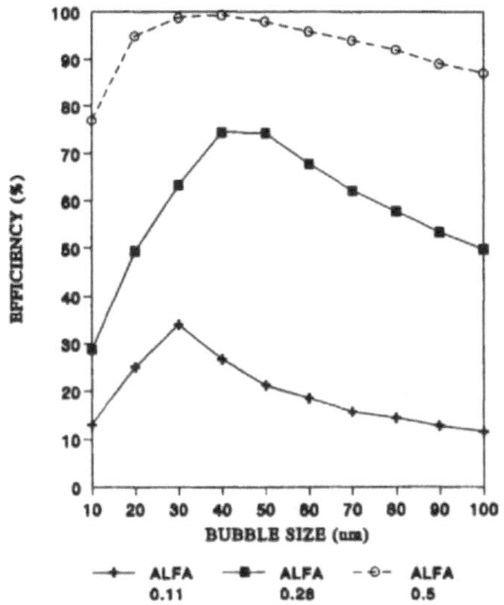

Fig. 7.8 Particle removal efficiency as predicted by the model as a function of bubble size, under conditions similar to those met in Dutch DAF plants circumstances [7] and assumed α_{pb}=0.5, d_p =30 μm, model water jar test, and pilot plant experimental conditions.

In the jar test the mean particle (floc) size measured by computer image analysis was 24 μm, and the mean bubble size 35.4 μm (bubble volume concentration $\phi_b=2.8*10^{-3}$ m^3/m^3), resulting in a calculated α_{pb} value of 0.11. In the pilot plant the mean particle (floc) size was 36.6 μm and the mean bubble size 62 μm (bubble volume concentration $\phi_b=3.2*10^{-3}$ m^3/m^3), resulting in a calculated α_{pb} of 0.28.

Fig. 7.8 suggests that the particle (floc) removal efficiency of 30% for the jar test and 66% for the pilot plant would increase if the bubble size distribution had a mean bubble size of only 30 μm and 40 μm, respectively. Similarly, under the jar test conditions a significantly lower removal efficiency value was achieved because of the relatively low α_{pb}.

Fig. 7.9 relates the α_{pb} value to the removal efficiency as calculated by the model for the three cases. Due to the exponential character of the model equation (eq. 15), α_{pb} rises steeply for high removal efficiencies (>90%). In the jar tests the theoretically maximum achievable particle (floc) removal efficiency (further referred to as Eff) would be 99%>Eff>90%, and α_{pb}=0.69 for a hypothetical 90% removal efficiency. In the pilot plant experiments this value was also 99%>Eff>90%, with α_{pb}=0.59 for the measured 90% removal efficiency. A 99.99% removal efficiency is predicted to be achieved only under the assumed circumstances [7] with α_{pb}=0.91. Comparison of the efficiency for the assumed [7] and the pilot plant experiments showed that a variation in raw water quality (e.g. temperature, NOM concentration and characteristics), and in coagulation, flocculation, and DAF conditions, can influence the (mean) bubble and particle sizes. The model suggests the latter as variables that substantially affect the DAF efficiency and α_{pb}.

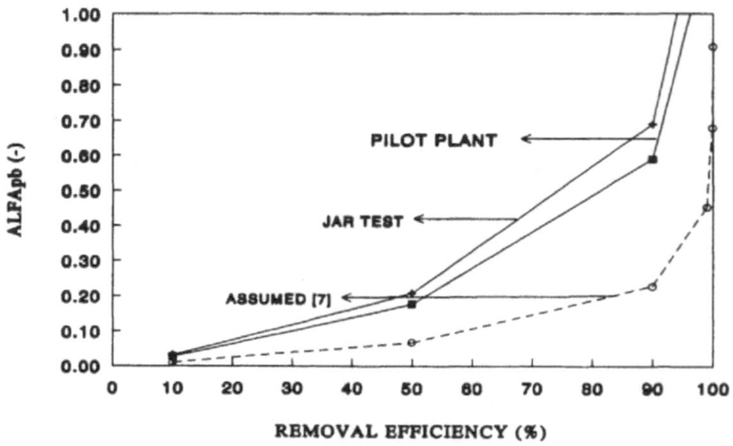

Fig. 7.9 Particle-bubble attachment efficiency as a function of the removal efficiency calculated by the model, for assumed [7] and experimental conditions.

On the other hand, the dependence of α_{pb} on raw water quality is difficult to predict. It has already been suggested [8] that flotation of natural waters is dependent on the presence of suitable organic impurities (e.g. humic acids). It has further been suggested that specific characteristics of the NOM (the existence of fine organic matrices or meshes of filaments which embed very small organic and inorganic colloids in natural surface waters [46, 47]) may positively affect treatment [18]. The suggested relation between the existence of these organic meshes and improved particle coagulation and agglomeration by bridging may also be extended for DAF, the same organic matter structures possibly serving as sites for bubble attachment and increased DAF particle removal efficiency. Thus, more accurate modelling of DAF would additionally require quantification and qualification of the NOM and its impact on α_{pb}.

b. DAF kinetic model : sensitivity analysis

The sensitivity analysis was carried out by using the model to calculate α_{pb} as a dependent variable (Fig. 7.10). Under the jar test conditions (3 mg Fe(III)/L, pH 8, T=20 °C, G_f=70 s^{-1}, t_f=15 min) the calculated α_{pb} value was 0.11. It was set as a reference level (100 % value in Fig. 7.10). The mean bubble size entered into the equation was altered in steps of 5% till +25% and -25%, starting from the original mean bubble size of 35 μm. Process efficiency and mean particle size were kept constant at their actually determined values. The procedure was then repeated in the same manner, but while keeping the mean bubble size constant (35 μm) the mean particle size was altered in 5% increments till +25% and -25%, starting from the mean particle size of 24 μm. The 25% limit was set so that the mean particle and bubble sizes remained within the mean size range recorded during the experiments.

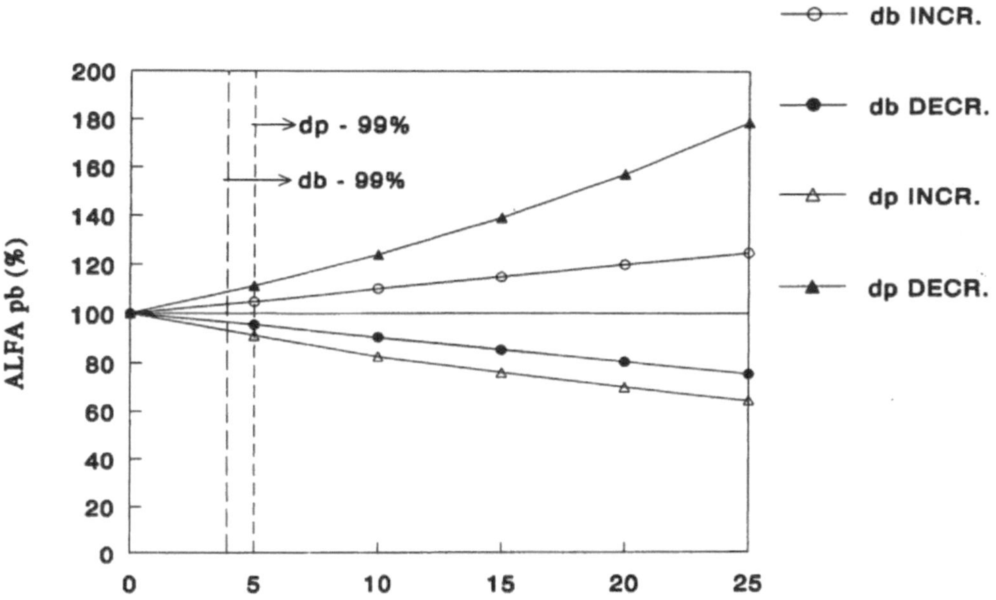

Fig. 7.10 Model based α_{pb} calculation for assumed particle and bubble size increase or decrease, under the assumption of constant process efficiency in a jar test experiment. Reference α_{pb}=0.11; db and dp bubble and particle (floc) size, respectively; 99% confidence limits for dp and db.

The model appeared to be more sensitive to change in particle size than in bubble size. It is most sensitive to a particle size decrease : a 25% decrease of particle size caused a 78% increase of α_{pb}. These results need to be treated with caution because the α_{pb} calculations are based on the assumption of constant process efficiency.

In practice, the particle size distribution and the mean particle size can be influenced by manipulating the coagulation/flocculation process. Similarly, our experience and that of others [43, 48] show that different recirculation ratio, saturator pressure and air recycle stream introduction influence the bubble size distribution in the DAF reaction zone.

In order to also investigate the influence on DAF efficiency of the bubble size distribution, we assumed a number of normal bubble size distributions. Based on the bubble size analysis discussed in Fig. 7.4 it may be assumed that a normal distribution would be reasonably representative. Modes of 30, 40 and 60 μm were chosen, with standard deviations of their populations lower (S = 3, 5 and 7) and higher (S=11 and 13) than the experimentally determined values of S=9.39 and of the mode of 40 μm (Fig. 7.11). The higher the standard deviation the 'wider' the bubble size distribution. For each set of assumed bubble size distributions grouped around the chosen bubble size modes, the theoretical model efficiency was calculated under the jar test experimental conditions of Fig. 7.3 (3 mg Fe(III)/L, G=70 s^{-1}, t=10 min, R=5%, P=500 kPa). In Fig. 7.12 the calculated values are presented versus the standard deviations of the assumed bubble size distributions.

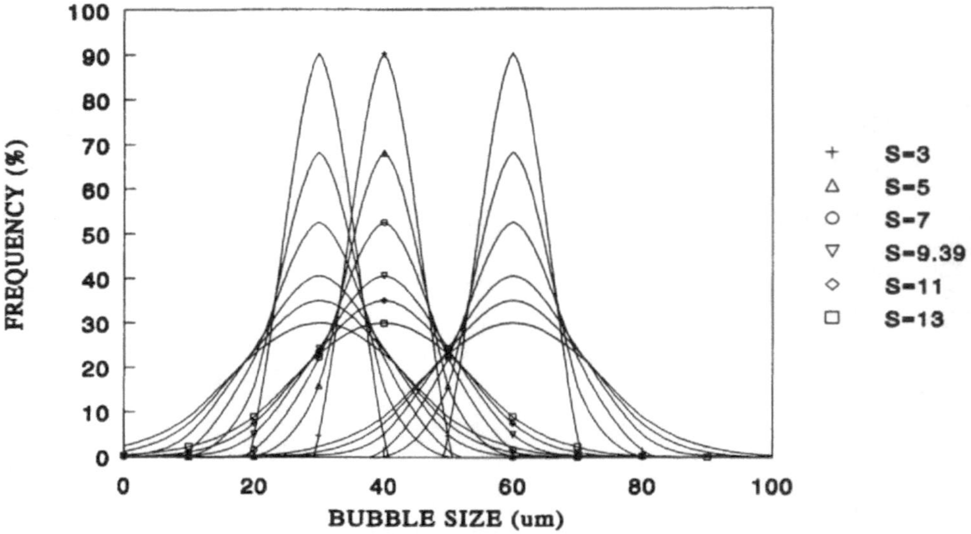

Fig. 7.11 Assumed normal bubble size distributions, with modes of 30, 40 and 60 μm and different standard deviation - S.

For bubble size distributions with modes of 30 μm and 40 μm a slight decrease in process efficiency was calculated with the 'wider' bubble size distributions. The results further suggest that the only noticeable improvement in efficiency would be caused by a very 'narrow' (S=3) bubble size distribution with a mode of 30 μm; this would increase the process efficiency however, by only 5%. On the other hand, for the bubble size distributions with the mode of 60 μm very little influence on efficiency was observed as a function of the width of the distribution. The average calculated efficiency at this mode size was approximately 30% lower compared to that of the smaller bubble size distribution modes.

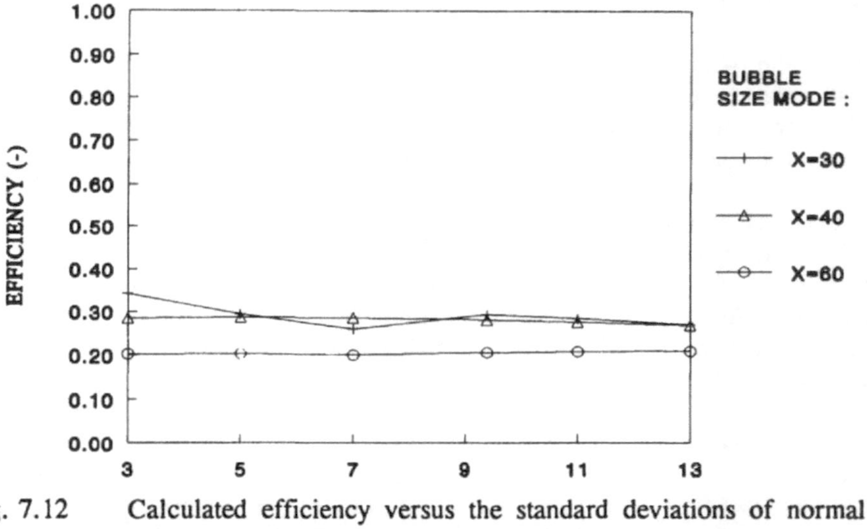

Fig. 7.12 Calculated efficiency versus the standard deviations of normal bubble size distributions (mode X=30, 40 and 60 μm).

c. Bubble size distribution and DAF efficiency

In the previous section, it was shown that DAF efficiency for the case discussed in Fig. 7.4 could be increased if a 'narrow' bubble size distribution (S=3) with a mode of 30 μm was produced. Commonly, bubble size distributions have been reported to have a mean bubble size of 40 μm and a maximum of 100 μm [26, 43]. The bubble size distribution is mainly influenced by the recirculation ratio R and the saturator pressure P. This section aims at assessing the impact of various R (R=500, 600 and 700 kPa) and P (P=5, 7 and 10%) combinations under bench-scale jar test conditions. The obtained bubble size distributions are given in Figs. 7.13, 7.14 and 7.15).

As noted by other research [21, 43, 48] a change in the saturator unit configuration can also affect the bubble size distribution. The distribution from the jar test results given in Fig. 7.4 significantly differed from the one in Fig. 7.13, though saturator pressure and recycle flow conditions were identical. However, the recycled flow in the former case was directly released into the main water 'stream', while in the latter case it was diverted towards the jar bottom outlet structure to avoid floc break-up due to turbulence and shear. Furthermore, the uneven length of the conduit between the saturator vessel and the nozzles, as well as partial or complete nozzle blockage, also affected the bubble size distributions. The design of these provisions is relatively standard. However, minor modifications or changes proved to be of considerable impact on the bubble size distributions. Thus, the analysis in Fig. 7.13, 7.14 and 7.15 were done for the same jar and after careful cleaning of the nozzles. The results are averages of duplicates.

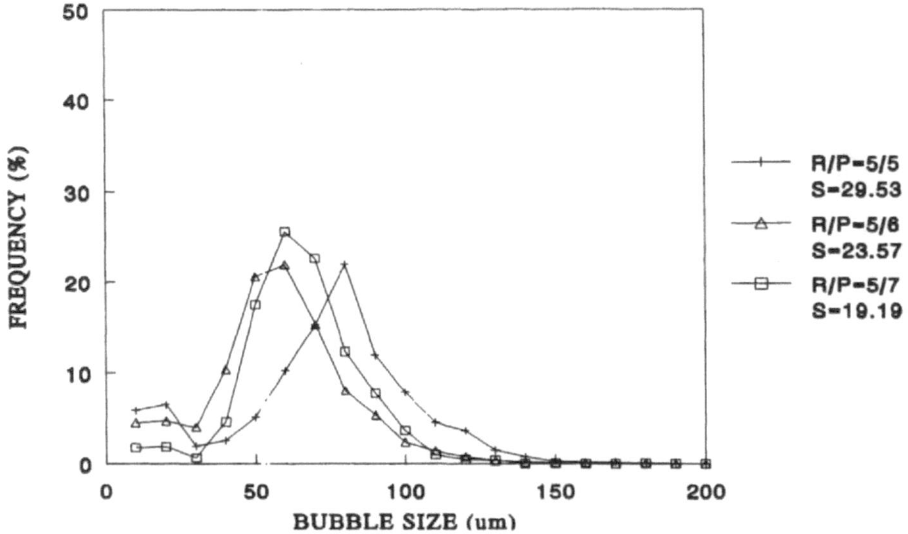

Fig. 7.13 Bubble size frequency distribution at constant recycle flow ratio R=5 % and different saturator pressure conditions (P=500, 600 and 700 kPa). S: standard deviation of obtained distributions.

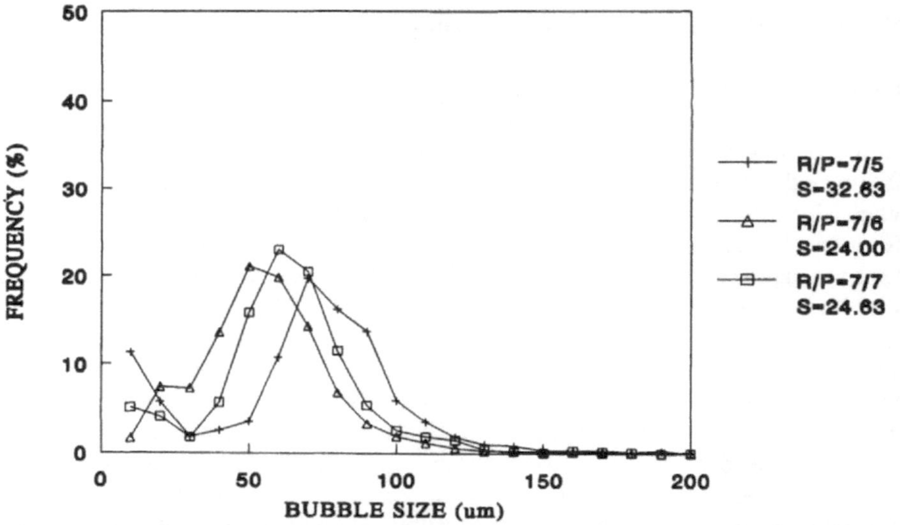

Fig. 7.14 Bubble size frequency distribution at constant recycle flow ratio R=7% and different saturator pressure conditions (P=500, 600 and 700 kPa). S: standard deviation of obtained distributions.

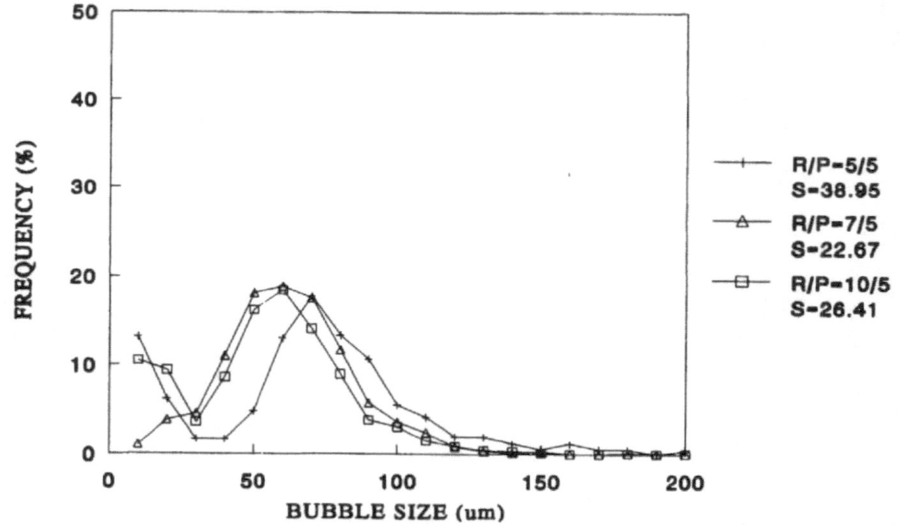

Fig. 7.15 Bubble size frequency distribution at constant recycle flow ratio R=10 % and different saturator pressure conditions (P=500, 600 and 700 kPa). S: standard deviation of obtained distributions.

The parameters which are varied in practice are the saturator pressure and the recycled flow. As observed in Figs. 7.13, 7.14 and 7.15, these process conditions affected the bubbles size distributions significantly, resulting in different mean bubble size (d_b=49-67 μm) and standard deviation (s=19.2-38.9).

Increase of the saturator pressure (500, 600 and 700 kPa) resulted in decrease of the mean bubble size for all recirculation ratios tested (5, 7 and 10%), similar to other research [48]. This was accompanied by 'narrower' more efficient bubble size distributions. An increase of the saturator pressure also led to an increase of bubble concentration. However, similar to other research for high saturator pressures (>300 kPa according to Takahashi et al. [48], in our case between 600 and 700 kPa), the bubble concentration differences were not very pronounced. Furthermore, it was visually confirmed during photo sample analysis that this increase of bubble concentration caused more frequent bubble collisions and coalescence. This resulted in larger mean bubble sizes for the higher pressure conditions than would be expected, as well as difficulties in quantifying the phenomenon of bubble concentration increase. This was most pronounced at the highest pressure of 700 kPa, where a large portion of the 20-40 μm bubbles fraction started forming aggregates of two or more bubbles which contributed to the larger bubble size fractions as well as increased the width of the distribution.

Raising the recirculation ratio from 5 to 7 and 10% for the same saturator pressure conditions, increased the bubble concentration and the width of the bubble distribution [48]. The effect on the mean bubble size distribution caused by rise of the recirculation ratio was most significant for the 700 kPa pressure (reduction of d_b=from 60 - 49 μm), while for P=500 and 600 kPa the differences were less pronounced.

In order to maintain a high N_b/N_p ratio and provide favourable DAF conditions, especially in case of high particle concentration, high values for the saturator pressure and recycled flow seem advisable and are often applied in practice. However, the insignificant gains in DAF efficiency differences measured as HIAC particle count, turbidity, residual coagulant and DOC removal both in bench-scale jar tests (Chapter 2) and pilot plant tests (Chapter 5), suggest that this may not be justified in some cases. Also the model calculations confirm that the changes would not significantly affect DAF efficiency. It was suggested that beside floc stability and size distribution, other factors such as the presence of NOM may also be an important factor.

7.4 CONCLUSIONS

The single collector collision efficiency DAF reaction zone model was considered for assessment under the assumption that in DAF drinking water treatment circumstances particle-bubble aggregation relies most significantly on the collision mechanism, and that laminar flow conditions exist around the bubbles (interception playing a major role compared to Brownian diffusion and sedimentation). By further assuming ideal plug up-flow conditions α_{pb} (particle-bubble collision efficiency) was calculated from the DAF kinetic model. Particle (floc) size distributions before and after the coagulation and flocculation, and after DAF were obtained by computer image analysis and used in the model calculations. Bubble size analysis was provided in the same manner for the model calculations. The particle (floc) density was calculated using the relative floc density concept. Incorporating the measured or calculated values under different process conditions yielded α_{pb}=0.04-0.75 for DAF jar test experiments, and α_{pb}=0.26-1.0 (1.79) for DAF pilot plant experiments.

The higher α_{pb} values for the DAF jar tests were accomplished under optimal coagulation/flocculation conditions (α_{pb}=0.39), and in case of enhanced coagulation conditions either due to application of cationic polyelectrolyte Superfloc C-573 as coagulant aid (α_{pb}=0.75), or in the case of a low coagulation pH 6 below the IEP=6.5-7 (α_{pb}=0.42). In the two latter cases, the relatively high α_{pb} values were related to improved particle coagulation due to adsorption coagulation. The accompanying particle (floc) removal efficiency (relative to the particle concentration after flocculation) was 50-70%, while the overall particle removal efficiency (relative to the raw water particle concentration) varied - it was lower or higher than the respective floc removal efficiency, depending on the coagulation/flocculation conditions. High(er) α_{pb} values do not necessarily come with high(er) overall DAF efficiency, suggesting that other factors which are not included in the model may influence the DAF efficiency.

A high flocculation G value of 100 s^{-1} had a negative effect on the α_{pb} value both at pH 7 and 9, compared to the same pH conditions at a flocculation G=50 s^{-1}. The use of polyelectrolyte shifted the particle (floc) size distribution to larger. The increased mean floc size was accompanied by decreased floc density and favourable flotation circumstances. The calculated corresponding α_{pb} values under bench-scale experimental conditions were as high as 0.75, while under pilot plant conditions a value of 1.0 (1.79) was calculated. The corresponding DAF particle (floc) removal efficiency was 56% and 83%, respectively.

The use of ozone as conditioner generally resulted in higher α_{pb} values than when Fe(III) was used as a coagulant without conditioning; this held both with or without the use of coagulant aid (α_{pb}=0.34 and 0.37 for combined ozone + Fe(III), and for ozone, Superfloc C-573 and Fe(III) treatment; α_{pb}=0.24 for Fe(III) treatment only. $KMnO_4$ conditioning under DAF jar test conditions resulted in low efficiency and accompanying α_{pb}=0.19 value.

Model calculations suggested that both for the jar tests and the pilot plant experiments efficiency could be increased by a bubble size distribution with a smaller mean size. Under these process conditions the model calculations suggested that it would be difficult to reach removal efficiencies >90%, as this requires generally high particle-bubble attachment efficiency ($\alpha_{pb}\geq0.5$).

Model sensitivity analysis under the assumption of constant process efficiency, showed higher sensitivity of the α_{pb} towards particle than towards bubble size change. The width of the bubble size distribution had little impact on process efficiency. This was supported by turbidity, particle count (HIAC), residual coagulant and DOC measurements. Model calculations suggested a slight efficiency improvement in the case of a bubble size distribution with a low standard deviation (S=3, "narrow" distribution) and low bubble size mode (30 μm), while an increase of the bubble size mode (60 μm) would decrease process efficiency by 30%.

With increase of saturator pressure bubble concentration (N_b) increased and mean bubble size was reduced (d_b). The increased bubble concentration at high saturator pressure increased the chances of bubble collision and resulted in their coalescence. The increase of the recycled flow ratio increased the standard deviation of the distribution, while the impact on the particle size was not so well defined. Model calculations confirmed the relatively low effect of the bubble size distribution on DAF process efficiency. The application of high recirculation ratios (>7%) and saturator pressures (>500 kPa) is not always fully justified. It may result in negligible or even adverse effects on process efficiency.

The single collector collision efficiency DAF kinetic model can be a valuable tool in the design of DAF plant facilities, as well as for DAF experimental design, since it defines a range of coagulation/flocculation and DAF process parameters that affect the DAF process efficiency. Physical verification of the model, however, largely relies on the accurate prediction of α_{pb}. Determination of this lump coefficient largely depends on the coagulation/flocculation process. The role of raw water quality factors affecting the coagulation/flocculation and DAF process efficiency, such as NOM presence, concentration and composition, the presence of surface active agents, etc. must also be considered. The modelling assumption of treating particle-bubble collision as the only relevant mechanism for their aggregation, while neglecting other proposed mechanisms such as entrapment of bubbles within a condensing network of flocs and growth of bubbles from nuclei within flocs is not fully justified. However, these mechanisms are complex and difficult to model, in view of their dependance of the water quality.

REFERENCES

1. Edzwald J.K. and Wingler B.J., 1990. Chemical and physical aspects of dissolved air flotation for the removal of algae - *J. Water SRT-Aqua*, **39**, pp. 24-35.
2. Edzwald J.K., 1993. Algae, bubbles, coagulants, and dissolved air flotation - *Water Science and Technology*, **27**, 10, pp. 67-81.
3. Zabel T., 1992. The advantages of dissolved air flotation for water treatment. *JAWWA*, **77**, pp. 42-46.
4. Janssens J.G., 1990. The application of dissolved air flotation in drinking water production, in particular for removal of algae. *DVGW Wasserfachlichen Aussprachetagung*, Essen, Germany.
5. Gregory R., 1997. Summary of general developments in DAF for water treatment since 1976. Proc. *CIWEM (Chartered Institution of Water Institution and Environmental Management) International Conference on Dissolved Air Flotation in Water Treatment - An Art or a Science?*, London, UK.
6. Malley J.P. and Edzwald J.K., 1991. Concepts for dissolved-air flotation treatment of drinking waters - *J. Water SRT-Aqua*, **40**, 1, pp. 7-17.
7. Schers G.J. and van Dijk J.C., 1992. Flotatie, de theorie en de praktijk. (Flotation, the theory and the practice) *H_2O*, **25**, 11, pp. 282-290
8. Kitchener J.A. and Gochin R.J., 1981. The mechanism of dissolved air flotation for potable water : basic analysis and proposal. *Wat. Res.*, **15**, pp. 585-590.
9. Fukushi K., Tambo N. and Matsui Y., 1994. A kinetic model for dissolved air flotation in water and wastewater treatment. Proc. *Joint specialised conference on Flotation Processes in Water and Sludge Treatment*, Orlando, FL, USA.
10. Edzwald J.K., 1997. Contact zone modelling and the role of pretreatment in dissolved air flotation performance. Proc. *CIWEM (Chartered Institution of Water Institution and Environmental Management) International Conference on Dissolved Air Flotation in Water Treatment - an Art or a Science?*, London, UK.
11. Han M, Dockko S. and Park C., 1997. Collision efficiency factor of bubble and particle (α_{pb}) in DAF. Proc. *CIWEM (Chartered Institution of Water Institution and*

Environmental Management) International Conference on Dissolved Air Flotation in Water Treatment - an Art or a Science?, London, UK.

12. Reay D. and Ratcliff G. A., 1973. Removal of fine particles from water by dispersed air flotation : effects of bubble size and particle size on collection efficiency. *The Canadian Journal of Chemical Engineering*, **51**, pp. 178-185.
13. Flint L.R. and Howarth W.J., 1971. The collision efficiency of small particles with air bubbles - *Chemical Engineering Science, Pergamon Press*, **26**, pp.1155-1168.
14. Bernhardt H. and Clasen J. 1991. Flocculation of micro-organisms. *J. Water SRT-Aqua*, **40**, 2, pp. 76-87.
15. Gregory J., 1993. The role of colloid interactions in solid-liquid separation. *Wat. Sci. Tech.* **27**, 10, pp. 1-17.
16. Malley J.P. and Edzwald J.K., 1991. Laboratory comparison of DAF with conventional treatment - *J. AWWA*, **83**, pp. 56-61.
17. Vlaški A., van Breemen A.N. and Alaerts G.J., 1995. Optimisation of coagulation conditions for the removal of cyanobacteria by dissolved air flotation or sedimentation, *J. Water SRT-Aqua*, **45**, 5, pp. 253-261.
18. Petruševski B., 1996. *Algae and particle removal in direct filtration of Biesbosch water.* Ph.D. thesis TU and IHE - Delft, published by A.A. Balkema, the Netherlands.
19. De Groot C.P.M. and van Breemen A., 1987. *Ontspanningsflotatie en de bereiding van drinkwater.* Mededeling van de Vakgroep Gezondheidstechniek & Waterbeheersing, Faculteit der Civiele Techniek, Technische Universiteit Delft, (*Dissolved flotation and the preparation of drinking water.* Internal Communication Report of the Sanitary Engineering and Water Management Group, of the Civil Engineering Faculty, Technical University Delft, the Netherlands).
20. Oldenziel D.M., 1979. *Bubble cavitation in relation to liquid quality.* Ph.D. thesis, Enschede, the Netherlands.
21. Haarhoff J. And Steinbach S., 1997. Towards maximal utilisation of air in dissolved air flotation. Proc. *CIWEM (Chartered Institution of Water Institution and Environmental Management) International Conference on Dissolved Air Flotation in Water Treatment - an Art or a Science?*, London, UK.
22. Steinbach S. And Haarhoff J., 1997. Air precipitation efficiency and its effects on the measurements of saturator efficiency. Proc. *CIWEM (Chartered Institution of Water Institution and Environmental Management) International Conference on Dissolved Air Flotation in Water Treatment - an Art or a Science?*, London, UK.
23. Tambo N., Igarashi T. and Kiyotsuka M., 1985. An electrophoretic study of air bubble attachment to aluminium clay/colour flocs. *J.JWWA*, **604**, pp. 2-6.
24. Okada K. and Akagi Y., 1987. Methods and apparatus to measure the ζ-potential of bubbles. *Jour. Chem. Eng. Japan*, **20**, pp. 11-15.
25. Ives K.J., 1995. The inside story of water treatment processes. *J.Environmental Engineer.* pp. 846-849.
26. Edzwald J.K., 1994. Principles and applications of dissolved air flotation - Keynote address, *Joint specialised conference on Flotation Processes in Water and Sludge Treatment*, Orlando, FL, USA.
27. Yao K.M., Habibian M.T. and O'Melia C.R., 1971. Water and waste water filtration: concepts and applications. *Environmental Science & Technology*, **5**, 11, pp.1105-1112.
28. Ives K.J., 1960. The significance of surface electric charge on algae in water

purification. *J. Biochem. Microbiol. Technol. Engineer.* 1, pp. 37-47.

29. Malley J.P., 1994. The use of selective and direct DAF for removal of particulate contaminants in drinking water treatment. Conference paper, *Joint specialised conference on Flotation Processes in Water and Sludge Treatment*, Orlando, FL, USA.
30. Nakamura T., Adachi Y. and Suzuki, M., 1993. Flotation and sedimentation of a single *Microcystis* floc collected from surface bloom. *Wat. Res.*, **27**, 6, 979-983.
31. Nakagawa M., Takamura Y. and Yagi O., 1987. Isolation and characterization of the slime from a cyanobacterium *Microcystis aeruginosa* K-3A. *Agric. Biol. Chem.* **51**, pp. 329-338.
32. Petruševski B., Vlaski A., van Breemen A. N. and Alaerts, G.J., 1993. Influence of algal species and cultivation conditions on algal removal in direct filtration. *Wat. Sci. Tech.*, **27**, pp. 211-220.
33. Roberts K.L., Weeter D.W. and Ball R.O., 1978. Dissolved air flotation performance. Proc., *33rd Industrial Waste Conference*, Purdue University,USA.
34. Fawcett N.S.J., 1997. The hydraulics of flotation tanks : computational modelling. Proc. *CIWEM (Chartered Institution of Water Institution and Environmental Management) International Conference on Dissolved Air Flotation in Water Treatment - an Art or a Science?*, London, UK.
35. Shawcross J., Tran T., Nickols D. And Ashe C.R., 1997. Pushing the envelope : dissolved air flotation at ultra-high rate. Proc. *CIWEM (Chartered Institution of Water Institution and Environmental Management) International Conference on Dissolved Air Flotation in Water Treatment - an Art or a Science?*, London, UK.
36. Breese S., 1997. Use of comutational fluid dynamics in the design and optimisation of DAF basins. Proc. *CIWEM (Chartered Institution of Water Institution and Environmental Management) International Conference on Dissolved Air Flotation in Water Treatment - an Art or a Science?*, London, UK.
37. O'Neill S. And Oddie G., 1997. Physical modelling study of the dissolved air flotation process. Proc. *CIWEM (Chartered Institution of Water Institution and Environmental Management) International Conference on Dissolved Air Flotation in Water Treatment - an Art or a Sience?*, London, UK.
38. Ta C.T. and Brignal W.J., 1997. Application of single phase computational fluid dynamics techniques to dissolved air flotation tank studies. Proc. *CIWEM (Chartered Institution of Water Institution and Environmental Management) International Conference on Dissolved Air Flotation in Water Treatment - an Art or a Science?*, London, UK.
39. Dahlquist J., 1997. The state of DAF development and applications to water treatment in Scandinavia. Proc. *CIWEM (Chartered Institution of Water Institution and Environmental Management) International Conference on Dissolved Air Flotation in Water Treatment - an Art or a Science?*, London, UK.
40. APHA, AWWA and WPCF, 1985. *Standard methods for examination of water and wastewater*, Am. Publ. Health Assoc., Washington D.C., USA.
41. Nederlands Normalisatie Instituut, 1982. NEN: 6460: Water - Determination of iron content by atomic absorption spectrometry (flame technique).
42. Nederlands Normalisatie Instituut, 1982. NEN: 6466: Water - Determination of manganese content by atomic absorption spectrometry (graphite furnace technique).

43. Rijk de S.E. and Graaf van der J.H.J.M. and Blanken den J.G., 1994. Bubble size in flotation thickening - *Water Research*, **28**, 2, 465-473.
44. Boller M. A. and Kavanaugh M. C., 1995. Particle characteristics and headloss increase in granular media filtration. *Wat. Res.*, **29**, 4, pp. 1139-1149.
45. Alaerts G.J. and Van Haute A., 1981. Stability of colloid types and optimal dosing in water flocculation. *Physicochemical Methods for Water and Wastewater Treatment*, Pawlowski, L. (eds.), Elsevier Publ., New York/Amsterdam.
46. Filella M., Buffle J. and Leppard G.G., 1993. Characterisation of submicron colloids in freshwater : Evidence for their bridging by organic structures. *Wat. Sci. Tech.*, **27**, 11, pp. 91-102.
47. Buffle J. and Leppard G.G., 1995. Characterization of aquatic colloids and macromolecules. Structure and behaviour of colloidal material. *Environmental Science and Technology*, **29**, 9, pp. 2169-2175.
48. Takahashi T., Miyahara T. and Mochizuki H., 1979. Fundamental study of bubble formation in dissolved air pressure flotation. *Journal of Chemical Engineering of Japan*, **12**, 4, 275-280.

APPENDIX 7.1

EXPERIMENTAL MODE		O_3 (mgO_3/ mgTOC)	$KMnO_4$ (mg/L)	Contact time(min)	Contact pH	Fe(III) (mg/L)	Poly. (mg/L)	Coag. G (s^{-1})	Coag. t (sec)	Coag. pH	Flocc. G (s^{-1})	Flocc. t (min)	R (%)	P (kPa)	α_{pb} (/)
BENCH-SCALE	A	/	/	/	/	3.0	/	1,000	30	8.0	10	10	7	500	0.39
	B	/	/	/	/	10.0	/	1,000	30	6.0	70	10	5	500	0.42
	C	/	/	/	/	3.0	0.5	+ 500	+ 30	8.0	70	10	7	500	0.75
	D	/	/	/	/	3.0	1.0	+ 500	+ 30	8.0	70	10	7	500	0.33
	E	0.2	/	/	7.5	3.0	/	1,000	30	7.5±0.2	10	10	7	500	0.34
	F	0.2	/	/	7.5	3.0	0.5	+ 500	30	7.5±0.2	10	10	7	500	0.44
	G	0.2	/	/	7.5	/	1.0	500	30	7.5±0.2	10	10	7	500	0.38
	H	/	0.7	30	8.0	5.0	/	1,000	30	8.0	70	15	7	500	0.19
PILOT PLANT	I	/	/	/	/	10.0	/	>3,500	<1	8.0	10	22	7	500	1.0
	J	/	/	/	/	10.0	/	>3,500	<1	8.0	50	22	7	500	0.37
	K	/	/	/	/	15.0	/	>3,500	<1	8.0	50/30	18/4	7	500	0.36
	L	/	/	/	/	15.0	/	>3,500	<1	8.0	50/30	18/4	7	500	0.28

Comment :

Poly. : Polyelectrolyte
Coag. : Coagulation
Flocc. : Flocculation
+ 500 : Coagulation G value in case of polyelectrolyte as coagulant aid, applied after 30 s at G of 1,000 s^{-1} for the $FeCl_3$
+ 30 : Coagulation time in case of polyelectrolyte as coagulant aid, applied after 30 s at G= 1,000 s^{-1} for the $FeCl_3$
50/30 and 18/4 : tapered flocculation with a G=50 s^{-1} and t=18 min in the first step and G=30 s^{-1} and t=4 min in the second flocculation step

Table 7.1 α_{pb} values obtained for the bench-scale and pilot plant experimental mode under different process conditions.

Chapter 8

EXTENDED SUMMARY AND CONCLUSIONS

8.1 Potential remedies for the algae problem

8.1.1 Reservoir water quality management

Restrictions on ground water utilisation render an increase in surface water use inevitable for drinking water supply. To protect the treatment process and the customer from excessive variability in the river water quality (incidental pollution events, or low discharge periods), and to improve the generally poor river water quality, usually reservoir storage is provided. These storage reservoirs are designed for detention times depending on the river water quality (expected season's and calamities pattern) and partly on the expected degree of 'self purification' during the storage period. Apart from natural purification processes, such as sedimentation, which occur in these impoundments, very often the water treatment process begins within or even before them. For example, Fe(II) application in the water intake for promoting phosphorus precipitation and the settling of suspended and colloidal matter (including algae) within the reservoir is current practice at the WRK III (Rhine-Kennemerland Water Transport Company) treatment plant in Andijk and at the DZH (Dune Water Supply Company South Holland) in the Andelse Meuse. The former also applies periodical reservoir mixing in its 20 m deep reservoir as a measure to avoid stratification, provoke light attenuation and create unfavourable algal growth conditions. On the other hand, the GWA (Amsterdam Water Supply) coagulates the water to cause sedimentation before the water enters Loenderveen Lake. Thus, apart from suspended and colloidal matter precipitation, phosphorus which is often the limiting algal growth nutrient is also precipitated.

According to the statistical analysis of the water quality of the five water resources conducted during this study there is a significant difference in their water quality, while no significant changes of the water quality occurred during the analysed periods. Statistical analysis further identified phosphorus as the key element which controls algae growth within the studied reservoirs (r^2=0.87 for log(chlorophyll-total) versus log (PO_4-total)).

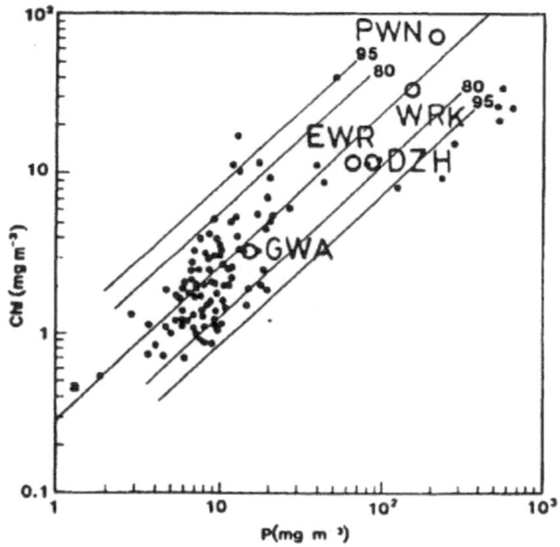

Fig. 8.1 Annual mean chlorophyll vs. P concentrations for the five impoundments, superimposed on a graph from Harris (Chapter 1, [23]) .

A similar correlation was obtained between log(chlorophyll-*a*) and log (PO_4-total) and pH (r^2=0.80). The transformation equation predicting the chlorophyll-*a* value on a year average basis is log(chlorophyll-*a*) = 0.52 log (PO_4-total) + 0.6 (pH) - 3.2. The superimposed data for the chlorophyll-*a* and the PO_4-total on a graph containing data for water impoundments elsewhere, fell in the 95% confidence limit of the respective regression line (Fig. 8.1). Accordingly, the approximate PO_4-total concentration which is probably to cause problems in terms of algae growth (expressed as chlorophyll-*a*) lies above 0.06 mg P/L.

Most of the reservoir water quality management measures have algae as their direct or indirect target. Based on the degree of success of the reservoir management, the algae population varies by concentration and by species composition. Expressed in chlorophyll-*a* concentrations, the algae concentration in the studied water resources varied from 1.6-8.4 µg/L for the Loenderveen Lake to 18-238 µg/L for the PWN (North Holland Water Supply Companies) reservoir as minimum and maximum yearly average, respectively (Table 1.1, Chapter 1). This suggests different water quality at different locations, oligotrophic for e.g. the Loenderveen Lake and highly eutrophic for e.g. the PWN reservoir; each condition is characterised by a specific algae population. The more eutrophied the water source analysed, the more restrictive was the composition of the algae population. The species encountered included the diatoms *Melosira* spp., *Stephanodiscus* spp., *Asterionella* spp., *Nitzschia* spp. and µ-diatoms, the cyanobacteria *Microcystis* spp., *Oscillatoria* spp., *Aphanizomenon* spp., *Anabaena* spp., the green algae *Scenedesmus* spp., *Coelastrum* spp., *Oocystis* spp., as well as *Cryptophyceae* spp. and *Crysophyceae* spp.

8.1.2 Algae and water treatment

a. Water treatment problems related to algae

The rationale behind the definition of the algae genera that are most relevant for the treatment was to identify the algae responsible for concentration peaks. The analysis pointed to cyanobacteria as the most frequent, versatile and periodically dominant species that regularly would occur in excess concentrations in all the analysed water resources. Further analyses pointed to *Microcystis aeruginosa* and *Oscillatoria agardhii* as the most relevant cyanobacteria.

Although of very different morphology and physiology, the abundant concentrations in which these two species are encountered during their blooms contribute to a range of typical water treatment problems. As a result, very often they end up in high concentrations in the plants' effluent. The single cells form of *M. aeruginosa* poses the greatest threat in this context, as its size and shape (oval shape, 3-10 µm diameter) make it difficult to remove. A comparable problem exists for pathogenic microorganisms (e.g. *Cryptosporidium* oocysts and *Giardia* cysts) that are of a similar size. The problems associated with the cyanobacteria also include toxicity to humans and animals.

Filtration, whether as rapid or slow sand filtration, commonly provides the ultimate barrier to prevent passing of particles, including algae (Table 1.5, Chapter 1). Some plants, such as the WRK III and PWN, practice pre- and post-straining (200 and 35 µm mesh size), while further treatment before final filtration includes sedimentation or dissolved air flotation (DAF). The algae related treatment problems (Table 1.6, Chapter 1) include increase of coagulant demand (up to

100% of the usual dose), increased filter backwash quantities, trihalomethane formation (only during periodical chlorination for prevention of mussels growth in pipe lines, which recently ceased to be applied for this reason), clogging of filters, algae passing through treatment in objectionable quantities, and consequent increase of MFI (Modified Fouling Index) and AOC (Assimilable Organic Carbon) values of the final filtrate. In order to cope with these problems, a number of measures are applied, including deepening of the reservoir (PWN) and change of treatment mode and ozone dosing (GWA), while some of the water supply companies are considering new and more efficient treatment technologies for their future extensions. In extreme situations raw water sources are abandoned and changed for better ones (Table 1.6, Chapter 1).

b. DAF, an increasingly attractive option for algae laden water treatment

Given the natural tendency of algae to float, DAF emerges as an attractive treatment technology to remove particles and flocs, as opposed to conventional sedimentation. It is characterised by high rate, high efficiency, low chemicals consumption, and high versatility and adaptability. However, existing discrepancies in literature and practice, suggest that DAF is a process worth of further research. The interest that DAF has attracted worldwide since the seventies has resulted in the construction of numerous plants in Scandinavia, the United Kingdom, Australia, South Africa, Canada, Japan, France and the Benelux (Section 1.3.2). The recent construction of the largest DAF facility in the world in Hong Kong ($1.2x10^6$ m^3/day), as well as the planned DAF facility for the city of Vancouver, Canada ($1.0x10^6$ m^3/day) confirm the interest for DAF. In the nineties and early next century most of the treatment plants in Western Europe require an increase of production capacity (in our case : at GWA, DZH and GWG) due to increased water consumption, or upgrading due to more stringent water quality regulations. This compounds our interest in DAF as a viable and suitable option for algae laden water treatment.

c. Coagulation/flocculation, a prerequisite for efficient DAF

The most important prerequisite for efficient DAF is sufficiently destabilised particles and floc material that would result in favourable floc-bubble attachment. Thus, an approach including the up-stream particle agglomeration (i.e. coagulation/flocculation) is necessary. Metal coagulants, especially Fe(III), are a more attractive option for Dutch water treatment circumstances than alum. Although the issue remains controversial, in this study polyelectrolyte coagulant aids were used as sole coagulant or in conjunction with Fe(III) to improve DAF performance (Chapter 3, 4 and 5). The emphasis was placed on the agglomeration phase, considered essential whether between particles in the coagulation/flocculation process or between particles (algae) and bubbles in the DAF process.

d. Algae conditioning, an option for enhancing DAF efficiency

Up-stream algae conditioning by oxidants such as chlorine, ozone, hydrogen peroxide, $KMnO_4$, or their combinations, is currently considered an option for improving the coagulation/flocculation process and hence the down-stream treatment, especially in cases where these chemicals are also used for disinfection. Due to the trihalomethane formation, the practice of chlorine disinfection in Dutch drinking water production is almost discontinued. The bromate formation has led to a cautious attitude of Dutch water utilities in its application of ozone. Ozone however, has been proven to considerably influence particle removal in several cases (Chapters 4 and 5). Another strong oxidant, $KMnO_4$, which is not compromised by the production of hazardous by-products, has also received attention as an alternative to ozone (Chapters 4 and 5). Both oxidants were considered here in the experimental context of metal coagulant, or metal coagulant and polyelectrolyte treatment.

e. DAF process kinetics

An essential aspect of the study here were the kinetics of the DAF process. The complexity of the DAF process and the numerous determining factors (raw water quality, pre-treatment and other process variables) have caused substantial controversy (Chapter 7). We opted to address these inconsistencies and critically assess one of the proposed DAF kinetic models, namely the single-collector collision efficiency contact zone model.

8.2 Methodology

8.2.1 Experimental modes

Algae laden water treatment research faces the challenge of having access to water with high algae concentrations representative of algae blooms. This problem is imposed by the periodicity of peak concentrations of relevant species, limited to relatively short periods of the year (late summer and early autumn in the case of *M. aeruginosa*). The possibility to concentrate algae by a tangential flow system as applied elsewhere, can be used for research where algae of particular interest are present in the water all throughout the year. This was not the case in any of the studied raw water resources here.

Therefore, model water was produced with laboratory cultured algae; this however imposed limits with regard to algae culturing capacity. Of the two cyanobacteria, *M. aeruginosa* and the *O. agardhii*, only the first one could be cultured in the laboratory without too many reproducible problems, and in repeatable quantities. Experiments with model water had clear drawbacks. Spiking water of well defined quality with cultured algae potentially biased results as a consequence of the introduction of chemicals (nutrients and complexing agents) that are constituents of algae growth media; and the inconsistent quality and quantity of the cultured algal cells (bacteria infestation and algal die-off, differences in nutrients concentration, and in algae growth stage). These effects were carefully assessed during the discussion of results.

A batch jar test experimental research set-up, provided with DAF and small-scale rapid sand filters, was used for laboratory model water research. The results formed the basis for the continuous-flow DAF pilot-plant research (Q_{max}=13 m^3/h), executed twice during the summer/autumn period of 1995 and 1996, at the site of the WRK III treatment plant, using directly reservoir water and naturally occurring cyanobacteria blooms with *M. aeruginosa* being the dominant species. The DAF efficiency trends recorded for the two experimental set-ups were in general well correlated. In all cases the reproducibility of the obtained results was regularly tested, and, when necessary, equipment was adjusted. However, the scale difference and the difference between the model and the reservoir raw water characteristics, resulted in differences in the absolute values of the DAF removal efficiency (see Section 8.7).

8.2.2 Analytical methods

The absence of an algae related drinking water standard led to adoption of a range of analytical techniques to assess process efficiency and interpret process mechanisms. Apart from conventionally applied lump parameters like turbidity, TOC/DOC and UV_{254nm}, the assessment was complemented with particle counting. Computer image analysis (Magiscan) was used for qualitative description of the particle (floc) size distribution after different treatment stages, in particular after flocculation when a HIAC particle count was inapplicable due to the high shear and the floc break-up in its particle sampling orifice. This analytical methodology was the basis for assessment of the DAF kinetic model. Inverted microscope inspection was also periodically carried out in order to verify particle count results, as well as for the observation of flocculated water. Other visual observation techniques included high resolution scanning electron microscopy (SEM), which was especially important for the assessment of process phenomena after conditioning with oxidants.

Alga electrostatic surface charge was measured through electrophoretic mobility (EM). Additional measurements included residual coagulant (total iron), residual Mn (total and dissolved Mn in case of permanganate conditioning), and BrO_3^- analysis in the case of ozone conditioning. Modified fouling index (MFI), indicative of the pore clogging capacity of the treated water, was measured during the pilot-plant research.

The use of sophisticated analytical technologies such as the computer image analysis and the SEM opened new avenues and provided more powerful research tools for the process analysis.

8.3 DAF versus sedimentation

Model water batch experiments confirmed that sedimentation or DAF alone is insufficient to cope with moderate to high initial concentrations of the *M. aeruginosa* species (≈10,000 cells/mL), confirming the need of the down-stream filtration step. Optimal DAF and sedimentation conditions resulted in insignificant particle (algae) removal efficiency differences, while sedimentation was more efficient in turbidity removal. This is partly a result from the three times higher coagulant dose applied for optimal particle removal in case of sedimentation (10 mg Fe(III)/L compared to 3 mg Fe(III)/L for DAF). Coagulant dose stoichiometry was established for DAF and the single celled cyanobacteria *M. aeruginosa* within the tested concentration range of 5,000-20,000 cells/mL (the latter concentration corresponding to approximately 50 μg/L chlorophyll-*a*). Further increase of algae concentration by 25 % did not raise the coagulant dose requirement, suggesting that a critical mass (concentration) of algae and coagulant has been reached allowing for full utilisation of the sweep coagulation. The accompanying sludge quantities emerging from the higher coagulant dose for sedimentation would result in additional costs and environmental inconveniences.

The high coagulant dosages required for the sedimentation resulted in floc volume distributions with increased volume of particles both in the smaller (<50 μm) and larger (>50 μm) particle size ranges. In addition to operational and literature data which suggest high(er) flocculation G values as favourable for DAF, a low flocculation G value of 10 s^{-1} resulted in similar and better algae removal efficiency than the value of G=70 s^{-1}; it resulted in a floc size distribution comprising more voluminous and lighter floc material favouring efficient DAF. Sedimentation required 30 min flocculation time at G=30 s^{-1}, compared to the 10 min flocculation time at G=10 s^{-1} required for DAF. Similarly, a longer detention time (60 min) was necessary for sedimentation, compared to 10 min for DAF. The combined interaction of the flocculation G and t values was found to be statistically significant for the DAF process efficiency, the effect of each separately of each other being insignificant. Although this implies the existence of an optimal Gt value region, this was not confirmed experimentally. Tapered flocculation with diminishing G in consecutive flocculation stages did not improve DAF efficiency.

Table 8.1.a and b address the potential benefits and drawbacks related to application of DAF and sedimentation, respectively. In practice, the choice between these two treatment technologies depends upon a range of water quality parameters, including algae concentration, turbidity, NOM presence and nature, colour, hardness, temperature, etc. It also depends upon environmental considerations regarding the quantity and quality of the produced sludge, as well as space and skilled labour force availability.

Table 8.1.a Comparative benefits and drawbacks associated with a DAF based treatment scheme (including filtration), in the removal of moderate to high concentrations of cyanobacteria ("algae").

DISSOLVED AIR FLOTATION (DAF)	
BENEFITS	**DRAWBACKS**
■ high algae removal efficiency, at low(er) coagulant consumption up to 50% less than sedimentation ■ lower sludge production (less environmental impact) ■ high(er) dry solids content of sludge (4-6% compared to 2-4% after sedimentation) - reduced water losses (1% compared to 3% in sedimentation) - consequent lower sludge treatment costs ■ high(er) loading rates of up to 25-30 m^3/m^2h - smaller unit space demands - lower construction cost ■ relatively short flocculation time (≥10(15)min) - low(er) construction cost ■ efficient at low flocculation energy input (G=10 s^{-1}) - low energy costs ■ efficient at cold temperatures, low sensitivity to temperature variations ■ robust but flexible process, reaches steady state after approximately 45 min	■ low(er) coagulant dose may result in less complete precipitation and less efficient sweep coagulation (relative to optimal sedimentation), due to partial organic matter-metal ion complexation - high(er) residual turbidity - high(er) residual coagulant concentration - high(er) down-stream filter effluent MFI values ■ possible bubble carry-over onto the filters at very high process rates, and filter operation disturbance ■ high(er) operational costs for recycled stream pumping and air saturation ■ skilled operating personnel required

Table 8.1.b Comparative benefits and drawbacks associated with a sedimentation based treatment scheme (including filtration), in the removal of moderate to high concentrations of cyanobacteria ("algae").

SEDIMENTATION	
BENEFITS	**DRAWBACKS**
■ efficient sweep coagulation conditions achieved at high(er) coagulant dose ■ additional benefit of improved particle (floc) sedimentation due to high particle concentration - good turbidity removal - low(er) residual coagulant values - low(er) down-stream filter effluent MFI values ■ efficient DOC (TOC) removal associated with the high inorganic coagulant dose ■ low flocculation energy input and costs ■ robust and easy to operate	■ moderate to high algae removal, at high to very high coagulant consumption ■ low loading rates of 1 - 2 m^3/m^2h - larger unit space demands - larger construction cost ■ low(er) solids content of the sludge (2-4% compared to 4-6% after DAF) - larger quantities of sludge - higher sludge treatment costs

The treatment technology choice should be based on pilot-plant and cost-benefit analysis, considering an integrated approach which includes at least up-stream agglomeration and down-stream filtration. Enhanced treatment options including e.g. the use of polyelectrolytes and/or algae conditioners should also be evaluated.

In this context, relatively low doses of cationic polyelectrolyte as coagulant aid, of 0.5 mg Superfloc C-573/L for DAF and 1.0 mg C-573/L for sedimentation increased process efficiency of both processes (without down-stream filtration considered) up to 99%. These results, together with the doubled removal efficiency achieved at coagulation pH below the iso-electrical point (IEP) (pH 6 vs. pH 8), implicate the importance of adsorption-coagulation and particle charge neutralisation for down-stream process efficiency.

The non-optimised down-stream bench-scale filtration step (one-layer H=0.2 m filter bed of sand d=0.8-1.25 mm commercial fraction, v=10 m/h, without considering the use of polyelectrolyte as coagulant aid) increased removal efficiency for both DAF + filtration to up to 90% and for sedimentation + filtration to up to 95%. The filtration efficiency was in general sensitive to the flocculation G value and the resultant floc size/density ratio, as well as to the residual coagulant level. The filtration efficiency increased with higher residual coagulant concentrations after DAF or sedimentation. The small scale of the filters may have given rise to filter wall effects, and together with the non-ripened state of the filter, both imparted a negative effect on filtration efficiency. Furthermore, in order to achieve more than the obtained 2 log algae removal on the raw water algal concentration of 10,000 cells/mL, would require an effluent particle concentration of 10-100 /mL, which is a figure that often approached the particle count of the demineralised water used for effluent sample dilution.

Optimized DAF performed similarly efficient as sedimentation in terms of particle (algae) removal, but was accompanied by cost savings, including lower flocculation energy and shorter flocculation time, shorter hydraulic retention time and lower sludge production. Cationic polyelectrolytes and low coagulation pH (below the iso-electrical pH 6.5-7) significantly improved DAF particle removal efficiency. The surface charge effects confirm the important role of well destabilised floc material for improved particle-bubble attachment and hence efficient DAF. Thus, improving the coagulation/flocculation process is the key for more efficient DAF.

8.4 Options for enhanced DAF

8.4.1 DAF with Fe(III) coagulant

DAF pilot-plant investigations on reservoir water characterized by a seasonal cyanobacteria bloom (*M. aeruginosa*) confirmed its efficiency, reliability and robustness for the removal of algae. The optimal coagulant dose for DAF regarding particle (≥2.75 μm) removal was significantly lower than the one applied in the full-scale WRK III sedimentation line (7-12 g Fe(III)/m^3 compared to 20-24 g Fe(III)/m^3). This resulted in the relatively high particle removal efficiency of 92-97.3% (between 1 and 2 log removal). The accompanying turbidity removal was lower and residual iron content was higher than the ones achieved by the full-scale sedimentation. Complexation of the iron coagulant by the organic matter created difficult to remove organo-iron complexes and restricted the amount of coagulant available for destabilisation of the colloidal and particular matter. The large coagulant doses applied prior to sedimentation created very favorable sweep coagulation and settling conditions, that caused better turbidity removal and lower residual coagulant levels.

Flocculation times longer than 15 minutes were found to be necessary for DAF in order to produce effluent of reasonable quality (≥90% particle removal after DAF). The contact opportunities for the destabilised particulate and colloidal matter needed to be more numerous, and required flocculation time longer than suggested by other sources, typically in the range of 5 min (Chapters 1 and 3). These values are probably typical for the characteristics of the organic and inorganic particulate and colloidal matter in this raw water, the water temperature (10-17 ° C), and the hydraulic conditions in the flocculation units. The flocculation energy input (G value) was found to be of less influence on DAF process efficiency; the highest particle removal was obtained at a low flocculation G=10 s^{-1} both in the jar test and pilot-plant experiments.

Both DAF jar test and pilot-plant results suggested that the variation of the saturator pressure (within the range of 500-700 kPa) and the recycled flow ratio (5-10% of the throughput) did not significantly affect the particle removal efficiency. This implicated that the resultant bubble-to-particle concentration ratio (N_b/N_p) was sufficiently large in all investigated cases to result in comparable DAF efficiency. Thus, the particle removal efficiency under the tested circumstances was predominantly affected by the coagulation/flocculation process.

The DAF effluent was subject to down-stream filtration in conditions reflecting the full-scale WRK III filtration. The DAF + filtration had a slightly lower algae (≥2 log) and turbidity (1-2 log) removal efficiency than the full-scale sedimentation + filtration. Most difficult to achieve was the target MFI value of 5 s/L^2, however, the full-scale plant also experienced difficulties in achieving this value due to the occurring algal bloom (mean MFI value of 8.83 s/L^2 and 7.2 s/L^2 for the 1995 and 1996 study periods, respectively, with a maximum value of 24 s/L^2). The combined DAF + filtration effluent quality for larger coagulant doses in the range of 15 g Fe(III)/m^3 in combination with 0.5 g Superfloc C-573/m^3 polyelectrolyte performed equally well as the full-scale sedimentation + filtration treatment process with respect to all parameters, including the MFI which fell below the 20 s/L^2 level. This suggests that further optimization of the DAF + filtration process scheme, notably of its coagulation/flocculation and filtration stages, has a good potential for treating heavily algae laden WRK III reservoir water.

8.4.2 Polyelectrolyte enhanced DAF

Cationic polyelectrolytes as coagulant aid in the dose range of approximately 1/10th of the optimal Fe(III) coagulant dose, significantly improved the DAF efficiency achieved by Fe(III) coagulant in the sweep coagulation mode, in case of model water bench scale experiments. The DAF efficiency was affected by the polyelectrolyte electric surface charge and molecular weight characteristics. It is suggested that the application of higher charged, and lower molecular weight polyelectrolytes such as Superfloc C-573 are more suitable than the lower charged, and longer chain polyelectrolyte Wisprofloc-P. Anionic (Superfloc A-100) and non-ionic (Wisprofloc-N) polyelectrolytes proved to be unsuitable for DAF.

The cationic polyelectrolytes Superfloc C-573 and Wisprofloc-P proved to be a viable option for treating algae laden water as sole coagulants. DAF jar test results suggested that similar (or higher) particle removal efficiency can be achieved, than the one achieved by the Fe(III) coagulant alone. The particle and turbidity removal again depended on the surface charge and weight characteristic of the polyelectrolyte; the stronger charged Superfloc C-573 again was more efficient. The practical implications of these findings for Dutch water treatment are limited, since polyelectrolytes are rarely used.

Pilot-plant application of the cationic polyelectrolytes 0.1-0.7 g/m^3 Superfloc C-573 and Wisprofloc-P, combined with 7 g Fe(III)/m^3, did not result in significant DAF efficiency improvement. Unlike for the DAF jar test experiments with model water, EM measurements did not show significant particle surface charge changes after flocculation. A limited number of DAF experiments (with WRK III reservoir water) suggested that higher polyelectrolyte and coagulant doses were needed to benefit from the polyelectrolyte application. This demand is asserted by the abundant colloidal matter which needs larger amounts of metal coagulant and/or polyelectrolyte under the neutral coagulation (pH 8) conditions, in order to be more efficiently destabilised and removed.

The final choice of polyelectrolyte, whether as a coagulant or coagulant aid, also depends on the raw water characteristics, notably the character and the concentration of suspended and colloidal particulate matter, the TOC/DOC content, and the presence of certain metals such as calcium. The study proved that cationic polyelectrolytes can offer an efficient and reliable option for DAF. However, the issue of polyelectrolyte residual should be of concern, as long-term consumption of water containing polyelectrolyte monomer residuals may involve an unknown degree of health hazard.

8.4.3 Ozone conditioning and DAF

Ozone conditioning of algae in bench scale DAF experiments did, under a range of raw water quality characteristics and ozone dosing parameters (e.g., 0.4-0.8 mg O_3/mg C, hardness-to-TOC ratio >25 mg $CaCO_3$/mg TOC, low bromide concentration, pH <7.5), statistically significantly increase DAF algae removal efficiency (by 30-40%). Under ozone conditioning, the initial particle count and the initial turbidity were found to statistically significantly affect DAF treatment efficiency, suggesting the importance of the seasonality. Although an increase of the ozone dose (from 0.2 to 0.5 and 0.9 mg O_3/mg C) modified the organic matter (algae, EOM and IOM) substantially (as verified by UV_{254nm} measurements), the accompanying DAF particle and turbidity removal efficiency did not vary significantly. Increase of coagulant dose for the same ozone treatment did not statistically significantly improve DAF process efficiency, although a positive trend was noted.

Bench scale DAF with combined ozone conditioning, and metal coagulant and (cationic) coagulant aid addition tended to further increase the process efficiency by 5%. The polyelectrolyte's charge and molecular weight characteristics played an important role with respect to process efficiency, in view of the possible polyelectrolyte modification by the ozone or hydroxyl radical.

Bench scale DAF with combined ozone conditioning and (cationic) polyelectrolyte application as sole coagulant resulted in high algae removal efficiency (>80%) coupled with the absence of problems related to metal coagulants application, namely the complexing of metal coagulants by organic matter and resultant high coagulant residuals and turbidity. In this case the turbidity removal was comparably higher (>70% compared to approximately 50%). This option, however, proved to be sensitive to ozone overdosing, which resulted in deteriorated effluent quality. This is suggested to be the result of a modification of the polyelectrolyte structure into smaller, less effective components of lower particle destabilisation capacity.

Ozone conditioning of model water and (bench scale) DAF, combined with metal coagulant application, metal coagulant plus cationic polyelectrolyte coagulant aid, and cationic polyelectrolyte coagulant, all outperformed the treatment option with metal coagulant only. Ozone, polyelectrolyte, and their interaction had a statistically significant positive effect on DAF treatment efficiency. Ozone and polyelectrolyte treatment had a statistically significant effect on the particle removal efficiency irrespective of the initial particle count, while initial turbidity and polyelectrolyte treatment had a statistically significant effect on the turbidity removal efficiency, irrespective of the ozone treatment; ozone treatment itself did not statistically significantly affect turbidity removal efficiency. No statistically significant efficiency differences were detected between the above treatment combinations.

Technical limitations of the pilot plant facility did not allow verification of the bench-scale findings on a larger scale with reservoir water. However, bromate analyses showed that for the relatively low ozone dose of 0.2 g O_3/g TOC (1.2 g O_3/m^3) at pH 7, which proved highly beneficial in the bench-scale model water experiments, the bromate levels were below the WHO provisional guideline value of 25 μg/L and approaching the European guideline of 10 μg/L.

8.4.4 $KMnO_4$ conditioning and DAF

$KMnO_4$ conditioning of algae in the context of DAF treatment was investigated for the first time in this study. The interest for this oxidant was based on recent positive experiences in its application as a conditioner prior to direct filtration removal of algae (Chapter 4 and 5). Furthermore, no hazardous by-products are formed. For the model water conditions the optimal permanganate dose was found to be in the range of 0.7 mg $KMnO_4$/L, coinciding with the visually determined permanganate demand (i.e. no pink colour observable after treatment). The occasional change of permanganate demand during the study on model water, and the relatively high variability of DAF efficiency results, implicated organic matter as the main cause for these phenomena. Namely, the concentration and the characteristics of the organic matter determined the permanganate demand, the chemical activity of the permanganate and the resulting DAF efficiency. Pilot-plant $KMnO_4$ conditioning of reservoir water rich in organic matter proved to be much more efficient and reliable in algae treatment than the DAF bench-scale work had suggested. Here, a dosage of 1 g $KMnO_4/m^3$ combined with the previously determined optimal coagulant dose of 10 g Fe(III)/m^3 tended to raise DAF removal efficiency compared to that of the coagulant only.

A major drawback of permanganate conditioning is the potentially high Mn residual concentration after DAF and filtration. The residual Mn originates either from overdosing and non-reduced permanganate (dissolved Mn that is non-removable by filtration), or from the negatively charged, colloidal, hydrous MnO_2 precipitate (in principle removable by filtration). The DAF effluent Mn and Fe residual concentrations here, were higher than their respective MAC values. The cationic Superfloc C-573 coagulant aid led to significantly better particle and turbidity removal than that achieved by the Fe (III) coagulant only, or by the combined $KMnO_4$ and iron coagulant treatment. The pilot plant combined treatment by 1 g $KMnO_4/m^3$, 10 g Fe(III)/m^3 and ≥0.5 g Superfloc C-573/m^3 also tended to further increase DAF efficiency. This treatment option was more efficient in removing particles ≥2.75 μm than the full-scale WRK. III sedimentation treatment. The accompanying residual Mn and Fe of the DAF + filtration process scheme were below the 50 μg/L (in the range of 35 μg/L) and 200 μg/L MAC, respectively.

The option of lowering the Mn residual after DAF by the reducing agent Na_2SO_3 to convert all Mn into presumably removable colloidal MnO_2 also proved potentially viable if local circumstances allow it (i.e. if the effect of Na_2SO_3 dosing on residual Na and S concentrations is acceptable). Although below their respective MAC values, the Mn and Fe residuals (of particle size <1 μm) were thought responsible for high MFI values (≥50 s/L^2). This suggests that in WRK III circumstances, where the effluent MFI is of primary importance, application of the $KMnO_4$/DAF/filtration process scheme would require strict optimization of the filtration part to remove the residual Mn and Fe colloids. Optimising filtration removal for the <1 μm particle size range (e.g. by a fine-sized garnet layer) is expected to be able to achieve this goal.

8.5 Process mechanisms involved in algae laden water treatment

8.5.1 *M. aeruginosa*, an alga representative for qualitative algae removal assessment

The great variety of algal species precludes treating algae as any other inorganic particle. Their typical morphological and physiological features, such as size, shape, buoyancy, flagella activity, cell wall characteristics, EOM availability, etc., provide for a wide range of responses to treatment. The initial research plan included two distinctly different cyanobacteria species, *M. aeruginosa* and *O. agardhii*. However, problems encountered during the laboratory culturing of the latter species prevented reliable experimental work. Thus, the research findings related to *M.*

aeruginosa and the discussed process mechanisms apply to many other algae in a qualitative manner only.

Table 8.2 in Appendix 8.1 synthesizes the DAF bench-scale and pilot-plant results and observations related to the process mechanisms in the different treatment stages. It specifies beneficial processes and processes which are regarded as potential risks. Fig. 8.2 represents the schematized DAF process and the characteristic mechanisms occurring in the case of $KMnO_4$ algae conditioning.

8.5.2 Coagulation/flocculation and DAF of algae

The coagulation process is essential for efficient algae removal regardless of the type of downstream floc removal process. The origins of algal stability are their electronegative surface charge, the steric effects due to the surrounding water layer, protruding parts of the cell, extracellular organic matter (EOM) adsorbed at the cell surface, and the algae motility. Overcoming the repulsive forces and steric effects is a prerequisite for their agglomeration and subsequent removal by sedimentation, DAF, filtration or any combination of these.

It has been proposed that the concentration of aquatic colloids influences the coagulant dose in an inversely proportional manner (Chapter 4), and that this phenomenon is observed only after a critical concentration of the algae has been reached. For spherically shaped algae, such as *M. aeruginosa*, at concentrations below the critical one, a stoichiometric relation between their concentration and the required coagulant dose may be expected to exist.

The coagulant dose and coagulation pH notably influenced the particle (floc) size distribution after flocculation. DAF at coagulation pH below the IEP, which generally depends on the ionic content of the water (in our case pH 6.5-7), outperformed DAF at 7<pH<9. Under low pH conditions, i.e. below IEP, predominantly positively charged iron-hydroxo species are formed, causing efficient charge neutralisation and adsorption-coagulation of the algae, if coagulant dose is low to medium (3-5 mg Fe(III)/L for the bench-scale model water experiments and 7-10 g Fe(III)/m^3 for the pilot plant reservoir water experiments). For model water conditions this was verified by EM measurements. For reservoir water conditions at low pH particle surface charge neutralisation was incomplete, but DAF was still more efficient compared to its performance at pH>7.

Efficient sedimentation required an increase in the frequency of the larger particle size fraction (>50 μm), but also an increase of overall particle concentration, achieved by the relatively large coagulant doses that caused ample precipitation (Chapters 3 and 6). For DAF, considerably lower coagulant doses were required (2-3 times lower) with the result of raising the >30 μm particle (floc) size fraction. This occurred most efficiently at low pH conditions (pH<IEP). The most efficient DAF removal of the larger size fractions in this case was accompanied by a substantial removal of the smaller particles (with a size similar to that of non-flocculated single cells *M. aeruginosa*), on account of the improved attachment between the negatively charged air bubbles and the neutral but non-flocculated particulate (algal) matter.

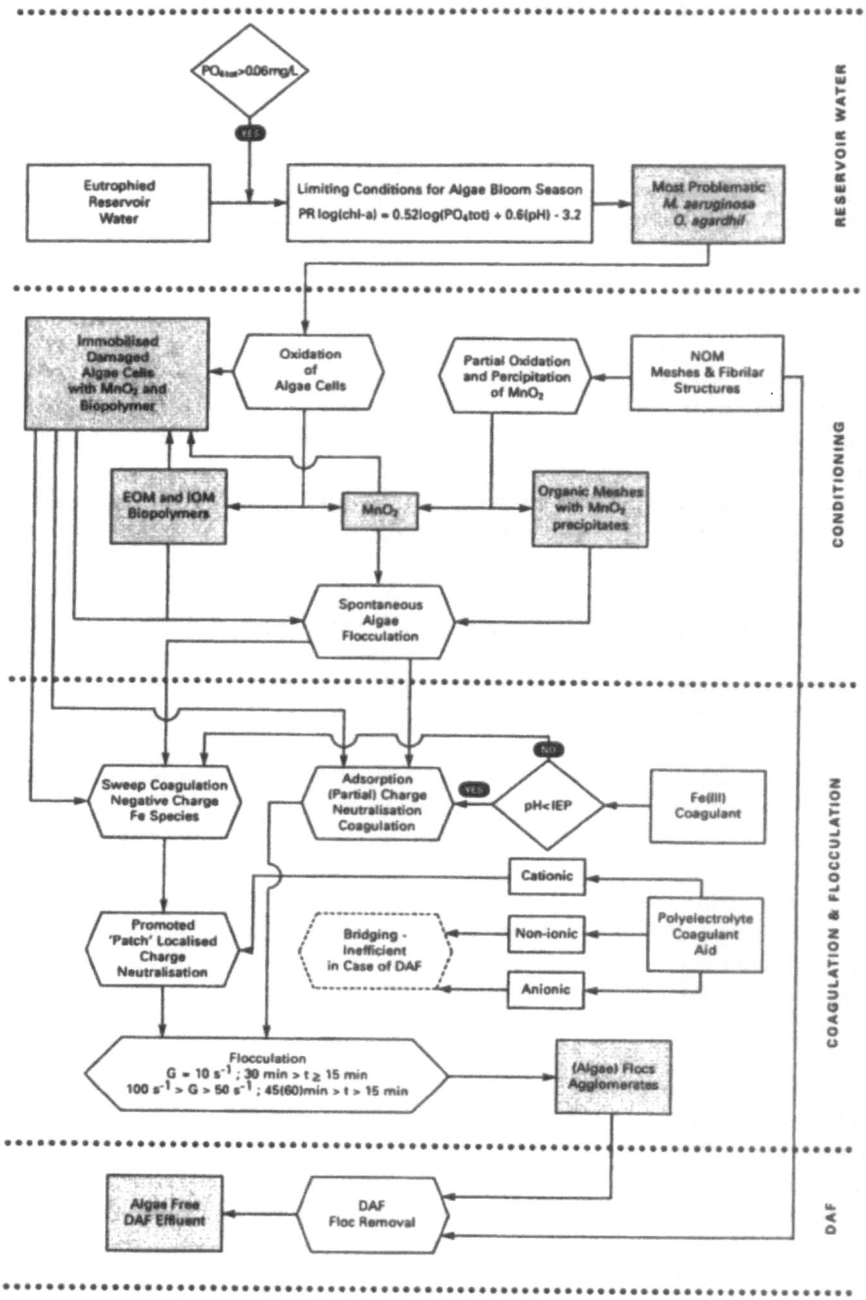

Fig. 8.2 Schematized process mechanisms occurring during different treatment phases of enhanced DAF, including the use of $KMnO_4$ algae conditioning, coagulation with Fe(III) coagulant at different pH and different (charge) polyelectrolyte coagulant aids.

The flocculation energy input was of considerable influence on the floc size distribution and hence on down-stream process efficiency. An increase of the energy input resulted in a decrease of the floc size and an increase of floc density (Chapters 3 and 6). Low to medium flocculation energy input of G=10-30 s^{-1} resulted in efficient sedimentation, however, the higher value resulted in a denser floc structure and higher particle removal efficiency. Increase of the larger particle size fraction at the expense of the smaller one was essential for efficient DAF.

The common opinion that high flocculation G values favour DAF by resulting in small, resistant flocs able to withstand the shear forces caused by the introduction of the pressurised recycled flow via nozzles or valves, was not confirmed. In bench. scale experiments using model water, a low flocculation G value of 10 s^{-1} contributed to the growth of the larger particle size fraction more significantly and resulted in efficient DAF, better than or comparable to the performance obtained at higher G value ≥ 70 s^{-1}. In pilot plant experiments with reservoir water all tested flocculation G values (G=10, 30 and 50 s^{-1}) resulted in a comparable effect on the particle volume distribution in the larger size range, and in similar particle and turbidity removal efficiency. An increase of the flocculation energy input is assumed to create better contact opportunities for particles of size <50 μm, resulting in slightly higher turbidity removal efficiency for G=50 s^{-1}. The lowest flocculation G of 10 s^{-1} resulted in flocs of lowest density and slightly more efficient particle removal.

8.5.3 Polyelectrolytes and DAF of algae

The improvement in DAF efficiency achieved at coagulation pH below the IEP is not always feasible because of the cost of pH adjustment. Similar process improvements were achieved at pH>IEP by the use of cationic polyelectrolytes as coagulant aid. The efficiency in DAF particle and turbidity removal, achieved by the two tested cationic polyelectrolytes (Wisprofloc-P and especially Superfloc C-573) was significantly higher than the one achieved by the metal coagulant only under sweep coagulation conditions. The adsorbed cationic polyelectrolyte assumes a localised 'patch'-like structure on the particles (algae) which results in a mosaic of patches of positively charged polyelectrolyte and negatively charged original particle surface, or iron-hydroxo species co-existing on the same particle. The patches could be observed under SEM (Fig. 8.3).

The bridging mechanism prevalent in the case of the non-ionic and anionic polyelectrolytes and under the circumstances of model water was of inferior efficiency compared to cationic polyelectrolytes associated with the charge neutralisation mechanism. The abundant Ca^{2+} present in the model water, which enhances adsorption, is thought to be the reason for slightly higher DAF efficiency in the case of the anionic polyelectrolyte Superfloc A-100, compared to the least efficient of all, the non-ionic polyelectrolyte Wisprofloc-N.

The application of the cationic coagulant aid (Superfloc C-573) caused more efficient flocculation in comparison to Fe(III) being used as a sole coagulant. This was reflected in the particle (floc) size distribution shifting further towards the larger particle size range, at the expense of a significant reduction of the smaller size fraction. An increase of the polyelectrolyte dose resulted in better flocculation of smaller size particles, however, the resulting larger size flocs were of lower volume compared to the case of the lower polyelectrolyte dose, suggesting that flocs were more dense (Chapters 3, 5 and 6).

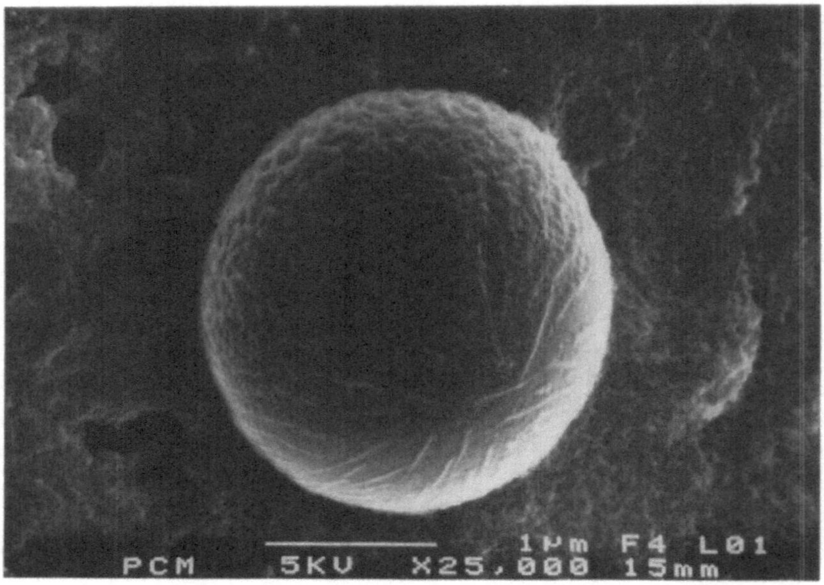

Fig. 8.3 Localised 'patches' of cationic polyelectrolyte (lower half of algae) and iron coagulant precipitates (upper half of algae). Bar length is 1 μm.

Cationic polyelectrolytes, especially Superfloc C-573, proved as efficient (sole) coagulants. Here, particle adsorption phenomena and particle charge neutralisation occurred between oppositely charged particles and cationic polyelectrolytes, as well as improved particle-bubble attachment. This again promoted cationic as a more efficient coagulant than the non-ionic and anionic polyelectrolytes with their associated bridging mechanisms.

8.5.4 Ozone conditioning of algae and DAF

Particle (algae) conditioning by the application of the oxidants ozone and $KMnO_4$ proved potentially beneficial for enhancing the coagulation and flocculation process in the context of more efficient down-stream DAF. Although of different chemistry, similar coagulation/flocculation improvement mechanisms are involved as a consequence of O_3 and $KMnO_4$ conditioning.

Ozone created stress conditions in the algal environment, and was responsible for EOM release and IOM leakage from damaged and ruptured algal cells. Algal cell rupture was verified by lower particle count after ozonation, as well as by an increase of accompanying UV_{254nm} absorption level. The latter are indicative of the presence of organic matter, some of which contains hydrocarbon rings in its structure, such as pigmented material from lysed algal cells. Furthermore, high resolution SEM (scanning electron microscopy) showed that at a dose of 2 mg O_3/L (0.57 mg O_3/mg C) significant changes of the algal cell wall occur, and that IOM starts leaking into the water (Fig. 4.15, and Figs. 8.4, 8.5 and 8.6).

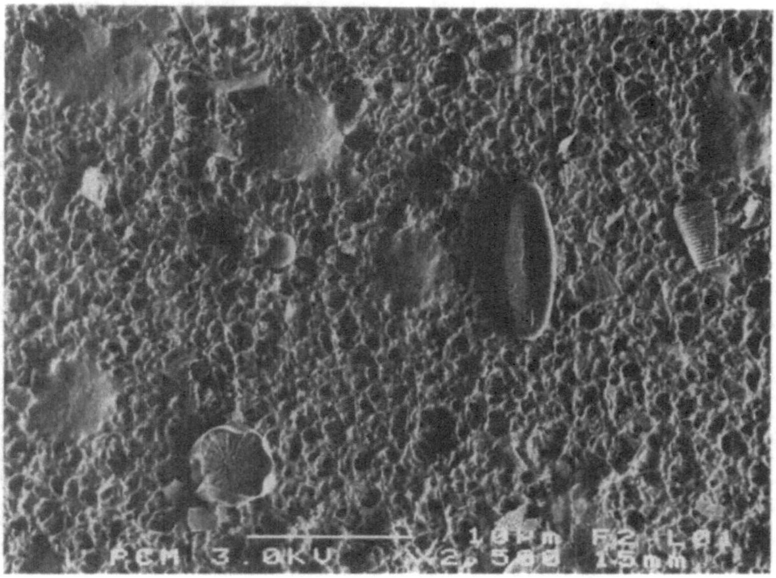

Fig. 8.4 SEM depicting the effect of ozone (2 mg O_3/L or 0.57 mg O_3 /mg C) on a combined cyanobacteria (*M. aeruginosa*) and diatoms (*Melosira* spp. and *Navicula* spp.) population (bar length is 10 μm).

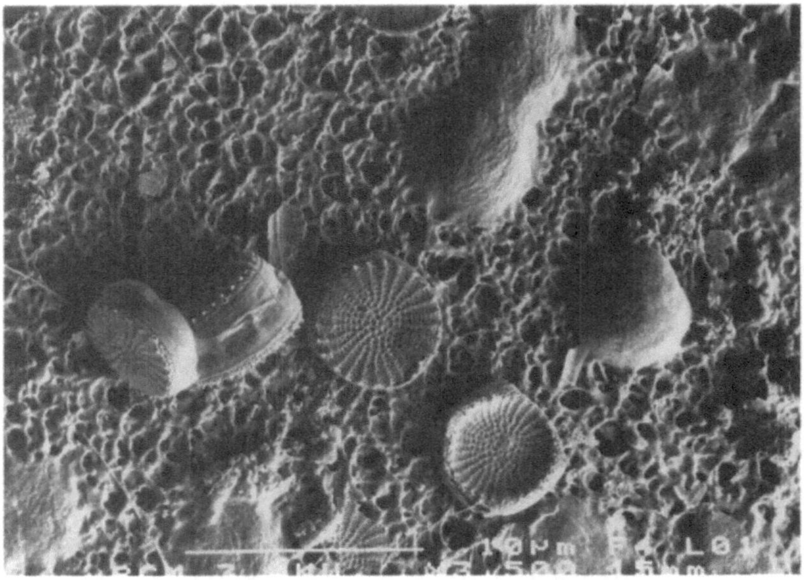

Fig. 8.5 SEM depicting the effect of ozone (2 mg O_3/L or 0.57 mg O_3 /mg C) on a combined cyanobacteria (*M. aeruginosa*) and diatoms (*Melosira* spp.) population (bar length is 10 μm).

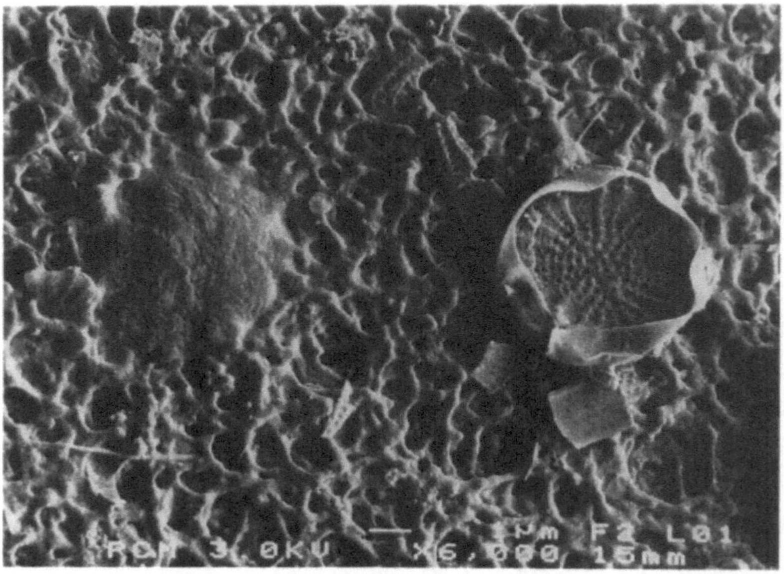

Fig. 8.6 SEM depicting the effect of ozone (2 mg O_3/L or 0.57 mg O_3/mg C) on a combined cyanobacteria (*M. aeruginosa*) and diatoms (*Melosira* spp.) population (bar length is 1 μm).

Comparison of the area in the close vicinity of the circular shaped *M. aeruginosa* algae cells with the area further away from the algae clearly suggests IOM leakage. A distinction should be made between the porous structure of the sample's base and the denser structure in the close vicinity of the algae originating from the EOM and IOM material. The released EOM and IOM act as natural coagulant aid and promote efficient flocculation. The occurrence of spontaneous microflocculation without any externally applied coagulant due to ozone application was also verified by particle counting and computer image analysis data (Chapter 4). The microflocculation effects were stimulated by an ozone induced decrease of the initial algal surface charge, as verified by EM measurements (Fig. 4.17).

Low ozone doses of 0.2 mg O_3/mg C (or 0.57 mg O_3 /L) reduced the initial particle count reflecting the modification of the algal cell wall layer, as well as the partial rupture and lysis of the algae ($UV_{254\ nm}$ measurements in Fig. 4.16). Larger ozone doses (0.5 mg O_3 /mg C) provided opportunity for a stronger oxidation and modification of the algal cells, resulting in further decrease of their concentration. This was accompanied by continued chemical modification by ozone of the released EOM and IOM into molecules of expectedly lower MW and higher acidic group content, and of a more hydrophilic character; this resulted in a negative impact on flocculation and DAF efficiency (Fig. 4.10).

The highest ozone dose (0.9 mg O_3/mg C) had the severest impact on the algal population in terms of the release of EOM and IOM. However, in this case the excess ozone further oxidised the released EOM and IOM into compounds of expected lower MW, which were thought to eventually result in the formation of difficult to remove organo-iron complexes, and increased residual coagulant in the effluent.

In this case it was accompanied by ozone induced increase of the coagulant demand, relative to the optimal coagulant (only) dose. Namely, the different ozone doses combined with the optimal Fe(III) coagulant dose resulted in lower efficiency than achieved by the same Fe(III) dose when it was used alone. This suggests that a critical organic matter concentration is needed in order to benefit from ozone conditioning. In our case, the rise of the initial single cells *M. aeruginosa* concentration from 5,000 to 10,000 cells/mL provided such a critical concentration and ozone benefitted the process efficiency. In our model water the main portion of the organic matter prone to ozone reaction and modification was in the form of algae and their EOM and IOM. It is assumed that flocculation improvement in the case of the higher algae concentration is caused by the provision of better bridging opportunities.

The spontaneous particle microflocculation served as basis for improved particle coagulation and flocculation by subsequent application of metal coagulant and polyelectrolyte. Different treatment combinations of ozone, iron and polyelectrolyte resulted in different particle (volume) distributions. The production of floc material in the larger particle size range in the case of ozone conditioning and iron coagulant resulted in improved DAF efficiency. Although the use of the Superfloc C-573 coagulant aid resulted in an opposite trend (floc size decrease and the density increased), DAF process efficiency increased. The cationic coagulant aid is thought to aid the particle flocculation by adsorption coagulation of particles which have already been partly destabilised by the ozone conditioning and the metal coagulant induced sweep coagulation. Thus, more process flexibility and reliability are provided for achieving higher process efficiency. An additional benefit in the case of DAF is the improvement of particle-bubble attachment efficiency due to the lowered negative surface charge of the floc material.

The use of Superfloc C-573 as a sole coagulant combined with ozone conditioned water demanded larger doses for efficient particle destabilisation (in comparison to being used as a coagulant aid), which however resulted in compact and dense floc material and good DAF removal efficiency. The cationic polyelectrolytes destabilise the ozone conditioned algae (of lower negative surface charge) efficiently by adsorption leading to partial charge neutralisation. The non-ionic and anionic charge character of the ozone induced algal EOM and IOM results in their adsorption and entrapment in the polyelectrolyte-floc material, also found to enhance flocculation. However, this scheme proved sensitive to ozone overdosing, which resulted in deteriorated effluent quality. The use of polyelectrolyte in the coagulant aid mode provided more process flexibility, since algae coagulation and flocculation do not solely depend on the polyelectrolyte unlike when it is used as a sole coagulant.

8.5.5 $KMnO_4$ conditioning of algae and DAF

The flocculation improvement induced by $KMnO_4$ is based on a range of process mechanisms which are similar to the ones encountered with ozone conditioning. This is related to the creation of stress conditions and algal EOM and IOM release, as verified by particle count, and $UV_{254\,nm}$ absorbance and DOC measurements. The released EOM and IOM act as coagulant aids and promote spontaneous particle microflocculation even before the coagulant addition, as observed by computer image analysis (Chapters 4 and 6).

Under normal pH conditions in water treatment permanganate reduction results in the formation of the colloidal (0.3-0.4 µm size) hydrous MnO_2 precipitate. The negatively charged MnO_2 exhibits high adsorption capacity. Furthermore, the overall particle concentration increases and better conditions for particle flocculation are created. Thus, the overall effect of permanganate application is based on a combination of its oxidative activity with inorganic and organic matter, and on its adsorptive capacity and activity of MnO_2. Although its attachment to algae slightly increased their negative surface charge potential, presumably creating unfavourable flocculation conditions, the permanganate addition tended to result in a DAF algae removal efficiency similar to or higher than that achieved by the iron coagulant only.

The application of cationic polyelectrolytes after $KMnO_4$ conditioning and with Fe(III) coagulant, resulted in improved DAF particle removal efficiency by 30-40%. This efficiency improvement is thought to be based on adsorption coagulation, which aided the removal of both the complexed Fe coagulant and the negatively charged MnO_2, lowering their residuals close to the respective MACs. Based on the same principles, the particle-bubble attachment was improved resulting in more efficient DAF.

8.6 DAF process kinetic modelling

8.6.1 Particle and bubble size analysis

Particle counting techniques can provide fast and reliable data for the assessment of treatment process efficiency, complementing traditional lump parameters such as turbidity. However, the HIAC particle count for effluent characterisation can be influenced by the residual coagulant by raising the count roughly proportionally to the coagulant dose (15-20% for the coagulant dose of 15 mg Fe(III)/L). Also, flocs may break up due to high shear forces during the passage of the sample through the recording chamber; this discredits its use for flocculated water characterisation. On the other hand, computer image analysis enabled qualitative as well as quantitative characterisation of the coagulation/flocculation process, as well as of bubble size. The acquired data were used to calculate the floc size - density ratio (P) and the absolute floc density (ρ_{fl}), used for DAF and sedimentation model calculations (Chapters 6 and 7).

8.6.2 DAF single collector collision efficiency reaction zone model

Based on image analysis data of flocculated and recycled stream samples, under model water and pilot plant circumstances, the following parameters were determined : mean particle size d_p=20-40 µm, and mean bubble size d_b<100 µm and predominantly in the range of 35-75 µm, depending on P and R. Figs. 8.7 and 8.8 represent typical particle and bubble size frequency distributions obtained under the stated experimental conditions.

Theoretical considerations related to the flow conditions around an air bubble of this size suggested that the DAF single collector collision efficiency reaction zone model was applicable. Namely, calculations showed that the application of laminar flow conditions in the bubble vicinity can be considered correct with reasonable accuracy. The model equation considers two coefficients : (i) η_T (total single collector collision efficiency coefficient) including a range of physical variables such as floc and bubble size, water temperature and viscosity; and (ii) the lump coefficient α_{pb} (the particle-bubble attachment efficiency), which is supposed to cover a range of physical and chemical aspects, including (particle and bubble) surface charge, ionic strength of the water, presence of surface active agents, etc. The predicting capacity of α_{pb} poses the greatest problems in the verification of the model and it has largely been based on assumptions rather than measurements.

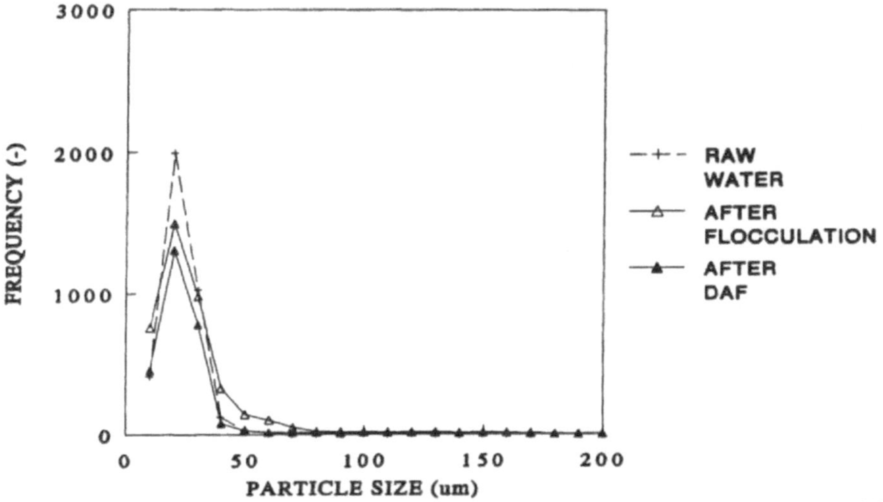

Fig. 8.7 Particle size frequency distributions after different treatment stages. Conditions : 3 mg Fe(III)/L, pH 8, G_f=70 s^{-1}, t_f=10 min, R=5%, P=500 kPa.

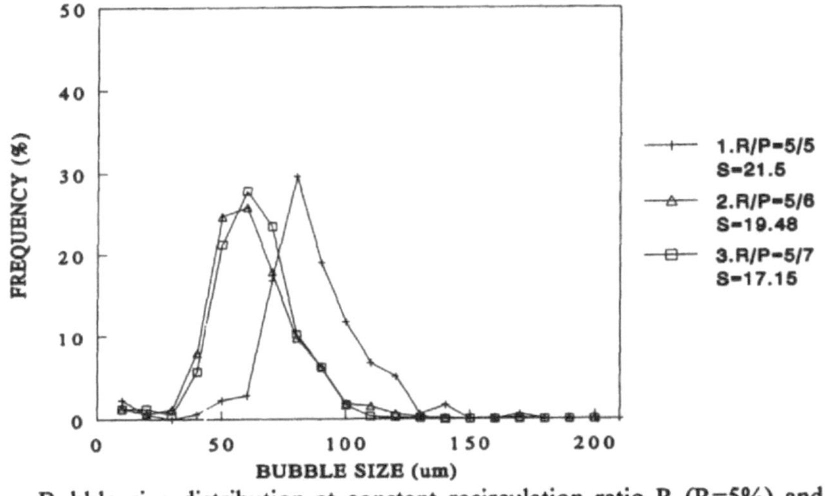

Fig. 8.8 Bubble size distribution at constant recirculation ratio R (R=5%) and varied saturator pressure P (P=500, 600 and 700 kPa). S=standard deviation.

8.6.3 Particle-bubble attachment efficiency α_{pb}

Under the assumption of ideal plug up-flow condition, a solution of the DAF kinetic model equation for α_{pb} demands the provision of a range of input data after different treatment stages. Based on experimentally obtained or calculated input values for relevant model variables under different process conditions (coagulant dose, coagulation pH, flocculation G value, coagulant aid, ozone or $KMnO_4$ conditioning) the α_{pb} was calculated to be in the range of 0.04-0.75 for DAF jar test conditions. The higher α_{pb} values range was accomplished under optimal coagulation/flocculation conditions for DAF (α_{pb}=0.39), and in the case of enhanced coagulation either by application of Superfloc C-573 as coagulant aid (α_{pb}=0.75) or under low coagulation pH 6 below the IEP of 6.5-7 (α_{pb}=0.42). In the two latter cases, the relatively high α_{pb} values are assumed to be related to improved particle coagulation due to adsorption coagulation. The accompanying particle (floc) removal efficiency (relative to the particle concentration after flocculation) was in the range of 50-70%, while the overall particle removal efficiency (relative to the raw water particle concentration) varied. It was lower or higher than the respective floc removal efficiency, depending on the coagulation/flocculation conditions. This suggests that high(er) α_{pb} values do not necessarily stand for high(er) overall DAF efficiency (Chapter 7).

A high flocculation G value of 100 s^{-1} had a negative effect on the α_{pb} both at pH 7 and 9, compared to a flocculation G=50 s^{-1} value. The use of ozone as conditioner generally resulted in higher α_{pb} values than when Fe(III) was used as a coagulant without conditioning, both with or without the use of polyelectrolyte coagulant aid (α_{pb}=0.34 and 0.37 for combined ozone + Fe(III), and ozone + Fe(III) + Superfloc C-573 treatment, respectively, versus α_{pb}=0.24 for Fe(III) based treatment only). $KMnO_4$ conditioning under DAF bench-scale conditions resulted in low efficiency and accompanying α_{pb}=0.19 value.

The significant difference in DAF efficiency between experiments on model water and pilot plant experiments on reservoir water, suggested unforeseen process determining factors. The assessment of raw water quality pointed at the critical role of organic matter. The relatively low TOC (≈2.5 mg C/L) and the TOC composition (partly created by the spiked algae) of the model water differed from the high reservoir water TOC (up to 15 mg C/L) and composition (mainly humic NOM). Previous TEM (transmission electron microscopy) observation of the structure of small colloids (size <100 nm) in water of similar quality to our reservoir water, suggested that fine organic matrices or meshes of filaments acted as nesting sites for such small colloids. These fibrilar organic structures may also be associated with occasionally observed particle coagulation improvement in the case of higher TOC content, associated with bridging phenomena. Furthermore, they have been suggested to be responsible for the improved particle removal in the case of permanganate conditioning, embedding algal EOM, IOM and MnO_2, and enhancing flocculation by bridging. We extend this speculation by suggesting that such organic meshes may be favourable for DAF, where the organic meshes form a gel-like structure which is efficiently removed by rising air bubbles.

In the case of reservoir water experiments, the high TOC content exerted additional coagulant demand, while no charge neutralisation took place neither at low coagulation pH nor when cationic coagulant aid was used. An increase of the coagulant and the polyelectrolyte dose, however, increased the process efficiency (the first being more significant). The use of polyelectrolyte shifted the particle (floc) size distribution to larger particle sizes and resulted in an increased mean floc size. The increased mean floc size was accompanied by decreased floc density and favourable flotation circumstances. The calculated corresponding α_{pb} values were in the range of 0.28-1.0, while the corresponding DAF particle (floc) removal efficiency was in the range of 45-85%.

Calculation suggested that both for the bench-scale and the pilot plant experiments, and for the obtained particle size distribution after coagulation and flocculation, an increased DAF efficiency can be achieved by obtaining a bubble size distribution with a smaller mean size. Furthermore, the model suggested that high removal efficiencies of >90% generally require high particle-bubble attachment efficiency ($\alpha_{pb} \geq 0.5$).

Model analysis showed a higher sensitivity of the α_{pb} particle-bubble attachment efficiency to particle size change than bubble size change. The standard deviation ("width" of the bubble size distribution curve) was of less influence on process efficiency than the mean bubble size. Calculations suggest higher process efficiency for bubble size distribution with low(er) mean bubble size and standard deviation. The model calculations were supported by turbidity, particle count, residual coagulant and DOC measurements.

Bubble size distributions under an increased saturator pressure (P) resulted in an increase of bubble concentration (N_b) and a reduction of mean bubble size (d_b). Higher bubble concentrations, however, increased chances of bubble collisions and resulted in their coalescence. Increase of the recycled flow ratio (R) increased the standard deviation on the bubble size distribution (S), and decreased the mean bubble size. The latter observation was most significant for the high saturator pressure (P=700 kPa). Model calculations confirmed the relatively low impact of the standard deviation on DAF process efficiency. Overall, model calculations and experimental work suggested that the application of high recirculation ratios >7% and saturator pressures >500 kPa are not always fully justified. It results in increased production cost with negligible or even adverse effects on process efficiency.

Physical verification of the single collector collision efficiency model largely relies on the prediction capacity for α_{pb}. Modelling of this lump coefficient depends on the ability to model the coagulation/flocculation process, the resultant particle (floc) stability and particle (floc) size distribution (and density). Charge neutralisation induced by low pH conditions (below the IEP) or addition of cationic coagulant aid significantly increased α_{pb} and DAF process efficiency. The same applies for the use of ozone as particle (algae) conditioner. Furthermore, the role of the raw water quality, especially the ionic strength of the water, the presence of surface active agents, as well as the suggested role of NOM, additionally constrain the predicting capacity for α_{pb}.

8.7 Raw water quality and process parameters affecting DAF efficiency

The conceptual treatment process selection diagramme proposed by Janssens and presented in Fig. 1.7 uses chlorpohyll-*a* and turbidity as the relevant raw water parameters which affect the choice of appropriate treatment technology. Although only tentative, this diagramme should not be considered as a sole tool for process selection. It appears it does not cover the range of raw water quality characteristics that have been proven here to affect a technology choice (e.g. frequency and duration of algal bloom episodes, water temperature, NOM presence and nature, etc.), nor does it consider cost-benefit aspects which often prove crucial in the selection process. Thus, proper treatment technology choice should be based on carefully designed pilot plant research and cost-benefit analysis.

The comparison between bench-scale and pilot plant DAF performance outlined the role of NOM. Fig. 8.9 summarises the obtained particle removal (>2.75 μm) efficiency ranges from bench-scale experiments with model water and from pilot plant experiments with reservoir water, for different process combinations (1. sedimentation + filtration and 2. DAF + filtration) and different process conditions (coagulant and coagulant aid dose, and oxidants conditioning). The results are given in perspective to the composition and the concentration of NOM in the model and in the reservoir water.

NOM concentration and composition affect the particle (algae) coagulation process, e.g. by the complexing of the metal coagulant and the subsequent additional coagulant and/or coagulant aid demand. On the other hand, it may also improve down-stream solid-liquid separation. The presence of NOM in the form of fine meshes composed of fibrilar organic structures may enhance particle flocculation and DAF. These structures may serve as sites for floc (as well as colloidal MnO_2) entrapment and embedment into a gel-like structure which is easier to remove by rising air bubbles than the solitary floc material that would exist in the relatively clean model water used in the case of DAF jar test experiments. This factor calls for further analysis and quantification. Fig. 8.10 schematizes the effect of the water quality parameters relevant for DAF treatment.

The coagulant type and dose are important process parameters. DAF is equally efficient as sedimentation in particle separation (of particles in the algae size range ≥2.75 μm), but operates best at 2 times lower coagulant doses; however, at the common pH of 7-8.5, the lower doses generally result in lower sweep coagulation efficiency, and hence somewhat lower turbidity removal and higher residual coagulant. Furthermore, the low doses appear less efficient in removing the colloidal particles which are largely responsible for the turbidity.

The pH plays a key role in the coagulation and floc formation. The adsorption-coagulation occurring at coagulation pH below the IEP benefitted the DAF efficiency. Although initially controversial, the role of cationic polyelectrolytes as coagulant aids prior to DAF, was proven similarly beneficial for DAF. The floc structure and charge characteristics favored floc-bubble attachment. They also lowered the Fe and MnO_2 residuals, both after DAF and after filtration.

The high flocculation energy input and resulting floc material that is shear resistant are commonly preferred for DAF. It is suggested that this holds more for relatively "clean" water (low particulate and colloidal matter concentration, low turbidity and low NOM concentration), where flocculation energy ($G \geq 50s^{-1}$) helps to provide sufficient contact opportunities. On the other hand, our study showed that lower flocculation energy (e.g. $G=10s^{-1}$) resulted in larger particle (floc) mean size of lower density (Chapter 3), which tended to favor DAF. Again, the presence of NOM fibrillar structures may play a positive role in the flocculation process, rendering the flocculation energy input less critical as compared to flocculation time.

ORGANIC MATTER ORIGIN AND TYPE	CORE TREATMENT
Reservoir water : ❒ TOC=5-15 mg C/L ❒ DOC=90-95% of TOC ❒ NOM characteristics : -comprised mostly of humic matter; existence of NOM fine matrices or meshes of filaments.	**1. Sedimentation + Filtration** **2. DAF + Filtration**
Model water : ❒ TOC=2.3-3.0 mg C/L ❒ DOC=2.0-2.7 of TOC ❒ NOM characteristics (after full treatment of Biesbosch water used for model water preparation, and after spiking it with algae see Chapter 5) : -reduced molecular weight and increased functional group content, thus lower precipitating tendency -possible lack of functional groups that would enable sorption onto metal (coagulant) oxide surfaces. -spiked algae and their EOM (and IOM in case of oxidants application) comprised the most significant portion of OM prone to exert a coagulant demand and react with oxidants.	**1. Sedimentation + Filtration** **2. DAF + Filtration**

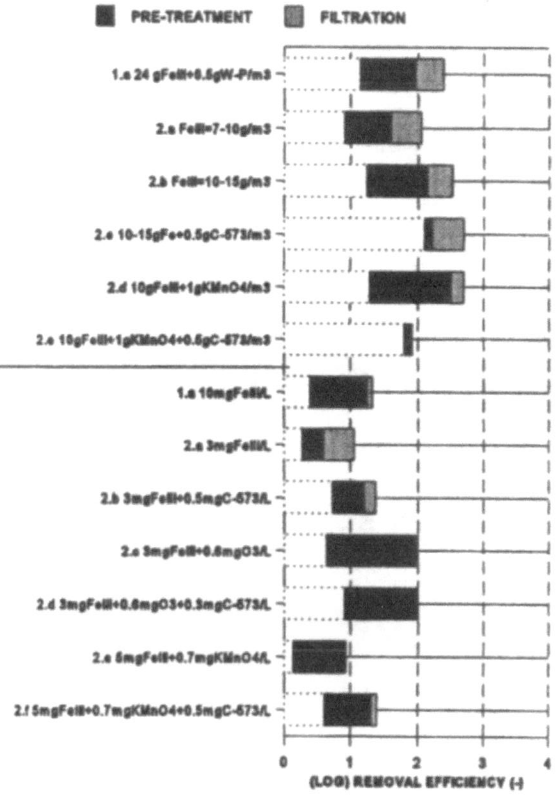

Fig. 8.9 Particle removal efficiency ranges obtained by bench-scale experiments with model water and pilot plant experiments with reservoir water, for different core treatment and different process conditions (coagulant and coagulant aid dose, and oxidants conditioning). Missing data for filtration in case of model water ozone conditioning (2.c and 2.d) and $KmnO_4$ conditioning (2.e), and in case of reservoir water $KMnO_4$ conditioning and polyelectrolyte coagulant aid (2.e) are the result of the lack of sufficient amount of experimental data. See Table 7.1 for other process conditions.

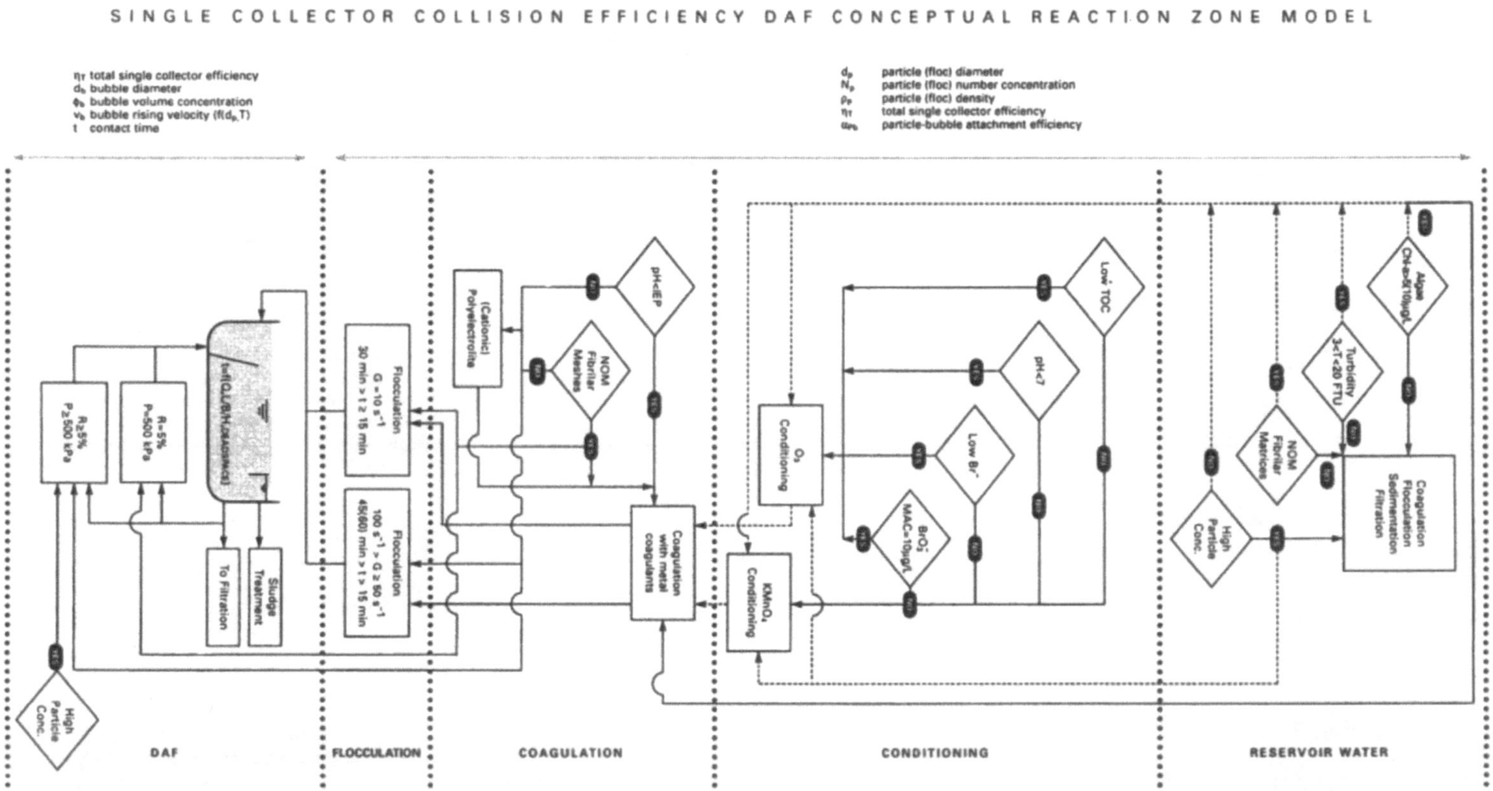

8.10 Schematized raw water quality and process parameters that influence (enhanced) DAF process efficiency and modelling. * Low Br^- is difficult to quantify, since even very low levels can result in BrO_3^- levels above the low 10 μg/L EU guideline. Thus, the level of BrO_3^- formation will primarily depend on the presence (concentration and characteristics) of organic matter and other ozone scavengers.

The benefit of short flocculation times (typically below 5 minutes) is the subject of dispute in literature. However, in our study an increase of flocculation time increased DAF efficiency; the flocculation time of 15 minutes was considered a good economic optimum. The necessary flocculation time is expected to increase inversely proportionally with water temperature. Larger flocculation energy input appeared to be of greater importance at the shorter flocculation times, implying the existence of an optimal flocculation Gt range. The unambiguous existence of this range, however, was not demonstrated by the bench-scale DAF experiments. Tapered flocculation did not benefit DAF.

Oxidants application in water treatment carries potential health risks. Ozone notably is related to bromate production, and its applicability depends on the possible change of the existing maximum allowable concentration (current EEC bromate MAC = 10 μg/L, newly proposed Dutch standard is 0.5 μg/L). To restrict bromate formation, the ozonation pH should preferably be ≤ 7, which also corresponds to pH conditions at which improved DAF efficiency was noted. The ozonation pH determines the reaction path; pH<7 results in direct oxidation by ozone, while pH>7 increases the decomposition rate and favors action of the resulting hydroxy radical.

Since the effect of particle conditioning by $KMnO_4$ is based on a combination of its oxidative activity and the adsorption capacity of the colloidal MnO_2, its eventual effect depends also on the composition and concentration of the organic matter, including the algae. Higher reactivity was exhibited at longer contact times and higher pH (>pH 8), which is consistent with theory and suggests higher down-stream removal efficiency should be obtained. However, this was not always the case, especially with regard to the pH conditions in the model water experiments. The adsorption-coagulation occurring at pH lower than the IEP (pH 6) outperformed the effect of the $KMnO_4$ conditioning at pH 8 (Chapter 4). Furthermore, even at very short contact times before coagulant dosing (less than 1 s) and at pH 8, comparably good results were achieved by the pilot plant. It is suggested that NOM provides embedment sites for colloidal and particulate matter, including the active MnO_2.

The conditioning, and the coagulation/flocculation, define parameters that affect DAF kinetics : particle (floc) mean size (d_p) and particle concentration (N_p) after coagulation/flocculation, and the single collector collision efficiency factor η_T. The recirculation ratio (R) and the saturator pressure (P) also affect the DAF efficiency via the mean bubble size (d_b), bubble volume concentration (ϕ_b), the single collector efficiency factor η_T, and the bubble rise velocity (U_b). Although of significant impact on the bubble size distribution, R and P proved to be of less overall importance for DAF efficiency (within the investigated range of R=5-10% and P=500-700 kPa) than the particle related characteristics (i.e. particle charge, size and density). The generally high bubble-to-particle concentration ratio and the beneficial role of the NOM fibrillar structures could largely explain this phenomenon.

8.8 Enhanced DAF, an attractive option for the treatment of algae laden water

Coping with algae bloom problems at short notice, and guided by the philosophy that 'the more algae are removed the better', is likely not to be the economically most feasible approach. Furthermore, surface water sources have proven to be continuously changing environments, resulting in short and long term periodical changes of their algae population which are not always easy to predict. The lack of a proper algae related water quality standard compounds the difficulties that treatment plant management face in operation. The establishment of such a standard would mark a period in which algae removal will be substantially easier to approach and prepare for at short notice.

DAF proved to be a viable and equally or more efficient alternative to conventional sedimentation for heavily algae laden water treatment, attaining over 2 log particle (algae) removal after DAF and filtration. This was achieved however, at lower chemicals consumption (up to 50% lower than for the sedimentation based process), at higher process loading rates, lower flocculation energy input and shorter flocculation times. It proved to be a robust, and yet a flexible process. The enhanced DAF treatment options, including different combinations of Fe(III) coagulant and the application of polyelectrolytes, ozone or $KMnO_4$, tended to further increase its efficiency as well as reliability.

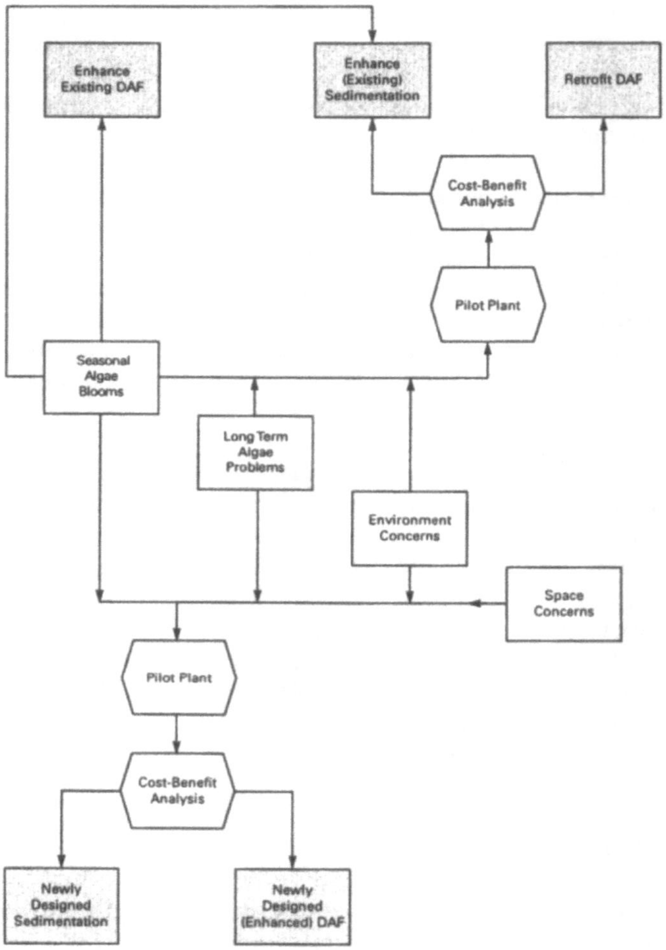

Fig. 8.11 Process treatment (mode) selection diagram for algae laden water treatment.

Fig. 8.11 schematizes the possible scenarios related to the choice of DAF or sedimentation. It considers algae problems and other treatment related issues, including environmental and available space for treatment unit construction, as the core issues of concern. On the other hand, the wide range of raw water quality and process characteristics that affect process efficiency calls for pilot plant experiments as basis for cost-benefit analysis of possible treatment options. This should enable the enhancement of existing DAF or sedimentation facilities possibly by some of the options considered in this study. The same applies for the design of new and efficient (enhanced) DAF or sedimentation facilities, and even retrofit DAF when existing sedimentation cannot cope with high algae loads.

Knowledge and experience with the history of the raw water quality is essential for coping with current and future algae problems. Experienced DAF or sedimentation plant operators can cope efficiently with the algae concentrations occurring during the longest part of the year. Seasonal algae blooms will still pose a particular threat. Importantly, periodical outbreaks of certain new algal species or comparable organisms are possible, as was the case with the recent *Cryptomonas* spp. bloom in the WRK III reservoir. Timely algae monitoring is an essential tool for anticipating the problem at short notice.

8.9 CONCLUSIONS

-Coping with high algal loads (algal blooms) in water treatment with the philosophy that "the more algae are removed, the better" is not always the most feasible and cost effective approach to the problem. An efficient solution to the problem requires an approach which combines simultaneous, and continuous water quality monitoring, (reservoir) water quality management and application of latest developments in drinking water treatment technology.

-The total phosphates concentration which is to probably result in troublesome algal blooms (>10 μg chlorophyll-*a*/L) in surface water impoundments lies >0.05 mg/L. The predominant algal species, however, will depend on a range of additional factors, including the water pH, the combined ratio of other nutrients, water temperature, light availability, the presence of zooplankton grazers, etc.

-The cyanobacteria in general, and *Microcystis aeruginosa* and *Oscillatoria agardhii* in particular, are the most relevant species in Dutch water treatment circumstances, regarding the amount of related treatment problems.

-Dissolved air flotation (DAF) is a viable treatment alternative to conventional sedimentation for the removal of particulate matter and algae in particular. It is an efficient, robust, reliable and flexible, high rate treatment technology which is to be highly considered in case of treatment design for waters with expected long or short term algae related problems.

-The coagulation/flocculation process plays and essential role in determining the down-stream process efficiency, irrespective of, however, defined by the applied process (DAF, sedimentation or filtration).

-Otimal DAF algae removal (≥2.75 μm) requires approximately 50% less coagulant than sedimentation, with considerable accompanying cost savings and positive environmental implications. This, however, is accompanied with less efficient sweep coagulation conditions (partly due to complexing of the metal coagulant with the organic matter) than achieved by the high coagulant doses usually applied in the sedimentation, and results in higher residual turbidity, residual coagulant and modified fouling index (MFI) values.

-Highest DAF efficiency can be achieved at a coagulation pH below the iso-electrical point (IEP), related to the positive effect of adsorption coagulation occurring under these pH circumstances, which aids the predominantly occurring sweep coagulation (dosage dependant). Under normal drinking water treatment pH conditions (pH 7-8.5), the applicability of this option for process enhancement is generally not feasible on long term, however, its application during short periods of algae blooms is an attractive option worthy of consideration.

-Although DAF operated efficiently at a high flocculation G value of 70 s^{-1} (which is generally related to the production of small, shear resistant flocs able to withstand the high shear from the introduction of the recycle flow), a low flocculation G value of 10 s^{-1} resulted in similar or higher process efficiency (the lower shear resistance being compensated by the larger size and lighter floc structure obtained at lower G values).

-Increase of flocculation time for the lower G value resulted in further increase of DAF efficiency, which was not the case for the high G value. The emerging suggestion of the possible existence of an optimal Gt value range could not be verified.

-The optimal flocculation time for DAF is significantly shorter than for sedimentation (more than twice). Under normally encountered drinking water treatment pH conditions of 7-8.5, however, flocculation times shorter than 15 min are not recommended, in view of the insufficient down-stream DAF efficiency (<90%).

-Tapered flocculation did not significantly affect DAF process efficiency. It is suggested that the total flocculation time plays a greater role in this respect than the tapering effect of diminishing energy input in consecutive flocculation stages.

-The application of cationic polyelectrolytes as coagulant aids significantly increased process efficiency in case of model water bench-scale experiments. This efficiency improvement is related to the promotion of adsorption coagulation. Efficiency improvements were absent in case of the non-ionic (Wisprofloc-N) and anionic (Superfloc A-100) polyelectrolytes, while efficiency differences also existed between the two cationic polyelectrolytes (Superfloc C-573 outperformed Wisprofloc-P). The latter is related to their charge and molecular weight characteristics.

-Similarly, the application of cationic polyelectrolytes as sole coagulant in DAF treatment proved efficient; Superfloc C-573 outperformed Wisprofloc-P. The application of cationic polyelectrolytes as sole coagulants is attractive due to the absence of metal coagulant residuals, however, the absence of a suitable residuals determination technique precludes their application, due to uncontrolled carcinogenic monomer residuals.

-The application of ozone as an algae conditioner can significantly raise DAF process efficiency at relatively low doses (0.2 mg O_3/mg TOC, or 0.5 mg O_3/L). This process enhancement option is, however, restricted due to bromate formation. Its applicability is dependent on raw water quality (e.g. ozonation pH preferably <7) and stringency of applied drinking water regulations. If the EU guideline of 10 μg BrO_3^-/L maximum allowable concentration (MAC) is considered, ozone algae conditioning may still be an attractive option.

-The application of $KMnO_4$ as an algae conditioner proved a potentially beneficial option for enhancing the algae coagulation/flocculation and subsequent removal by DAF, compounded by the absence of health hazardous by-products.

-Both, ozone and $KMnO_4$ resulted in significant modification of the algal cell wall structure, release of extra-cellular organic matter (EOM) and inrtra-cellular organic matter (IOM) leakage from lysed algal cells, and the promotion of spontaneous microflocculation, even before the coagulant was added.

-The application of cationic polyelectrolytes in the combined Fe(III) coagulant-ozone or $KMnO_4$ conditioning treatment scheme, can further increase DAF algae removal efficiency. Furthermore, it can reduce the residual iron (organo-iron complexes) and the colloidal MnO_2, below their MAC values.

-The flocculation process affects DAF or sedimentation efficiency via the resultant floc size - density relation. The relative floc density (in relation to the original density of particles incorporated in the floc) decreased with increasing floc size. The floc size increased as a consequence of decreasing pH, decreasing G values (as single stage or as part of the tapered flocculation concept), ozone or $KMnO_4$ conditioning (spontaneous microflocculation occurred) and cationic polyelectrolyte coagulant aid application. Increase of the polyelectrolyte coagulant aid dose, however, increased the relative floc density.

-Comparison of bench-scale experiments with model water and pilot plant experiments with reservoir water implicates that the efficiency of the coagulation/flocculation and down-stream process (DAF or sedimentation) is significantly influenced by the natural organic matter concentration and composition. The process efficiency recorded in the reservoir water experiments (high organic matter content, mostly humic matter) was significantly higher than the one recorded in the model water experiments (low organic matter content, already treated by ozone in full scale treatment).

-The high NOM content of the reservoir water asserted a high coagulant demand in order to meet the stringent MFI value of 5 s/L^2. For a given coagulant dose, the impact of the flocculation G value and the application of cationic polyelectrolyte coagulant aid were less significant for the DAF process efficiency compared to the model water circumstances. The flocculation time was of greater significance for process efficiency, while contrary to model water experiments no charge neutralisation occurred in case of the cationic polyelectrolyte application.

-The existence of NOM organic matter meshes and fibrilar structures is suggested partly responsible for the higher efficiency of the coagulation/flocculation process in case of the reservoir water experiments. The organic matter meshes also served as sites for floc and MnO_2 entrapment. Furthermore, they favoured efficient DAF, providing a gel-like structure which is efficiently removed by the rising air bubbles.

-The solution of the particle removal efficiency equation of the single collector collision efficiency DAF kinetic model yielded α_{pb}=0.04-0.75 for the jar test experiments and α_{pb}=0.26-1.0 (1.79) for the pilot plant DAF experiments. The higher range of values were accomplished under optimal jar test coagulation/flocculation conditions (α_{pb}=0.39), and in case of enhanced coagulation due to coagulation pH<IEP (α_{pb}=0.42) or due to application of cationic polyelectrolyte as coagulant aid (α_{pb}=0.75); for pilot plant conditions the application of coagulant doses of 10-15 g Fe(III)/m^3, alone or in combination with 0.5 g Superfloc C-573/m^3 resulted in high α_{pb}=0.26-1.0 values.

-High DAF particle removal efficiencies (>90%) are related to the accomplishment of high particle-bubble attachment efficiency (α_{pb} values >0.5).

-The model showed higher sensitivity of the α_{pb} for particle floc size than bubble size. Modification of the bubble size distribution by variation of the recirculation ratio and saturator pressure has little impact on process efficiency, suggesting that high recirculation ratios (>7%) and saturator pressures (>500 kPa) are not always justified.

-The single collector collision efficiency DAF kinetic model can be a valuable tool in the design of DAF facilities and experimental design. However, the role of water quality factors which affect the coagulation/flocculation process and DAF, such as NOM concentration and composition, must also be considered. In this context, the basic assumption of the model that particle-bubble agglomeration occurs solely due to their collision is not fully justified.

WRK III RESERVOIR WATER

- turbidity ≥ 5 (10) FTU
- particle (algae) count of size ≥ 2.75 μm after microstraining N = 10-20 x 10^3/mL
- chlorophyll-$a_{(mean)}$ (01.09.95-15.11.95) = 28.8μg/L (01.09.96-15.11.96) = 29.3μg/L

MICROSTRAINING & INITIAL STAGE PUMPING

BENEFICIAL PROCESSES

- straining - removal of zooplankton
- straining - removal of large, filamentous and colonial species (e.g. *O. agardhii* and *M. aeruginosa)*

POTENTIAL RISKS

- fast blocking of the mesh pores by filamentous and colonial algal species (e.g. *O. agardhii* and *M. aeruginosa*), resulting in more frequent washing and servicing during algae blooms
- break-up of alga colonies (e.g. of *M. aeruginosa*) into difficult to remove single cells
- growth of and interference by $CaCO_3$ crystals

OZONE CONDITIONING

BENEFICIAL PROCESSES

- disinfection
- DOC (and AOC) oxidation and improved removal in subsequent flocculation stage of smaller aromatic compounds
- break-up of organo metal complexes, in-situ coagulant production
- reaction with organic matter on particle surface and reduced particle stability
- increased sorption of ozonated NOM on hydrolysed metal hydroxide flocs,
- increased Ca complexation with NOM
- improved taste and odour
- improved GAC operation and adsorption of organic matter, including pesticides
- possibly reduced coagulant demand (Fe), depending on organic matter structure and reaction mechamism with ozone (conflicting literature data)
- zooplankton and plankton conditioner :
  - immobilisation of motile algae and organisms, easier floc entrapment
  - reaction with functional groups on algal surface → no risk for increased Fe residual before filter
  - reduction of algal surface charge
  - excretion of EOM and IOM → production of biopolyelctrolyte coagulant aids
  - initial spontaneous algae flocculation
- continuing reactions of NOM with ozonides, organic peroxides, and organic free radicals, after O_3 has disappeared

POTENTIAL RISKS

- bromate formation
- break-up of colonial alga species (e.g. *M. aeruginosa*) into difficult to remove single cells
- cell lysis and IOM leakage
  - rise of AOC
  - rise of taste and odour problems if taste and odour causing algae are present
  - rise of (cyanobacteria related) toxicity
  - rise of THM potential
- possible increase of residual Fe (depending on the ozone imposed modifications of the NOM and the consequent impact on Fe-NOM complexation)
- possible negative effect on polymer structure (break-up of polymer chains)
- possible need for pH correction before ozone application (rapid decomposition, as well as increased bromate formation at pH>7.5)

$KMnO_4$ CONDITIONING

BENEFICIAL PROCESSES

- (mild) disinfection power at common water treatment doses (max ≈ 3-4 g/m^3)
- DOC (and AOC) oxidation
- adsorption of oxidised DOC onto MnO_2↓ precipitate
- break-up of organic-metal complexes → in-situ coagulant production
- accelerated flocculation through MnO_2 precipitation → increased particle concentration
- improved taste and odour
- reduction of THM formation potential
- zooplankton and plankton conditioner:
  - immobilisation of motile algae and organisms, easier floc entrapment
  - excretion of EOM and IOM → production of biopolymer coagulant aids
  - initial spontaneous algae flocculation

POTENTIAL RISKS

- increased residual Mn values (MnO_2 and dissolved Mn)
  - increased residual turbidity
  - increased MFI before filter
  - pink colour (dissolved Mn)
- break-up of colonial alga species (e.g. *M. aeruginosa*) into difficult to remove single cells
- cell lysis and IOM leakage
  - rise of AOC
  - possible rise of taste and odour problems (depending on KMnO4 taste and odour compounds reaction products)
  - rise of (cyanobacteria related) toxicity
  - possible rise of THM (conflicting results in literature)

Appendix 8.1

Fe(III) COAGULATION

BENEFICIAL PROCESSES	POTENTIAL RISKS
■ particulate and colloidal matter sweep and/or adsorption coagulation (pH - IEP and dose dependent) ■ conversion of soluble NOM molecules into insoluble metal-organic reaction products ■ initial microflocculation ■ in sweep coagulation Fe precipitates adsorb dissolved NOM	■ higher chance of increased Fe residuals in case of DAF due to generally lower coagulant demand and partial organic matter-Fe complexation - increase of turbidity before filter - increase of MFI

(CATIONIC) POLYELECTROLITE

BENEFICIAL PROCESSES	POTENTIAL RISKS
■ enhanced adsorption coagulation ■ improved agglomeration of : - algae - MnO_2 precipitates - colloidal $Fe(OH)3$ ■ improved floc structure: larger mean flocs size and denser flocs ■ improved particle (floc)-bubble attachment in DAF due to charge related phenomena (attraction between positively charged sites on flocs and negatively charged air bubbles)	■ scarce experience on measuring polyelectrolyte residuals ■ potentially hazardous residual monomers

DISSOLVED AIR FLOTATION (DAF)

BENEFICIAL PROCESSES	POTENTIAL RISKS
■ particle (floc) removal by rising air micro- bubbles (≤ 100 μm) via: - entrapment of bubbles within a condensing network of flocs - bubble growth from nuclei within flocs - adhesion of flocs upon collision with bubbles (dominant in 'common' drinking water circumstances) ■ formation of sludge with high dry solids content (4-6%)	■ possible air bubbles carry-over onto filters → disturbed filter operation and deteriorated filter effluent quality

$DAF_{EFFLUENT}$ (no conditioning)
- turbidity: 0.5 - 0.9 FTU
- particle (algae) count of size ≥ 2.75 μm: N = 200 - 900/mL (i.e. < 2 log removal)
- residual Fe: 0.7 - 1.2 mg/L

$DAF_{EFFLUENT}$ ($KMnO_4$ or O_3 conditioning)
- turbidity: 0.4 - 1.0 FTU
- particle (algae) count of size ≥ 2.75 μm: N = 100 - 300/mL (i.e. ≤ 2 log removal)
- residual Fe: 0.5 - 0.7 mg/L
- residual Mn: 0.12 - 0.17 mg/L
- BrO_3^- (pH7, ~ 1 mg O_3/L): 10 μg/L

MULTI MEDIA RAPID SAND FILTRATION (NON-OPTIMISED)

BENEFICIAL PROCESSES	POTENTIAL RISKS
■ additional particle in-bed flocculation ■ partial residual permanganate reduction ■ particle (floc) removal	■ filter optimisation is needed to be done in a combined DAF-filtration treatment mode, to achieve higher efficiency

FILTRATE (no conditioning)
- turbidity: 0.1 - 0.2 FTU
- particle (algae) count of size ≥ 2.75 μm: N = 150 + 300/mL (i.e. < 2 log removal)
- residual Fe: 0.05 - 0.30 mg/L
- MFI_{min}: ~ 19 s/L²

FILTRATE ($KMnO_4$ or O_3 conditioning)
- turbidity: 0.15 - 0.25 FTU
- particle (algae) count of size ≥ 2.75 μm: N = 30 - 150/mL (i.e. ≤ 2 log removal)
- residual Fe: 0.1 - 0.15 mg/L
- residual Mn: 0.030 - 0.08 mg/L
- MFI_{min}: ~ 50 s/L²

Table 8.2 Schematized process mechanisms, including beneficial processes and potential risks occurring during (enhanced) DAF treatment of algae laden water.

Samenvatting

Vlaški A., (1997). Verwijdering van *Microcystis aeruginosa* door toepassing van z.g. Dissolved Air Flotation - Mogelijkheden voor een verbeterde werking en modellering van het proces. Proefschrift, International Institute for Infrastructure, Hydraulic and Environmental Engineering (IHE)/Technische Universiteit Delft, Delft, Nederland, 254 pagina's.

De mate van eutrofiëring van het ruwe water dat door de drinkwaterbedrijven wordt gezuiverd, is aan sterke schommelingen onderhevig. Fosfor is hierbij de limitierde nutrient gebleken. Concentraties van meer dan 0.05 mg P/L zijn potentiel problematisch in de zin van algengroei (>10 μg/L). De predominant algenpopulaties zijn afhankelijk van een scala van aanvulende factoren zoals pH, temperatuur, de verhouding met andere nutrienten, licht intensiteit, etc. Als gevolg van de fluctuaties in algenpopulaties (van gemiddeld 8.4 μg/L chlorophyll-*a* voor het Loenderveense Plas tot gemiddeld 238 μg/L chlorophyll-*a* voor het PWN-bekken) treden zuiveringsproblemen op.Uit veld-onderzoek bleek dat cyanobacteriën in het algemeen, maar vooral *Microcystis aeruginosa* en *Oscillatoria agradhii* soorten, de hardnekkige soorten zijn en de meeste zuiveringsproblemen gaven. De Nederlandse ervaringen zijn over het algemeen representatief voor geïndustrialiseerde landen waar eutrofiëring gewoonlijk resulteert in een opbloei en hieraan gekoppeld in zuiveringsproblemen. Deze problemen bestonden uit een tot 100% verhoogde vlokmiddelbehoefte, substantieel verhoogde hoeveelheden filterspoelwater, trihalomethaanvorming (gedurende korte perioden van onderbrekingen in chlorering voor contrôle van mosselgroei), (oppervlakte) verstopping van filters, het door algen passeren van de behandelingsinstallatie in bezwaarlijke hoeveelheden, een verhoging van de MFI (Membrane Fouling Index) en de AOC (Assimilable Organic Carbon) waarden, enz. De ook voorkomende eéncellige *M. aeruginosa* is qua grootte (3-10 μm) en geaardheid representatief voor de pathogene *Cryptosporidium* oocysten en de *Giardia* cysten, die in de behandeling zijn bijzonder moeilijk te verwijderen.

Deze studie richtte zich op de verwijdering van de cyanobacterie *M. aeruginosa* door conventionele en meer geavanceerde behandelingsmethoden, die beschikbaar zijn of die als haalbaar beschouwd worden voor de kwaliteit van ruw water en de procesomstandigheden in Nederland. Zuiveringsmethoden die in de betrokken bedrijven toegepast worden omvatten sedimentatie-filtratie en de meer geavanceerde DAF (Dissolved Air Flotation)-filtratie. Het coagulatie/flocculatie-proces is van doorslaggevend betekenis voor de efficiëntie van voornoemde zuiveringstrappen, onafhankelijk van het zuiveringssysteem. De mogelijkheid van algenconditionering door oxidatiemiddelen zoals ozon of $KMnO_4$, werd als aantrekkelijk beschouwd om (tijdelijk of continu) de efficiëntie van de algenverwijdering te vergroten, alhoewel deze methode in Nederland niet toegepast wordt. Daarom richtte deze studie zich op een onderzoek naar een reeks zuiveringsmogelijkheden voor de verwijdering van *M. aeruginosa*, met de nadruk op DAF. Naast de efficiëntie van het proces werd aandacht besteed aan de procesmechanismen.

In deze studie worden twee experimentele opstellingen gebruikt: te weten een op laboratorium schaal ontwikkeld voor zowel DAF als sedimentatie, en een semi-technische DAF proefopstelling (Purac, Zweden). Voor de eerste opstelling werd modelwater gebruikt dat bereid was uit water afkomstig uit de Biesbosch reservoirs waaran een laboratoriumculture van *M. aeruginosa* (beginconcentratie van ≈ 10.000 cellen/mL) was toegevoegd. De standaardtemperatuur van de experimenten was 20°C en de coagulatie snelheidsgradient G was 1,000 s^{-1} voor een duur van 30 s bij een pH 8. Andere coagulatie-omstandigheden werden ook getest (pH 4, 6, 7, 8 en 9). De flocculatie G en de tijd werden gevarieerd (G_f= 10, 30, 50, 70, 100 en 120 s^{-1} en t_f = 5, 10, 15, 25, 30 en 35 min.). De vlokmiddel doses waren 0 - 15 mg Fe(III)/L, terwijl de kationische polyelectrolieten Superfloc C-573 en Wisprofloc-P, het non-ionische Wisprofloc-N, en het anionische Superfloc A-100 zijn getest als enige vlokhulpmiddelen. Ozon werd gebruikt als een algen-conditioner in de dosierbereik van 0.48-2.16 mg O_3/L of 0.2-0.9 mg O_3/mg TOC bij een pH 7.5. $KMnO_4$ behandeling werd getest in het dosierbereik van 0-2 mg/L bij een pH 8 waarbij de optimale dosis werd afgeleid met behulp van een visuele determinatie techniek. De $KMnO_4$-contacttijd werd gevarieerd tussen 0 en 30 min. Na de DAF behandeling (capaciteit installatie 4-6 m^3/h) werd het water over een op-waarts bedreven drielagensnelzandfilter gevoerd (v=10 m/h). Beide behandelingen zijn gesitueerd in de WRK III zuiveringsinstallatie in Andijk, Nederland. Het resultaat werd vergeleken met het zuiveringsrendement van WRK III zuiveringssysteem dat bestaat uit flocculatie, lamellenbezinking en - zandfiltratie. Analysetechnieken om doelmatigheid van het proces in te schatten bestonden uit bepaling van troebelheid, het restijzergehalte, TOC/DOC, UV absorptie bij 254 nm en het restmangaangehalte. Bovendien werden meer geavanceerde methoden toegepast bestaande uit bromaat en MFI-metingen, telling van deeltjes (in het afmetingenbereik van d>2.75 μm, en in een beperkt aantal gevallen d>0.3 μm) en computer beeldanalyse. Deze laatste methode biedt, gekoppeld met een hoge resolutie SEM (Scanning Electron Microscopy), waardevolle informatie om de proces mechanismen te schatten.

Voor de behandeling van algenrijk water (o.m. *M. aeruginosa*) is DAF een doelmatig alternatief voor sedimentatie gebleken. Zo resulteerde DAF behandeling in een gelijk of beter rendement van algenverwijdering dan het sedimentatieproces (71% vergeleken met 87% voor bekerproef-testomstandigheden en 96.3% vergeleken met 95.6 % voor de proefinstallatie omstandigheden), terwijl zij tegelijkertijd een 50 % lagere vlokmiddeldosis vereisten dan de sedimentatie (3 mg Fe(III)/L vergeleken met 10 mg Fe(III)/L in de laboratoriuminstelling en 7-12 g Fe(III)/m^3 vergeleken met 20-24 g Fe(III)/m ^{3} + 0.2-0.5 g Wisprofloc-P/m ^{3} in de proefinstallatie). Bovendien had het geproduceerde slib een hoger droge stofgehalte. Andere gunstige aspecten waren onder andere de relatief korte flocculatietijd (echter, niet korter dan 15 min, vergeleken met >30 min voor sedimentatie) en de lage hoeveelheid energie nodig voor de flocculatie (G = 10 s^{-1} vergeleken met G = 30 s^{-1} voor sedimentatie), zowel als de hoge belasting van het proces (over het geheel leidend tot 5-6 keer minder procesruimte). Bovendien, het DAF-filtratie behandelingsschema resulteerde consequent in een hoge (>2 log) efficiëntie in algenverwijdering. Echter, de relatief lage vlokmiddelbehoefte voor DAF zorgde voor minder efficiënte omstandigheden dan bij sedimentatie, wat resulteerde in vorming van organo-Fe complexen en dientengevolge hogere ijzergehalten en resttroebeling van het flotaat. Dit werd opgelost door toepassing van kationische vlokhulpmiddelen, of door instelling van een pH die lager ligt dan de pH van het IEP (Iso-Electrical Point). De verbetering in het rendement van de deeltjesverwijdering (algen) was duidelijker bij de experimenten met modelwater (20-40%, en maar 1-2% in experimenten met de proefinstallatie). Dit is te wijten aan de bevordering van adsorptie coagulatie van deeltjes (algen) wanneer de lading geneutraliseerd is. Bovendien, de geobserveerde verschijnselen van neutralisatie van de lading resulteerden in een verbeterde hechting tussen deeltje-bel en dus een efficiëntere DAF. Non-ionische en anionische polyelectrolyten werden minder efficiënt bevonden, gedeeltelijk vanwege de lage deeltjesconcentratie (lagere vlokmiddeldosis) en de negatieve lading op de luchtbellen.

Het rendement van de algenverwijdering met de proefinstallatie ligt 20-30 % hoger dan de bekerproeven met modelwater. Waarschijnlijk was de NOM (Natural Organic Matter) concentratie en samenstelling de oorzaak van deze verschillen, vooral omdat het aanwezig is in de vorm van netwerken van organisch materiaal en in de vorm van draadvormige strukturen. Hoewel het complexeren van de metaal vlokmiddelen door NOM een extra vlokmiddelbehoefte geeft, zullen deze strukturen uiteindelijk coagulatie bevorderen, omdat zij kunnen dienen als plaatsen voor vlokgroei en hechting en ook een gel-achtige struktuur bieden zodat deze gemakkelijker is te verwijderen door de flotatie.

Op laboratorium-schaal bleken relatief lage doses ozon (0.2-0.5 mg O_3/mg TOC of 0.6-1.5 mg O_3/L) gebruikt als algenconditioner de DAF efficiëntie significant (40-50%) te verhogen (2 log verwijdering). Gecombineerd met kationische polyelectrolyten werd de DAF efficiëntie nog 5 % beter. De toepassing van de lagere dosis van 0.2 mg O_3/mg C bij pH 7 resulteerde in bromaat niveau's onder de 10 μg/L MAC (Maximum Allowable Concentration), zoals voorgeschreven door de EU, alhoewel dit resultaat nog niet bevestigd is in de praktijk op semi-technische schaal. De voortgaande verbetering van de verwijdering van organische stof door aktief-koolfiltratie zou het belang van deze behandeling vergroten, speciaal als de ozon al gebruikt wordt voor andere doeleinden binnen de zuivering (bijv. desinfectie) en alleen als de MAC-waarde onder de huidige voorgeschreven concentratie blijft.

Het gebruik van $KMnO_4$ als een algen conditioner leidde ook tot een verhoging van de DAF-proces efficiëntie, hetgeen resulteerde in een gelijkwaardig of beter rendement dan in het geval van de conventionele sedimentatie technische schaal, in enkele gevallen zelfs in een betere prestatie dan in het geval van technische schaal sedimentatie + filtratie wat betreft de deeltjesverwijdering (d > 2.75 μm). Echter, als gevolg daarvan was de restijzer concentratie (organo-Fe-complexen) en Mn (over het algemeen MnO_2) hoger, wat het aantal deeltjes van colloidale afmeting (<0.5 μm) en de troebelheid deed toenemen. Deze situatie verbeterde toen de kationische polymeren geïntroduceerd werden als vlokhulpmiddelen, wat significant de efficiëntie van het DAF proces verbeterde (over het algemeen met 5%) door reductie in Fe en Mn-concentraties onder hun respectievelijke MAC-waarden. Maar deze mogelijkheid verlaagde niet de concentratie van het begeleidende filtraat MFI tot de gewenste waarde van 5 s/L^2. De laagste MFI-waarde die bereikt werd door de niet geoptimaliseerde filtratie (binnen het DAF + filtratie schema) was ongeveer 20 s/L^2. Dit suggereert de behoefte aan optimalizeren van de filtratiestap om geheel de conditionerende werking van $KMnO_4$ te gebruiken.

De verbeterde algenverwijdering in het geval van het conditioneren door een oxidatiemiddel is waarschijnlijk veroorzaakt door een aantal processen. Beweeglijke algencellen worden geïmmobiliseerd door oxidatie van delen van de buitenste cellaag, waardoor hun stofwisselingsprocessen verstoord worden. Dit gaat gepaard met struktuurwijzigingen in de algen, resulterend in EOM (Extra-cellular Organic Matter) uitscheiding, gedeeltelijke algen ontbinding en lekkage van IOM (Intra-cellular Organic Matter). De EOM en de IOM werken als natuurlijke vlokhulpmiddelen die een spontane micro-flocculatie tot gevolg hebben zelfs voordat het vlokmiddel toegevoegd wordt. Oxidatie en verwijdering van de organische beschermlaag om de deeltjes, en ook oxidatie van NOM, resulteren in gunstigere flocculatie-omstandigheden door de produktie van in-situ vlokmiddel, dat anders tot een complex zou zijn gevormd en niet beschikbaar zou zijn voor coagulatie. In het geval van conditioneren met $KMnO_4$ verhoogt het gevormde MnO_2 de deeltjesconcentratie en vergroot derhalve de vlokvormingssnelheid.

Verschillen in de verkregen deeltjesgrootteverdelingen en deeltjesvolumeverdelingen vormden de basis voor een discussie over de invloed van procesomstandigheden. Deze verdelingen maakten ook een proefondervindelijke berekening mogelijk van de relatieve en absolute vlokdichtheid, die een input-variabele is in het beschouwde DAF-model. Andere input-variabelen zoals de gemiddelde deeltjes (vlok) en belgrootte, alsook aantallen en volume-concentraties van deeltjes en bellen werden tegelijkertijd gemeten of berekend. De berekende waarden voor de hechtingsefficiëntie van deeltje aan luchtbel, α_{pb}, ondersteunen de geponeerde stelling betreffende de positieve invloed van kationische vlokhulpmiddelen en conditioners. De resultaten suggereren dat onder optimale flotatie omstandigheden bijna de helft van de botsingen tussen deeltjes (vlokken) en bellen resulteert in hechting. De recirculatie verhouding (5-10%) en de verzadigingsdruk (500-700 kPa) bleken minder essentieel te zijn voor de proces-efficiëntie, alhoewel zij de belgrootte aanzienlijk beïnvloeden. Derhalve eist het coagulatie/flocculatie-proces als de meest bepalende factor van het DAF-proces alle aandacht. Alhoewel moelijk te modelleren is de invloed van de NOM-concentratie en karakteristiken op de hechtingefficiëntie van deetje en luchtbel van groot belang. Op deze manier is een vereenvoudigde model benadering dat deeltjes-luchtbellen opeenlopig allen het resultaat van hun botsing is, niet geheel gerechtwaardig.

DAF blijkt een efficiënt, robuust en flexibel proces en is zeer geschikt als alternatief voor sedimentatie bij de zuivering van met algen belast water. De lagere investeringskosten en het lagere chemicaliën verbruik zullen vaak de hogere energie- en onderhoudskosten compenseren. De efficiëntie inzake algenverwijdering maakt deze vorm van flotatie aantrekkelijk als voorbehandeling voor directe filtratie of membraanfiltratie. Waar het al wordt toegepast kan de efficiëntie tijdelijk vergroot worden (met name gedurende algenbloei) door toepassing van met name kationische vlokhulpmiddelen, oxidatiemiddelen, of een combinatie van beiden. De toepassing van $KMnO_4$ lijkt bijzonder aantrekkelijk, omdat geen enkele bezwaarlijke bijprodukten ontstaan. De uiteindelijke keuze van de behandelingstechnologie dient eventueel na een onderzoek met een proefinstallatie en een kosten-baten analyse gemaakt worden.

Curriculum Vitae

Aleksandar Vlaški was born in Skopje, Macedonia on November 14, 1958. After attending primary school and gymnasium in Skopje, he received a B.Sc. degree in civil engineering in 1984 from the University of Cyril and Metodij in Skopje.

In 1984 he started his professional career as an assistant design engineer in the hydro-department of the "EMO-Institute for Energetics" in Skopje, a scientific and consulting daughter institution of the largest electronics company in Macedonia EMO-Ohrid. In 1987 he joined the Sanitary engineering department, where he formally stayed until 1992, advancing to a position of independent design engineer, with full responsibility for the design of different sanitary engineering projects. He was involved on a number of national and international projects of various nature: hydro-power plants and energetics, regional and rural water supply and distributions, regional and rural wastewater collection, and drinking water, wastewater and industrial wastewater treatment. He performed a variety of engineering tasks, including hydraulic and process technology design, site engineering, project management and finally project acquisition.

In 1989 he came to the Netherlands to the International Institute for Infrastructural, Hydraulic and Environmental Engineering (IHE), Delft, where he obtained a post-graduate diploma in sanitary engineering in 1990. In 1991 he returned to IHE and conducted his M.Sc. research on a joined project between IHE and the Technical University (TU), Delft, dealing with 'Aggregation of algae and direct filtration'. After seven months he concluded his research by obtaining his M.Sc. Degree with distinction in Environmental Engineering from IHE Delft.

From January 1993 until October 1997 he worked on what started as a joint research project between the IHE and TU Delft titled 'Relevance and removal of cyanobacteria in water treatment'. After two years of work at the Civil Engineering Faculty, Laboratory for Sanitary Engineering, and two and a half years of work at the IHE Delft, he concluded the research with a Ph.D. thesis titled '*Microcystis aeruginosa* removal by dissolved air flotation (DAF) - options for enhanced process operation and kinetic modelling'.

In November 1997 he joined the engineering and consulting company Tauw Milieu, Deventer, at the position of a drinking water treatment process technologist.

IHE
DELFT

The aim of the International Institute for Infrastructural, Hydraulic and Environmental Engineering, IHE Delft, is the development and transfer of scientific knowledge and technological know-how in the fields of transport, water and the environment.

Therefore, IHE organizes regular 12 and 18 month postgraduate courses which lead to a Masters Degree. IHE also has a PhD-programme based on research, which can be executed partly in the home country. Moreover, IHE organizes short tailor-made and regular non-degree courses in The Netherlands as well as abroad, and takes part in projects in various countries to develop local educational training and research facilities.

International Institute for
Infrastructural, Hydraulic and
Environmental Engineering

P.O. Box 3015
2601 DA Delft
The Netherlands

Tel.: +31 15 2151715
Fax: +31 15 2122921
E-mail: ihe@ihe.nl
Internet: http://www.ihe.nl